Supported by

(Environmental Records Centre for Cornwall and the Isles of Scilly)

© 2017 Rosemary E Parslow and Ian J Bennallick

The views expressed in this book are those of the writers. All rights reserved. No part of the publication may be reproduced, stored in a retrieval system, or transmitted, in any form or by means electronic, mechanical, photocopying, recording or otherwise, without prior permission of the publishers.

Distribution maps used in this publication are based upon Ordnance Survey material with the permission of Ordnance Survey on behalf of the Controller of Her Majesty's Stationery Office. © Crown copyright. Unauthorised reproduction infringes Crown copyright and may lead to prosecution or civil proceedings. Cornwall Council, 100049047, 2015.

Designer: Glynn Bennallick BA (Hons)
             pgbennallick@gmail.com

Publisher: Parslow Press

Printer: Short Run Press
             shortrunpress.co.uk

ISBN 978-1-5272-0483-6

The NEW Flora of the Isles of Scilly

ROSEMARY PARSLOW
IAN BENNALLICK

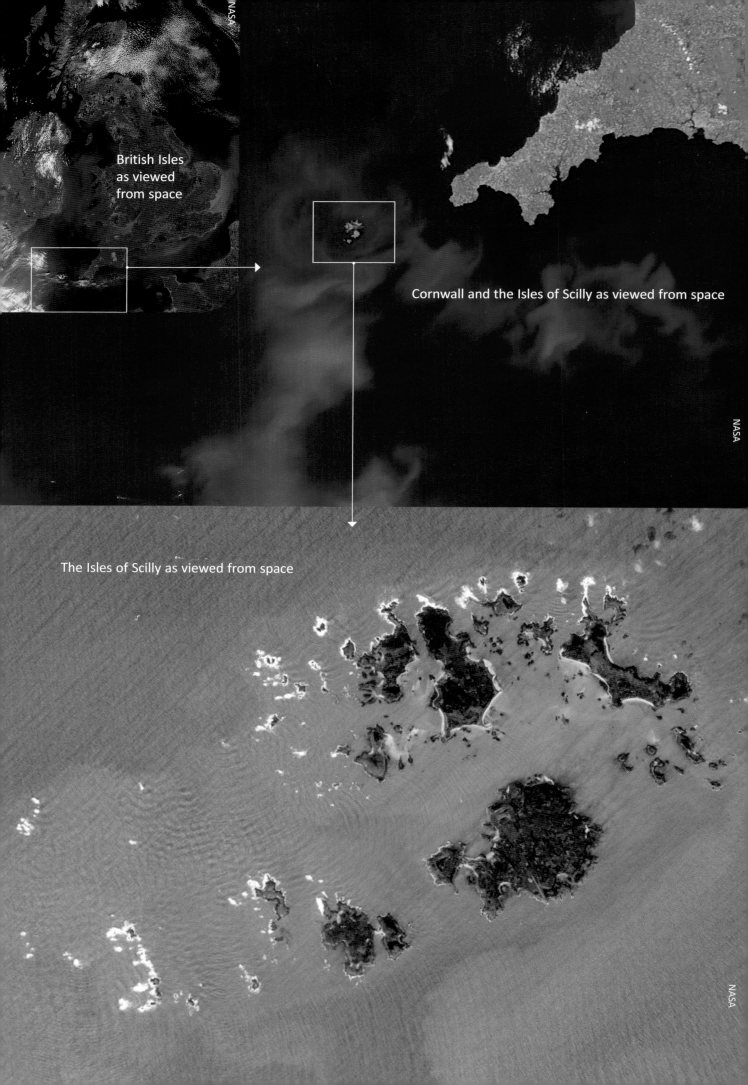

Contents

page	
7	**Contents**
8	**A New Flora for the Isles of Scilly**
10	**Map of the Isles of Scilly**
12	**Acknowledgements**
13	**Introduction**
14	Geographical position of the Isles of Scilly
16	Early history
19	The history of botanical recording on Scilly
23	Recording for The New Flora of the Isles of Scilly
26	The Islands
47	The origins of the alien flora
58	Plant habitats
80	Nature conservation in the Isles of Scilly
84	Visiting the Isles of Scilly
86	**Species accounts**
90	**PTERIDOPHYTES** — *Ferns and fern allies*
90	*Lycophytes* — *Clubmosses*
90	*Eusporangiate ferns* — *Adder's-tongues and Moonworts*
93	*Calamophytes* — *Horsetails*
94	*Leptosporangiate Ferns* — *True Ferns*
108	**GYMNOSPERMS** — **Conifers**
109	**ANGIOSPERMS** — **Flowering Plants**
109	*Pre-dicots* — *Primitive Angiosperms*
110	*Eu-dicots* — *True Dicotyledons*
427	*Monocots* — *Monocotyledons*
514	**Garden escapes and species needing confirmation**
516	**Bibliography and useful references**
522	**Glossary of terms**
524	**List of recorders, photographers, artists and contributors**
526	**Gazetteer**
530	**Index to species**

A NEW FLORA

The Isles of Scilly lie in the Atlantic Ocean 35 km (28 miles) off the south-west tip of Cornwall. The islands are something of a Mecca for botanists due to their unique flora which is unlike anywhere else in Britain and Ireland. This book attempts to describe the current (to 2016) state of our knowledge of the flora of the islands.

The introductory chapters will endeavour to give the background to this Flora, how it started and the history of plant recording on Scilly. Also the important habitats and an account of all the islands and their geography.

The main body of the text is the species accounts which attempts to include all those plants that occur on the Isles of Scilly or have been recorded there in the past, their distribution, status and history. Distribution maps are included in the text with dots covering two date classes, records pre-2000 and post-1999. Photographs of individual plants and their habitats are incorporated into the text where appropriate. The majority of the photographs have been taken in Scilly, with a few from Cornwall or elsewhere.

A bibliography, acknowledgements and list of contributors are also included.

For many years the Flora written by J.E.'Ted' Lousley (1971) was the bible for any botanist visiting the Isles of Scilly. Now 45 years since its publication an update is well overdue. Throughout this Flora reference will be made to Lousley's Flora to record and comment on any changes that have occurred. I took over as BSBI vice-county recorder for Scilly in 1983. At that time the only records inherited were print-outs from the Biological Record Centre at Monks Wood. Later I was able to borrow and copy the very valuable card index

FOR THE ISLES OF SCILLY

compiled by Peter Clough when he was the Head Gardener on Tresco in the late 1970s to 1981. At first, like everyone else, records were kept on a card index: later I moved on to a desktop computer eventually using the recording package and database MapMate.

The majority of the records on the database were entered post 1990. Initially most were from my visits as at that time there was only a handful of regular contributors. Since about 2000 the number of other recorders has increased, with visitors who send in their holiday lists and Scilly residents as well as staff and volunteers at the Isles of Scilly Wildlife Trust (IoSWT) and members of the Isles of Scilly Bird Group (IoSBG).

Compiling the Flora was started in the early 2000s but did not progress until after 2008. In 2013 Ian Bennallick joined as co-author, sorting out the distribution maps and getting the Flora ready for publication. The cut-off date for the flora is the end of 2016 which means we will not be able to report on the long-term effects of the extreme weather experienced during the early part of 2014. Although the storms caused great changes to some coasts, coastal plants are resilient and we expect them to recover eventually. Surprisingly Viola kitaibeliana Dwarf Pansy on Bryher survived, although its site on Teän has been buried under deep sand. The apparent total loss of Rumex rupestris Shore Dock was more of a shock, only time will tell if this loss is permanent. Of even more significance long-term may be the continuing downturn in bulb farming on Scilly which is leading to rare arable weeds becoming rarer or eventually lost altogether from the flora.

Rosemary Parslow

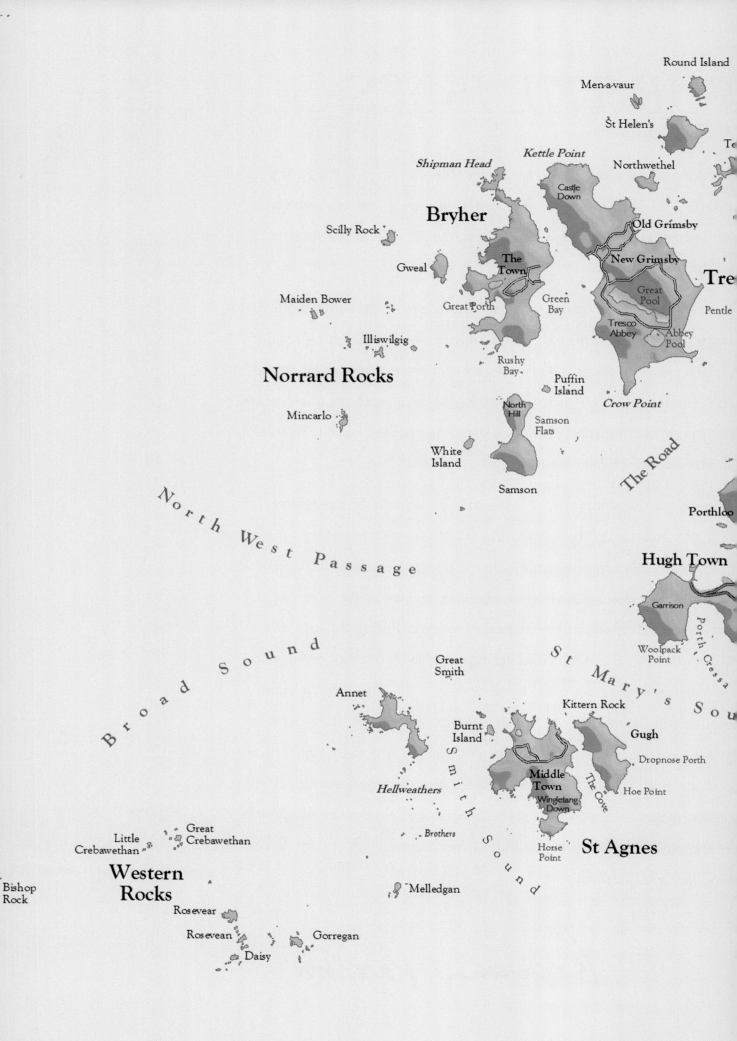

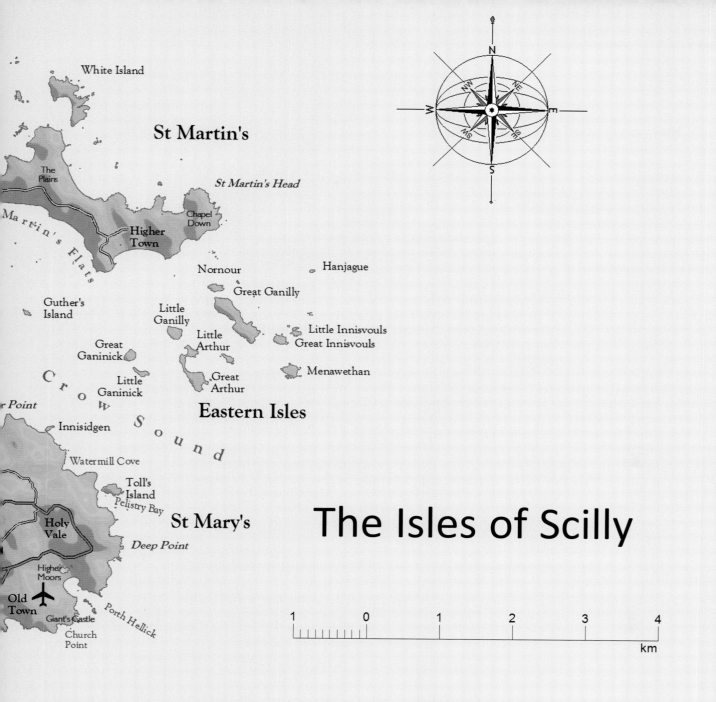

The Isles of Scilly

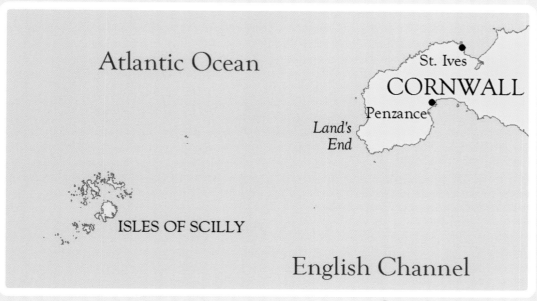

Acknowledgements

It is impossible to thank everyone individually who has contributed in any way to this Flora, but a list of those who sent in plant records is included on pages 524 - 5. Without them this book would not have been possible. If anyone has been inadvertently omitted or not credited we apologise.

Many Scillonians have contributed directly or indirectly to this book. They include IoSWT staff and volunteers, boatmen, farmers who allowed access to their fields, members of the IoSBG, islanders who gave me accommodation, sent photos or passed on information about plants or helped in many other ways.

Some islanders merit special mention, in no special order; Lesley Knight, Anne & Mike Gurr, David Mawer, Liz & Jim Askins, Wendy Hick, Keith Low, Fran Hicks, Adrian & Mandy Pearce, Martin Goodey, Mike Brown, Darren Hart, Penny Rodgers, Jaclyn Pearson, Ren & Jo Hathway, Will Wagstaff, Bob Dawson, Christine Blackwell, Carmen Stevens, Vicky Heaney, Katharine Sawyer, Melanie Woodcock and Celia Sison.

The following also contributed in many ways:

Staff from Natural England (and predecessors NCC and EN) with whom we worked in Scilly at various times, these included Lynne Farrell, Jeremy Clitherow, Cyril Nicholas, Jules Webber and Lone Mouritsen. Members or staff of BSBI, BPS, BCG, NHM, RSPB and visitors with Wildlife Travel and other natural history holiday companies. Also members of the Parslow family and Adrian Colston.

Additionally we have benefitted from the expertise of a number of botanists including David Allen (for Rubus), Chris Preston, Lynne Farrell, Alan Silverside, Geoffrey Kitchener, Rosaline Murphy, Ray Stephenson, Chris Page, Eric Clement, Roger Maskew, Mark Spencer, Fred Rumsey, David Pearman, Colin French and Gwynn Ellis (index).

ERCCIS for help with producing the distribution maps: Gary Lewis, Martin Goodall, Emma Weller, Alan Rowland and Tamara Weeks.

A number of photographers gave generously of their photographs, far more than we could use, those we have used are acknowledged in the captions.

The topographical map of the islands was produced by Hanno Koch, Latitude Cartography. IoSBG gave permission to use the outline maps of the inhabited islands from their annual Bird Report and Natural History Review. David Mawer is responsible for the habitat maps. The late Prof. Charles Thomas allowed me to modify maps from his book on The Drowned Landscape.

Thanks to Keith Hyatt and Daisy Walker for reading the text, and making many useful comments. Any errors however are ours.

My co-author Ian Bennallick was also responsible for converting the species records into distribution maps and for working with his brother Glynn Bennallick who designed the book.

Organisations or groups shown above

BCG	Botanical Cornwall Group
BPS	British Pteridological Society
BSBI	Botanical Society of Britain and Ireland
EN	English Nature
ERCCIS	Environmental Records Centre for Cornwall and the Isles of Scilly
IoSBG	Isles of Scilly Bird Group
IoSWT	Isles of Scilly Wildlife Trust
NCC	Nature Conservancy Council
NHM	Natural History Museum
RSPB	Royal Society for the Protection of Birds
WFS	Wild Flower Society

INTRODUCTION

page

14 Geographical position of the Isles of Scilly

16 Early history

19 The history of botanical recording on Scilly

23 Recording for The New Flora of the Isles of Scilly

26 The Islands

47 The origins of the alien flora

58 Plant habitats

80 Nature conservation in the Isles of Scilly

84 Visiting the Isles of Scilly

Geographical position of the Isles of Scilly

The Isles of Scilly is a small and remote archipelago well away from the coast of Cornwall. Their comparative isolation has resulted in a number of differences in the flora and habitats compared to that of the Cornish mainland. Although Scilly is included in the Watsonian botanical recording area of VC1 with West Cornwall, this is a constant source of irritation to Scilly botanists as it masks the distinctiveness of the Scillonian flora and leads to many misconceptions about what occurs and what is typical. Briefly Scilly was allotted its own vice-county number 114, but this does not seem to have caught on.

Geology, soils and climate

St Mary's granite "heads"

Compared with the Cornish peninsula the geology of the Isles of Scilly is relatively simple. The OS geological map shows the main rock is granite with areas of blown sand and a few very small patches of other formations. Scilly is what remains of one of the granite bosses or cupolas that formed part of the Cornubian batholith that extended from the high land of Dartmoor in a series of domes the length of the peninsula ending up with Scilly as the penultimate cupola (a final one, Haig Fras, lies under the sea nearly one hundred kilometres further west). At one time Scilly and Cornwall would have been covered by killas, a soft, shale-like rock which was pushed up from below by the pressure of the granite. Eventually the killas became eroded and on Scilly are now only found as an altered rock called tormalised schist of which a small amount exists on St Martin's.

Granite on Scilly is classified into two main types, called simply G1 and G2. The G1 forms the outer rim of the archipelago; it is coarse-grained granite. G2 has a finer structure and is mostly in the centre of the islands as well as often obtruding into the older G1 so there is no obvious cut-off between the two types. The Scilly granites are very beautiful with large crystals of feldspar, quartz and shiny mica. Characteristic veins of white or coloured quartz and black tourmaline form dykes running through the granite. Usually these details can only be seen in rocks on the coast or beach pebbles where the surfaces are not obscured by lichens.

Scilly granite

A feature of the landscape is the granite tors, especially around the coast. These have been subject to horizontal and vertical weathering to such as an extent they have been naturally sculpted into extreme shapes. Some of the carns often resemble animals or other structures and have been given whimsical descriptive names! In recent years *Pittosporum crassifolium* Pittosporum bushes have invaded the crevices in these carns where they flourish.

Pulpit rock

Another important feature on the islands is the material called 'ram' or 'rabb'. This is a claylike substance formed from decomposed granite and is found below the granite rock. Large pockets of ram have in the past been quarried to use as mortar and for roads. Around the coasts there is usually a shelf of ram which can act as an impermeable layer where cliff edge seepages emerge or as a platform for coastal plants.

Carreg Dhu garden in old ram pit

At one time is was believed the glaciers had missed Scilly completely, but it is now known that the glaciers just caught the northern edge of the islands leaving a legacy of different rocks, including flint which was utilised by the prehistoric people who later lived on Scilly.

The soils on Scilly are mostly acidic, derived from wind-blown silt, loess and blown sand. In a very few places there are lenses of slightly more calcareous sand associated with shell deposits. Where land has been farmed the soils have been modified by the use of fertiliser and lime, also in the past seaweed was frequently used as manure.

Climate has a significant effect on the flora of the Isles of Scilly being generally mild with an average temperature of 11.8°C. The summers are slightly cooler than on the Cornish mainland but the winter temperatures are milder with average lows of 6°C. Frosts tend to be light and brief and snow is rare. The coldest winter recorded recently was in 1987, with lows of -7.2°C on 13th January and 23cm of snow the previous day. This was an unusual event that killed many of the tender plants and numbers of large trees as well as miles of *Pittosporum* hedging. Average rainfall is 850 - 900mm, slightly less than Cornwall, although the frequency of sea fogs means the humidity is high. The clear, unpolluted air is reflected in the rich lichen flora. It may be that the bright reds and purples of some plants are related to high UV levels.

As small, low islands lying far out at sea without benefit of shelter from any larger landmass, the islands are especially vulnerable to storms and gales. Storms are

Closed path due to erosion

common and every few years there are more extreme gales and storm surges that are significantly more severe than in 2014. These cause much damage and erosion to the coast, send salt spray and even waves inland, flood lower land, blow down trees and tear rocks from the shore and hurl them into the fields. It seems these storms are becoming more frequent or certainly more violent. The latest series of severe storms, culminating with the worst on 14th February 2014 was very destructive and has changed the configuration of many bays and washed away coastal plants or buried them under sand. The damage caused by the salt was also very evident with trees and heathland plants near the coast all burnt and discoloured. Of course these are natural events and many of the plants will return or re-colonise in time. But each storm event takes back a little more land and so the gradual submergence of the islands continues.

Early history

The Isles of Scilly were believed to have separated from the Cornish mainland during the late Glacial period, about 12,000 years ago. There is no evidence that there were people living on Scilly during the Palaeolithic period, although the find of a possible 'penknife point' held at the Royal Cornwall Museum suggests the islands were being visited during that period. The landscape would have been tundra – very different from today.

During the Mesolithic period Scilly was a separate land mass from the mainland. It covered about 17 km x 8.5 km which extended from what is now the Western Rocks to the Eastern Isles and from Shipman Head on Bryher to Peninnis Head on St Mary's. The land mass would have been one large island that as the ice sheet melted and the seas rose was cut into two or three parts. The larger island that formed encompassed the northern part of Scilly with what would eventually form the present islands of Bryher, Tresco, St Mary's, St Martin's and the Eastern Isles. The rest of the land became what would be the smaller islands of St Agnes, Annet and the Western Rocks. This separation of St Agnes from the rest of the archipelago by deeper water may explain some slight differences in the flora there from the rest of Scilly.

Shoreline before submergence

Some clues to the environment of Mesolithic Scilly are based on pollen evidence from studies on St Mary's and St Martin's. An article by Charles Johns in 'The Archaeologist' (2011) describing the 'Lyonesse Project' gives a brief summary of vegetation changes shown by pollen in sub-marine samples. These were described more fully when the report of the project was published (Charman et al., 2012). From the samples it was deduced that woodland habitats were present throughout the period with oak-ash and hazel woodland followed by some birch scrub. This has been interpreted as small-scale scrub regeneration after woodland clearance. Later there is evidence of mixed broadleaved woodland that included oak, birch, lime, hazel, holly, alder and willow. It seems there were rapid/or slower changes with the vegetation changing from herb-rich grassland, through colonisation by oak, birch and hazel to woodland. Charcoal found at the transition to overlying peat is associated with an increase in birch and decline in oak which suggests forest clearance by fire, perhaps to grow crops. There was a later recolonisation by oak woodland and then by willow, possibly an indication that the area was becoming much wetter. The final stage shows open marsh dominated by sedges, then grasses and increasing salinity, possibly from sea-level rise. Radiocarbon* and OSL** ages from Crab's Ledge and Bathinghouse Porth, Tresco indicate a saltmarsh environment by c1200 cal BC. There are also marshland indicators from sub-tidal samples from between St Martin's and Nornour with low background woodland levels later in the Mesolithic. Interestingly one of the saltmarsh species that occurred was *Triglochin maritima* Sea Arrowgrass a plant no longer found on Scilly.

Signs of Neolithic people living in Scilly are few, just some pottery from the entrance to Bant's Carn on St Mary's and East Porth on Samson, plus a few stone axes and flint arrowheads. But an excavation directed by Dr Duncan Garrow at Old Quay, St Martin's in 2014 as part of the 'stepping stones'*** project discovered a typical Neolithic occupation site but most exciting a haul of 50 tiny microlith flints of a type from the Mesolithic (period before Neolithic) of a type from Belgium and Northern France. This is convincing evidence of people sailing

Bant's Carn

*** Radiocarbon dating**
Plants and animals absorb the radioactive carbon isotope 14C from the atmosphere during their lifetime by photosynthesis and respiration. When the plant or animal dies it can no longer take carbon from the atmosphere to replace the carbon 14C that was being lost when it was alive. 14C is depleted at a constant rate which can be measured so that the date at which the organism died can be calculated.
Dates are usually shown as calibrated years before present (cal BP) or calibrated years (cal BC/AD). 'Present' in this case is taken to be 1950 AD.

**** Optically stimulated luminescence (OSL)** *dating is quite a recent technique. Under natural conditions quartz and feldspar found in sediments such as sand or silt absorb ionizing radiation (Uranium, Thorium and Potassium) at a constant rate. When exposed to sunlight this stops and the electrons are released. OSL dating calculates the amount of radiation that was absorbed between the time the sediment was first buried and the later exposure to light under laboratory conditions.*

between Belgium, France and Scilly. Whether Scilly was permanently settled at this time is not clear although there is evidence of some initial clearance of land for arable crops. But the further woodland regeneration may imply these were only temporary habitations.

By the Bronze Age there is evidence of considerable human activity on Scilly as shown by the large numbers of Bronze Age remains in the islands. Many of their habitations and field walls have now vanished under the sea but the remains of some walls, hut circles and

Obadiah's Barrow, Gugh

other structures are still visible at low tide or in aerial photographs. On land there are still plentiful remains of buildings, walls and tombs; at least 80 entrance graves have been found, although many structures have been robbed out by later inhabitants. Evidence of the life style of the people can be gleaned from materials found in these sites as well as from the numerous middens that have been excavated. The inhabitants were now farming the land, growing cereal crops and building stone houses. Cultivation of crops such as cereals and grazing of farm stock would have been associated with changes in the plant communities on the land. Unfortunately there are few clues to what species of plants were already growing on the islands before the land was farmed and what species had been introduced, either deliberately or accidentally. There is evidence of traffic between the islands and mainland Britain and Ireland, so plant seeds would have come in with grain, or with imported goods. Some of the seeds and other plant remains found in the remains of a hut (Late Bronze/ Early Iron Age) at West Porth on Samson and other sites are of common arable weeds such as *Spergula arvensis* Corn Spurrey, *Urtica urens* Small Nettle, *Fallopia convolvulus* Black Bindweed and *Chenopodium rubrum* Red Goosefoot, all of which are still present on Scilly (Ratcliffe and Straker, 1996). We do not have any evidence whether any of the Lusitanian or Mediterranean species now part of the Scillonian flora arrived at this time in the history of the islands or were brought in much later.

From the archaeological material we know most of the habitats that are present today were there in the Bronze Age. That there was heathland can be deduced from remains of typical plants such as *Calluna vulgaris* Ling and *Erica cinerea* Bell Heather, *Potentilla erecta* Tormentil, *Danthonia decumbens* Heath Grass, *Ulex* spp. gorse and *Pteridium aquilinum* Bracken. The evidence of arable weeds such as those found at West Porth (see above) also some grassland species are evidence farming was taking place as well as plants from freshwater wetland habitats. Other plants that occurred at this time include salt marsh species such as *Suaeda maritima* Annual Sea-blite,

Nornour hut remains

*** **The Stepping Stones project** was directed by Duncan Garrow (Liverpool) and Fraser Sturt (Southampton) and funded by the Arts and Humanities Research Council. They also worked with Historic Environment, Cornwall Council, and in Scilly, the Isles of Scilly Museum. The project aimed to answer key questions about how and why the Neolithic arrived in and around Britain and Ireland at that time. They analysed all of the late Mesolithic and early Neolithic sites within the western seaways zone, carrying out computer modelling of the sea around that time, and excavated three key Early Neolithic sites including Old Quay in Scilly.

Salicornia sp. glasswort and possibly *Limonium* spp. sea-lavender. There is now no typical saltmarsh in Scilly so what are now the tidal flats between the islands were probably then low-lying areas of saltings and pools which may have been grazed except at the highest spring tides.

Although the Romans did not occupy the Isles of Scilly their influence would have been felt and some artefacts that are clearly Roman have been found. Even more significant was the maritime shrine on the tiny island of Nornour where an extraordinary quantity of coins, brooches, figurines and other votive objects were found. There is also what is thought to be a small harbour. So the island was visited by many outsiders who came in from the sea.

Halangy Village

Perhaps the most interesting historical period, botanically speaking, came later, from the fifth century onwards when there is evidence of goods reaching Scilly that had originated much further afield, even as far as the Mediterranean and North Africa (Thomas, 1985). Certainly some of the arable weeds we now know could have easily arrived alongside the cereals and other imports which originated in the Mediterranean and Asia, although sadly there is no actual evidence to support this. If such species did arrive on Scilly they would have found climate and conditions that enabled them to flourish. Lousley (1971) suggested that many plants followed the same route up the Atlantic coasts from the Mediterranean as the Bronze Age people who settled on Scilly and the Channel Isles, although there is no archaeological evidence.

The submergence

The present day configuration of the islands was reached as the sea rose and submerged the lower-lying land. The first mention of the inundation was made by William Borlase who saw the submerged walls on Samson flats (Borlase, 1756), he also suggested the islands of Tresco, Bryher and Samson had once been all one island. In 1926 the archaeologist O.G.S. Crawford followed up Borlase's observations and with the Scillonian photographer

Ancient walls, Samson

Alexander Gibson investigated the walls on Samson flats during a low spring tide. Later he published his observations and photographs of the exposed walls (Crawford, 1927). When the submergence occurred is not known, although there are different models of when it might have been. Charles Thomas suggested the land area of Ennor (the former name for the island before it separated into the individual islands) was still present at low water until the eleventh century, finally separating in the beginning of the sixteenth century (Thomas, 1985). More recent research based on studies of intertidal peat deposits, has suggested the initial separation of Ennor at high tide into smaller islands was much earlier from about 1,000 years BC. Whenever the inundation took place, clear evidence of the former, larger land area can be seen when the lowest equinoctial tides expose the sand flats between Samson, Bryher and Tresco and across to St Martin's. Under the shallow sea between

St Martin's Flats

these islands are remnants of stone walls, hut circles and other evidence of farming and habitation. There are also tree remains that are either buried under sand or on the sea bed. That the sea level is still rising can be seen by comparing old photographs of the coast with those of the present day. The demise of the populations of strictly shore-based, strandline plants such as *Rumex rupestris* Shore Dock more recently is also evidence of the continuing erosion of coastal habitat.

Historical Context

Although some clues to early habitats and vegetation of Scilly may be surmised from pollen and archaeological material, later evidence from written records is sparse. Few visitors to the islands have mentioned natural history in their accounts and at worse they have been sometimes misleading. Much is made of the treeless state of the islands and certainly there is minimal evidence of any woodland surviving into recent times. Just a few places may have a link to former woodland, such as references to a now vanished 'Wilderness' area near what is now the Valhalla on Tresco, that was described by Richard Lynch, the Curator of Cambridge University Botanic Gardens in the 1870s as 'a grove of Cornish Elm, alders and sycamore' (Lynch in King, 1985; Lousley, 1971). There have also been a number of finds of tree trunks and roots either buried under the sand or seen by divers under the sea. Under the present day plantation woodland of Abbey Wood are a few Ancient Woodland Indicator* plant species that are thought to be a link to the former woodland. For example, *Veronica montana* Wood Speedwell and *Carex sylvatica* Wood Sedge. Lousley also found a few old oak trees on Tresco that could have been native (although they appear to have now gone). Prof. Charles Thomas produced a map in his book *Exploration of a Drowned Landscape* (Thomas, 1985) based on the species accounts in Lousley's Flora showing the probable distribution of the former woodland in prehistoric Scilly. Since then we have additional records which have further refined the purported boundaries of this former woodland.

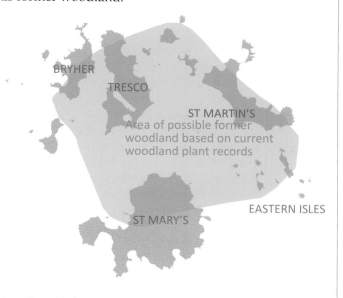

Area of possible former woodland

Although we know barley and other cereals were cultivated from the Bronze Age onwards (Ratcliffe and Straker, 1996) we do not know what plants may have been deliberately introduced to Scilly over the centuries although some would have come in with goods and as accidental introductions, including seeds. Monks from Tavistock Abbey who came to Tresco in the twelfth century are credited with the introduction of useful plants such as herbs, elder and also narcissus.

AWI species refer to plants that were present in woodland that has existed continuously since 1600 or before.

The history of botanical recording on Scilly

Even after the start of a regular mail boat service in 1827 there were not very many people who were prepared to make the difficult sea crossing to Scilly. The small packet boats that made the passage usually took eight to nine hours on a 'good passage', at other times the passengers could be shut in the tiny crowded cabin for thirty-six to forty-eight hours in rough seas (Woodley, 1822). Despite this a few brave souls visited Scilly; among the most frequent visitors were friends of Augustus Smith and subsequently those of the Dorrien Smiths. Usually these were for shooting parties, although some were also interested in the progress of the Abbey gardens and some took note of wild flowers. It was not until 1937 that the first flights to the islands began. There is now a regular ferry, RMV Scillonian from Penzance to St Mary's as well as Skybus flights by fixed wing aircraft. Previously there was a popular helicopter service between the islands and Penzance which was withdrawn in 2012 after 48 years. Attempts to reintroduce the service are currently under discussion.

RMV *Scillonian III*

Skybus

In his 1971 Flora, 'Ted' Lousley gives an account of the early botanists who visited the islands from 1813. One of the first was Sir William Hooker, the first Director of the Royal Botanic Gardens at Kew, who spent nine days visiting the larger islands but left scant account of his visit and appeared to note very little. Lousley dismisses this as 'a great opportunity lost'. After Hooker things appear to have improved and a number of other botanists visited the islands; Lousley comments on the species they recorded and his opinion on their validity. However none of the botanists who visited Scilly in the early days left comprehensive accounts. From 1852 there were quite a few who made the journey to Scilly. John Ralfs paid several visits, and although a good botanist he was having severe problems with his sight and Lousley rejected some of his records as misidentifications. Despite that, Lousley did accept thirty-four new records from his account. In 1862 Frederick Townsend spent ten days on Scilly and recorded 348 species (Townsend, 1864). Unfortunately Lousley comments that there are twenty-one certain errors, owing to 'haste and carelessness', perhaps also due to the short visit and being unwell after the 'boisterous' crossing. Some of the errors were later altered by Townsend in his own copy of the published manuscript. Lousley reserves his greatest praise for the Misses Millett who spent five weeks in the islands in June and July in 1882 or thereabouts. These two sisters recorded 144 flowering plants and sixteen ferns, unfortunately without localities, but assumed to be mostly from St Mary's. From the second half of the century many more botanists visited Scilly and produced lists which included new species (and also more errors according to Lousley). An exception was Sir William Wright Smith who wrote two excellent and accurate papers on the flora after his visit in 1906 (Smith, 1907 and 1909). Unfortunately the references to the Isles of Scilly in Davey's *Flora of Cornwall* (Davey, 1909) which were based on records from a number of sources (Davey never got to Scilly) were unfortunately often unconfirmed or dubious. A list of all the botanists who visited Scilly between 1750 and 1940 are tabled by Lousley, with an additional short account of his contemporaries who contributed many records that he incorporated. In the list of Herbaria he consulted there is reference in his Flora to the collection of 260 species of plants collected by Miss Gwen Dorrien-Smith and her niece Ann Dorrien-Smith between 1922 and 1940, which were verified by Edinburgh Royal Botanic Garden.

Lousley paid many visits to the islands from September 1936 and in all months from March to September. Often he was accompanied by other botanists. When he last visited the islands is not recorded, but he certainly went there in 1972 (Ron Parker pers. comm.) and he was still active just before his death. As so often happened at that time he was not too particular about recording detailed localities so unfortunately some potentially interesting records are just too vague to be useful now. Even in the Flora accounts he frequently assigns a record to an island without locality, particularly with the smaller islands such as the individual Eastern Isles, which he often lumped together. Similarly Northwethel and other islands nearby are all included as 'the St Helen's Group'.

During his involvement with Scilly, Lousley kept a series of notebooks and also wrote reports for the Nature Conservancy (Council). Some of his notebooks, photos and other material are now stored in the archives of the Isles of Scilly Museum. Sometimes it has proved difficult to interpret his findings – clearly the notes were for his personal use and not intended for use by his successors (I fear my successors might well have the same complaint about my notebooks)! I did eventually get copies of three of the four reports he wrote for the Nature Conservancy, unfortunately only as black and white copies, so the colour notes of the originals are lost.

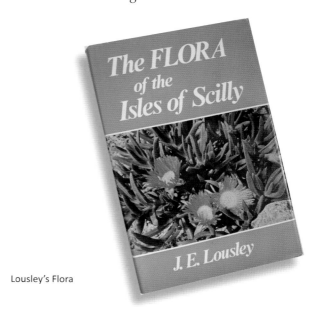

Lousley's Flora

A note published in 1939 (Lousley, 1939) sets out his intention of publishing an account of the flora of the Isles of Scilly starting with some of the information he had already collected, and this was followed by a second paper a year later (Lousley, 1940). So although Lousley's Flora was not published until 1971 much of the work had been done before the Second World War (1939-1945), when the manuscript was put on one side until he was able to visit the islands again. After 1971 there was something of a hiatus in the study and recording of plants in the islands. We know some botanists sent Lousley further records and comments after his Flora

Lousley's Dock monograph

was published, which were thought lost, except for a few letters in the archives of the Isles of Scilly Museum, until 2016 when an annotated copy of the Flora came to light with some 1971-5 records. One botanist, Dr David Allen, kindly sent me a duplicate copy of the records he had sent at that time that I had not previously seen.

Reference to Lousley's contribution to the study of the alien flora of Britain, his study of *Rumex*, which clarified much about that genus, and his huge herbarium were the subject of a number of papers. This included McClintock (1977) on his contribution to the study of alien plants and Jury (1977) on his herbarium which was presented to Reading University. It was the little book on *Docks and Knotweeds* by Lousley and Kent (1981) that alerted me to the confusion between *Rumex rupestris* and *R. crispus* in the Isles of Scilly that had led to the misrecording of *R. crispus* for the rare *R. rupestris* and had masked the serious decline in that species.

Clare Harvey was my predecessor as BSBI vice-county recorder for the Isles of Scilly, she moved to St Mary's in 1969 with her husband Leslie (L.A. Harvey, former Professor of Zoology at Exeter). They made a considerable contribution to local interest in natural history on the islands, especially wild flowers through the group of (mainly) ladies who organised a flower table in the Museum on St Mary's. Each week throughout spring and summer they collected fresh specimens for the display. Clare was very knowledgeable about the islands and especially their wild flowers; she was also an expert on seaweeds. Although she updated the plant checklist (Harvey and Lousley, 1993) that Lousley had produced for the Isles of Scilly Museum, sadly she kept no written records herself. Everything Clare knew was in her head and other than a few published notes such as those in the local magazine *Scilly Day to Day* and a few 'pink cards' of rarities sent to the BSBI, she left very little written material; so when she died in 1996 aged 92 most of what she knew died with her.

From 1983, when I took over as recorder from Clare, I visited her regularly and tried to make notes of what she could tell me. This process, usually over a glass or two of whisky was not ideal, although enjoyable! Then as Clare became older, and especially after Leslie died, she would send me specimens to identify – especially grasses or sedges which she did not like – and brief, tantalising notes about them. The most extraordinary example of Clare's exchanges was the small plastic box originally used to house a toothbrush in which she sent me a plant she did not recognise. When the dried-up plant was extracted and carefully unfolded after being soaked in warm water I was astounded to find it was almost a metre long *Cirsium arvense* var. *integrifolium* – the variety of Creeping Thistle that has no spines! See photo on page 362.

The wild flower display in the Museum continued to be maintained for a number of years after Clare Harvey's death but eventually the few remaining participants decided to retire. They continued to keep up their interest in the flowers, one, Julia Ottery even published a book of her wild flower paintings (Ottery, 1996) and several since have continued to send me many useful records.

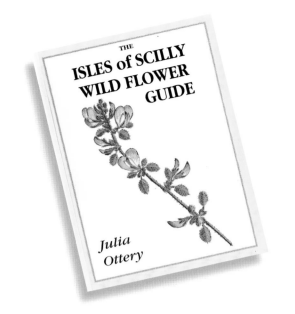

Another of the flower ladies, Lesley Knight, wrote a checklist of Scilly wildflowers in 1996. Both Julia Ottery and Lesley Knight for many years contributed regular notes on wild flowers to *Scilly Now and Then*, the local free magazine.

Another contributer to the interest in wild flowers on Scilly was the photographer Frank Gibson. He was the fourth of several generations of a family of

Frank Gibson's book

Gibson and Hunt's book

photographers mainly famed for their photographs of shipwrecks. Frank, with David Hunt (already well-known as the 'Scilly Birdman') published several booklets of photographs of wild flowers that proved very popular and useful, especially to visitors to Scilly. Will Wagstaff, who later took over David Hunt's wildlife tours in the islands, also includes some of the rarer and unusual plants in his programme of talks and guided walks.

Books with Scilly plant records included the *Review of the Cornish Flora, 1980* (Margetts & David, 1981), *The Cornish Flora Supplement, 1981-1990* (Margetts & Spurgin, 1991) and in the *Flora of Cornwall* (French, Murphy & Atkinson, 1999). Other publications have included the wild flowers of the Isles of Scilly, but none were, or were intended to be as comprehensive as Lousley. Jean Paton wrote the *Wild flowers of Cornwall and the Isles of Scilly* (1968), Patrick Coulcher wrote a natural history of Scilly entitled *The Sun Islands* (1999), Pat Sargeant and Adrian Spalding produced a booklet on *Scilly's Wildlife Heritage* (2000), the broadcaster Andrew Cooper produced *The Secret Nature of the Isles of Scilly* (2006) to accompany his television series and more recently David Chapman published a little paperback book of photographs *Wildflowers of Cornwall and the Isles of Scilly* (2006). I included plants in my natural history (2007) and produced a checklist for the Isles of Scilly Museum (2009a) and a booklet on arable plants for the AONB (2010a). More recently Liz Askins has had a wild flower 'spot' on Radio Scilly.

Parslow's Checklist

Paton's book

Group of visiting botanists on Tresco, May 2011

Recording for The New Flora of the Isles of Scilly

Generally getting around the inhabited islands presents few problems, at least between Easter and October when there are regular 'tripper' boats between the main islands. Although most farmers do not allow access to their land without permission, there is an extensive network of permissive paths. Also, most of the land tenanted by the Isles of Scilly Wildlife Trust (IoSWT), which covers most of the heaths and coast (except on Tresco), is open access. Using binoculars can be surprisingly useful if it is not possible to get into a field!

'Tripper' boats & *Scillonian*, St Mary's Harbour

At times I have been employed to survey much of the arable farmland for DEFRA and NCC, and also all the IoSWT land when writing a Management Plan for them. This enabled me to get to all of the inhabited islands and most of the uninhabited ones. This has been very useful in collecting records and other information used in this Flora.

Arable plants booklet

Getting to the uninhabited islands can be challenging; not only is it physically difficult to land on many of the smaller islands but weather and tides have to be taken into account. At times we have had to land on all the uninhabited islands, starting in 1984 to look for *Rumex rupestris*, the rare Shore Dock; this requires checking the seeds of every dock plant and walking all the shores. In the past when the Nature Conservancy Council (NCC) had their own boat and a boatman (the redoubtable Cyril Nicholas) on Scilly, it was sometimes possible to get put onto an uninhabited island for a few hours. Since English Nature (the successor body to NCC) gave up their boat it has occasionally been possible to get out on the IoSWT boat (while they had one) if it fitted in with their work. Unfortunately without regular use of a boat there has been a considerable decrease in data from the less accessible uninhabited islands in recent years. In 2013 and 2014 some of the uninhabited islands were revisited with Natural England officers.

Checking Shore Dock

A number of other groups or individuals have carried out botanical or ecological work on the islands. Cornwall Ecological Consultants (CeC), the consultancy arm of Cornwall Wildlife Trust has been employed to carry out surveys on a number of occasions and many of their

Alison Paul (from NHM) recording ferns

records have been made available to us. At times there have been visits by members of the Botanical Society of Britain and Ireland (BSBI), Wild Flower Society (WFS), British Pteridological Society (BPS) and Natural History Museum (NHM), these have increased the information we have on the Scilly flora.

In the late 1980s a programme of rat-baiting was started on some of the uninhabited islands to protect the important populations of seabirds from rat predation. The success of the programme led to a more ambitious programme being focussed from 2014 on St Agnes and Gugh. Although rats are known to eat seeds of some plants, any changes to vegetation that occur are being monitored. It will be interesting to see if the rats were having any impact on vegetation as well as on the birds.

Validation and verification of records

A phenomenon that Lousley describes in his Flora is the 'midsummer madness' that seems to overtake some of even the most experienced and reliable botanists when they visit the Isles of Scilly! This leads to the recording of plants unknown on Scilly. As Lousley noted, records

"Midsummer Madness" at Peninnis

of mountain flowers are easily rejected, but it is when common mainland plants that are not found on Scilly are recorded, it is much more problematic. These records can be very difficult to check and have occasionally been included in published accounts or databases. However it is difficult to be prescriptive as sometimes a suspect record has later proved to be genuine. Where possible, confirmation of unusual species is made by voucher specimens or photographs. One of the disadvantages I have had of being a non-resident recorder is being unable to go immediately to check the plants in the field. Digital photography and email have been a great advance in this respect as it can often be possible to confirm the identification remotely. Thanks are due to those kind friends on Scilly who have gone out to try and check a record or have spent their spare time searching for locations of unusual plants. Additionally the use of hand-held GPS units has been another improvement as it allows recorders to give more accurate grid references.

Anne and Mike Gurr at *Spergularia bocconei* site

The species accounts – what to include

Deciding what to include in a new Flora of the Isles of Scilly has not been straightforward. There are so many aliens and garden escapes on the islands that it was tempting not to tackle them at all, however, many are now established members of the flora and cannot be ignored. Every year at least one more introduced plant is found to have made itself at home outside the confines of Tresco Abbey Gardens, or another garden escape becomes established in the wild, or an accidental arrival begins to invade new habitats. Some Scillonian gardeners and farmers are quite cavalier about disposing of garden waste or unwanted bulbs into hedge banks or over the cliffs! Many of these plants survive and can subsequently spread.

One of the main problems with the alien plants is identification. I have received essential help in this from specialists who have a particular interest in them, such as Eric Clement. There is also a group of plant 'twitchers' for whom alien plants are their special quarry and who follow in the footsteps of the late Alan Underhill and his friends. Prof. Clive Stace included several more Scilly aliens in the third edition of his Flora (Stace, 2010). Dr Mark Spencer has also been able to correct my identification of some species and is taking an interest in groups such as *Aeonium*. Despite this, there are still some plants that have not been identified satisfactorily, for example some of the succulents.

Narcissus throw-outs at Halangy

In the species accounts most plants that have been recorded on Scilly get an entry. Some casuals and garden escapes are included even where there is no evidence that they have become established, on the grounds that some at least may do so eventually. Additionally, plants that were included in Lousley's Flora (Lousley, 1971) and are apparently no longer present in the islands get an entry for consistency, with a comment on the reason for their loss if it is known. Some unconfirmed records have been included, as they may be confirmed in future. Just a small number of records have been rejected as errors or misidentifications, although unfortunately some had got into national databases before we discovered our mistakes.

In the case of rare species we have in most cases given a locality or grid reference. It seems important for the future to know where plants grew, keeping localities secret is counterproductive. One of the very few exceptions to this are all the exact locations for *Ophioglossum lusitanicum*, the main sites are well known and act as 'honey-pots', so it is unnecessary for most people to search for other sites. Many years ago someone dug up a large piece of turf from an *Ophioglossum* site, an odd thing to do, it is doubtful the ferns would survive. A gazetteer of all the *Ophioglossum* sites is lodged with Natural England and with IoSWT. The location of the hybrid fern *Asplenium* x *sarniense* was also kept quiet as it was felt vulnerable, but it eventually died. The location of a colony of *Ornithopus pinnatus* that grows on private land (where it was accidentally introduced) has been omitted from the species account and map at the landowner's request.

Most trees, the hedge shrubs and plants that are clearly planted, or are grown as crops, have not been mapped although they will always get an entry in the systematic list. Plants found growing on walls, or on road verges where their provenance is often unclear, have also often been given the benefit of the doubt so are included.

Asplenium x *sarniense*

Unfortunately some of the 'critical groups' such as brambles and dandelions have not been tackled adequately or at all. Fortunately in the case of brambles the work and advice of David Allen has been indispensable in producing that section in the species account. This Flora is essentially a work in progress and hopefully will encourage others to continue to visit Scilly and fill those gaps. It seemed more important to publish the material already collected, rather than hope to cover everything before publishing. Another incentive has been that although I had managed to get to all the uninhabited islands at least once while recorder, this is getting more and more difficult. Sadly my days of leaping from small boats onto seaweed-covered rocks are now long past!

Landing at Great Arthur

The maps and photographs

Dot maps of almost all species are included, with localities included in the text as named sites or by grid references. The dots represent hectare (100 m x 100 m) records and for two year classes – before 2000 and after 1999. Native plants are indicated by green dots and aliens by red. In some cases plants that are natives will be mapped as aliens when they are introductions to or not native to Scilly. Inevitably some plants, especially ones on verges, in fields or near habitations may have gone from a particular locality so the maps give a good idea of a species distribution but are essentially just a snapshot in time.

Many people have kindly provided photographs that have been used in the book. For some species, especially some of the more photogenic, I have been spoilt for choice. It has been important to try to include photographs of some of the alien plants that are not easily identified or are not included in the usual field guides. Not every species in the account has a photograph, but most will have a map.

The Islands

The Isles of Scilly cover a very small land area, just 1,641 ha (4,054 acres); all the islands would fit easily into just over three 10 km x 10 km squares with quite a lot of sea left over. Over a hundred islands, islets and rocks make up the archipelago, but only five islands are now inhabited; St Mary's, and the 'off islands' of Tresco, St Martin's, St Agnes (including Gugh) and Bryher. In the past more islands were inhabited, either with settled communities, such as on Samson and Teän, or by hermits, or those at

RMV *Scillonian* at St Mary's

The islands from space

Old Town

Hugh Town from Porthcressa

the quarantine station on St Helen's. Some of the smaller islands were used for summer grazing or as temporary lodging by kelp burners in about the 1700s. Few of the smaller islands have a reliable fresh water supply so this would have limited their use for long-term habitation. There is an intriguing comment by Leland (in Kay, 1963) whose 'Itinerary' was compiled during 1535-43, 'there be 140 islettes of Scilly, that bere grass exceeding good Pasture for catail'. This obviously is an exaggeration as there are now far less than 140 islands that have had any vegetation present, let alone grass enough to be grazed! Leland also refers to pigs and corn on St Mary's, which is one of the earliest written references to farming on the islands.

THE INHABITED ISLANDS

St Mary's

St Mary's is the largest island in the group, c. 4 km by 3 km in extent and some 650 ha. Most of the human population of c. 2000 (which doubles in the summer) live here, the majority in Hugh Town the 'capital' of the islands. Hugh Town is where the Isles of Scilly Council and the Duchy of Cornwall offices are based, with most of the shops and business premises, hotels, pubs, guest houses and many homes. There is also the harbour from which the main boat traffic comes and goes, including the passenger ferry RMV *Scillonian* and the *Gry Maritha* that carries freight to the islands from Penzance. In 2017 the *Gry Maritha* is due to be replaced by the *Mali Rose*.

Scilly is owned by the Duchy of Cornwall and almost all premises are leasehold except for Hugh Town where they are freehold. The former main township of Old Town (now just a hamlet), lies just over the hill, on the opposite side of the island from Hugh Town. At Old Town is a tiny church and churchyard and the remains of Ennor Castle, recently cleared of trees.

Even in the built-up parts of Hugh Town there are a surprising number of plants to be found. *Asplenium marinum* Sea Spleenwort and other ferns are abundant on walls, and pavement cracks are home to *Polycarpon tetraphyllum* Four-leaved Allseed, *Erodium maritimum* Sea Stork's-bill and other tiny plants. Stands of *Diplotaxis muralis* Annual Wall-rocket, *Briza maxima* Greater Quaking-grass and many garden escapes can be found on street corners in any suitable niche. The steep hill up to the Garrison Arch is a good hunting ground; besides

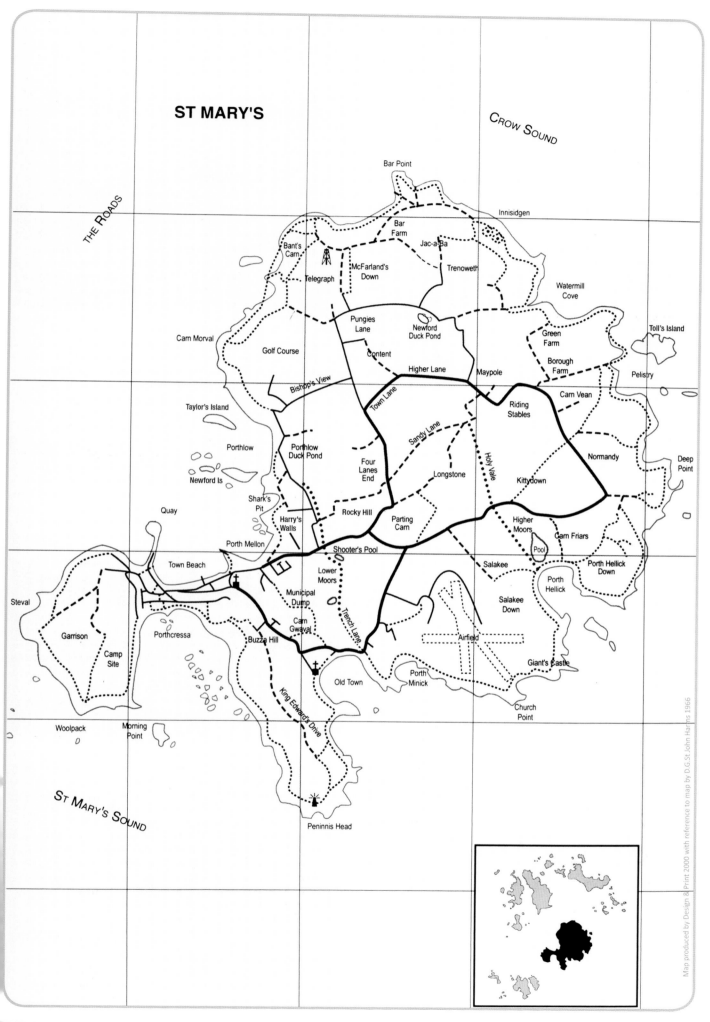

a host of multi-coloured *Pericallis hybrida* Cineraria established there, *Lotus subbiflorus* Hairy Bird's-foot-trefoil sometimes grows beside the steps as does *Erodium moschatum* Musk Stork's-bill. One of the best shows of *Euphorbia portlandica* Portland Spurge grows on top of a wall at Carn Thomas and at the foot of the same wall it is sometimes possible to find *Orobanche minor* Common Broomrape, apparently parasitic on *Gazania sp.* there.

Airport on St Mary's

Also on St Mary's is the airport which operates six days a week bringing passengers from the mainland by fixed-wing aircraft and, until 2012, helicopters before the service was withdrawn. There are approximately seven miles of quite busy roads on the island. These with the coastal path and the permissive footpaths are the best way to get around the island. All land on Scilly is private; there is no public footpath network or common land. Farmers generally do not like trespassers although some set out temporary paths across their land, especially in the autumn, for the benefit of the visiting birdwatchers. Otherwise it is advisable to ask permission before going into fields. As most of the coastal heath is leased to the Isles of Scilly Wildlife Trust it is generally open access. There are two nature trails, one through Higher

Coastline from the air

Moors and Holy Vale, and one through Lower Moors to Porthloo. There are numerous archaeological sites on St Mary's including many entrance graves including those at Bant's Carn, Innisidgen and Porth Hellick Down. The ancient settlement on Halangy Down has an interesting heathland flora as do Harry's Walls and the Garrison. Many of the more important monuments are in the guardianship of English Heritage and the grassland is kept mown, which is good for some plants.

Porth Hellick

Around the coast the landscape is predominantly windswept heath and rocky headlands. There are also two small islands just off Porthloo (or Porthlow); Newford Island which has planted shrubs around a garden, and Taylor's Island which is rocky with coastal grassland. Another larger island, Toll's Island, is in Pelistry Bay on the north-east of St Mary's. It is easily reached at low tide and has an interesting mixture of heathland and coastal grassland. There are a number of kelp pits, fortifications and other evidence of former use.

Newford Island

Taylor's Island

Toll's Island

Ophioglossum azoricum Small Adderstongue has been recorded here in the past. Around the coast are places where large granite 'carns' dominate the landscape and provide a base for small plants and lichens, and less beneficially, footholds for *Pittosporum* and *Coprosma repens* bushes. The heathland on St Mary's is in places quite spectacular 'waved' heath that has been shaped by the wind. In some places *Ulex gallii* can be found. *Scilla verna* Spring Squill is spreading along the coast near Trenear Point where the Wildlife Trust have been managing the heath and near Mount Todden is the best place for *Plantago maritima* Sea Plantain on Scilly. Between the headlands are a number of bays, some sandy, some rocky, with some degraded sand dunes and low cliffs. Peninnis Head is one of the larger headlands, with farmland along the top and heathland to the south. Protecting the harbour is another headland, the Garrison, with Star Castle and defensive granite walls extending almost the whole way round. Enclosed by the walls is the campsite and farmland as well as more heathland and planted trees. All these areas have botanical interest from the plants growing on top of the Garrison walls to the unusual aliens on Buzza Hill on the edge of Peninnis.

The only streams are also found on St Mary's, although just the one from Holy Vale that flows through Higher Moors to the sea at Porth Hellick is worthy of the name, the others through Lower Moors and Watermill are little more that ditches. Water from the Higher Moors stream and the well near Lower Moors is abstracted for the public water supply. It is mixed with water from the desalination plant on the coast near Mount Todden.

Porth Hellick pool

There are two main areas of wetland on the island, Higher Moors with the Holy Vale, and Lower Moors and the low-lying fields between there and Porthloo. Both of the Moors have substantial areas of reed or reed and sea rush swamp and some open water. There is also a small pool and more reeds near Porthloo and twin pools at Newford – Argy Moor. These last two are home to large flocks of tame ducks and have little botanical interest.

Porthloo duckpond

Holy Vale includes the main area of deciduous woodland on St Mary's, with smaller amounts scattered around the island, most of which was originally planted as windbreaks. The windbreaks are usually composed of elms or conifers. Especially in the north of the island there are substantial *Pinus radiata* Monterey Pine plantations that have now reached maturity. Although not native to the islands their unique silhouettes are a distinctive feature in the landscape.

Monterey pines

About half of the land area of St Mary's is farmland with more than 300 ha in bulbs and other arable crops as well as pasture (ERCCIS, 2013). The bulbfields are often small, 0.5 ha or even less, although not usually as small as those on the other inhabited islands where bulbs are grown (except Tresco). Surrounding the fields or 'squares' are high evergreen hedges that protect the crop from the strong, salt-laden winter winds. The bulbfields are of great significance for their unique arable flora. Grass and larger arable fields usually have boundaries of granite walls, locally called 'hedges', sometimes with shrub hedges as well. The evergreen hedges are called

Seaview Farm

'fences' locally; the main species involved include the alien shrubs *Pittosporum crassifolium, Coprosma repens, Olearia traversii* and *Euonymus japonicus*. Most grasslands on the island have had some kind of agricultural improvement or reseeding. Just a few fields exist that may be considered unimproved, they are acidic and typically not very species-rich. Otherwise there are a few grassland remnants of some interest scattered around the island. There are also coastal grasslands, grassland within heaths, and the mown grassland of cricket fields and around Ancient Monuments. Quite often some of the latter may be home to *Spiranthes spiralis* Autumn Lady's-tresses.

Halangy Village

Harry's Walls at St Mary's

THE 'OFF ISLANDS'

The other inhabited islands are generally referred to as the 'off islands'. Most of them have fewer than 100 inhabitants, only the larger island of Tresco having about 150 inhabitants.

Tresco

Tresco is the largest of the 'off islands', some 3 km from south to north and 1.5 km wide, about 297 ha in area. It is also rather different from the other islands as it is run as one estate by the Dorrien Smith family who lease it from the Duchy. As the main business of the Estate is tourism many of the dwellings are now time-share or holiday lets with the management of the built areas reflecting their use as a holiday venue. About a quarter of the island is under farmland, both arable and pasture. The north of the island rises up to an area of rocky ground and exposed heath vegetation; the south is sandy with dune heath, dune grassland and sand dunes. There are two lakes that take up a significant amount of the southern half of the island. The larger one, Great Pool, almost cuts the island in half. The main habitations and quays are situated at New Grimsby on the west coast opposite Bryher, and Old Grimsby on the east side. The famous Abbey Gardens occupy a large area south of the Great Pool surrounded by sheltering plantation woodland. Planted and garden escapes have spread into the woodland, Abbey Hill and along the paths and walls of the Abbey.

New Grimsby at spring tide

The woodland that is such a feature on Tresco is mainly dominated by alien species; some towering *Pinus radiata* and *Eucalyptus* and other trees with an under-planting of Rhododendrons and other exotic shrubs. It is the ground flora that is of most interest botanically. Among all the grasses and the native and introduced ferns that flourish here are a few Ancient Woodland Indicator species that provide a link with the former woodland cover cleared by the early farmers. Species recorded include *Veronica montana* Wood Speedwell, *Carex divulsa* Grey Sedge, *Carex sylvatica* Wood-sedge, *Carex remota* Remote Sedge and *Lysimachia nemorum* Yellow Pimpernel.

Great Pool was formed from former peat digging and is now a freshwater lake. It is linked to the sea by a leat at the west end with dunes at the eastern end, so some salt incursion is possible. The margins of the pool are

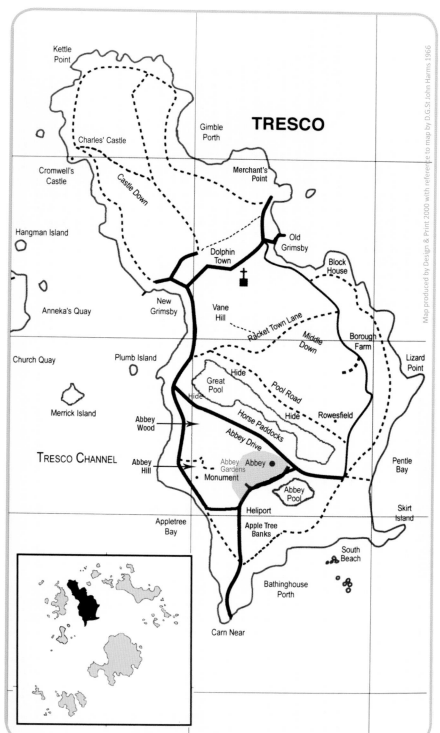

fringed by reeds and willows. The edges of both pools have a band of wetland vegetation which includes *Osmunda regalis* Royal Fern, *Typha latifolia* Bulrush, *Iris pseudacorus* Yellow Iris, *Lycopus europaeus* Gypsywort, *Ranunculus flammula* Lesser Spearwort and *Hydrocotyle vulgaris* Marsh Pennywort. In the reeds almost under the boardwalk to the David Hunt bird hide, is a stand of *Oenanthe fistulosa* Tubular Water-dropwort at one of its few sites on Scilly. Abbey Pool is smaller and almost round in shape. It is surrounded by exotic plantings and heathland. On the muddy edge of Abbey Pool grows an interesting form of *Ranunculus flammula* Lesser Spearwort with very tiny parts – quite different from the plants growing elsewhere on Scilly. In drought conditions the submerged *Elatine hexandra* Six-stamened Waterwort may be revealed – but this happens rarely.

Tresco Abbey and Pool

In the north of the island is Castle Down, a large area of heath (much of it an extreme form of 'waved heath'), many eroded braided pathways and exposed rocks. Many of the largest rocks and boulders that originally covered the heathland were blown up and removed in Augustus Smith's time, probably for building stone.

Walking in Tresco woodland

Great Pool

Castle Down

There are two castles; the ruins of King Charles Castle sit on top of the hill overlooking Tresco Channel, while below on coastal rocks stands Cromwell's Castle. Both castles have a few plants growing in among the masonry, notably *Erodium maritimun* Sea Stork's-bill in quantity.

Charles and Cromwell's Castles, Tresco

The rocky, heathy heights of Castle Down are in direct contrast to the south of the island which is low lying and dominated by dune systems and long stretches of white sandy bays. The dunes are mostly degraded and have an unusual appearance due to all the alien plants that are naturalised there. In late summer some dunes appear dominated by blue and white *Agapanthus praecox* ssp. *orientalis* African Lily, with spiky clumps of *Kniphofia* spp. Red-Hot Pokers, *Watsonia borbonica* Bugle-lily, *Ochagavia carnea* Tresco Rhodostachys, *Fascicularia bicolor* Rhodostachys and many other alien plants.

Pentle Bay

Agapanthus in the dunes

Tresco fields

The centre of Tresco is relatively protected by higher ground and shelterbelts so compared with the other 'off islands' most cultivated fields are large. A proportion is grassland, pasture for the herd of beef cattle kept by the Estate. Former bulbfields to the north of Great Pool have now either been put down to grass, game crops or quinoa.

Bryher

Bryher is the smaller twin of Tresco at 134 ha. The islands are only separated by the narrow Tresco Channel. There are several small hills on the island (Samson Hill, Heathy Hill, Gweal Hill and Watch Hill), a central core of farmland (mostly grassland), a large hotel complex, and a saline lagoon – Bryher Pool – plus small hamlets at Northward and The Town. On the north-west is the famed Hell Bay, well-known for spectacular storm scenes when big seas fetch into the bay. At the northern tip of the island the land rises up to a rocky promontory covered in coastal heath at Shipman Head Down, and beyond, separated from the main island by a deep cleft is Shipman's Head. In the south of the island is an area of grassland, sand dunes, dune grassland and sandy Rushy Bay.

Bryher

Much of the heathland on Bryher is home to the delightful little *Scilla verna* Spring Squill which is more frequent here than elsewhere on Scilly. Tiny plants such as *Arenaria serpyllifolia* Thyme-leaved Sandwort, dwarf forms of *Erodium cicutarium* Common Stork's-bill and other miniaturised plants are found in the sandy grassland at Rushy Bay. Also here are *Ophioglossum azoricum* Small Adder's-tongue growing in almost pure

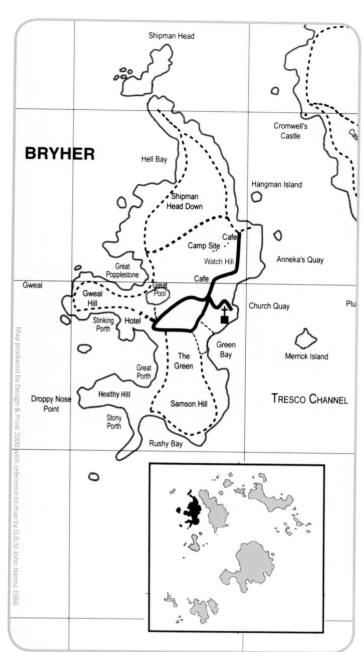

Breaking waves on Bryher

townsendii which seems particularly common on Bryher. Gweal Hill on the west of the island has some fine heathland on its landward slopes and *Armeria maritima* Thrift dominating the seaward side. The spread of *Scilla verna*, *Euphrasia* spp. Eyebright and *Ophioglossum azoricum* here is attributed to the reintroduction of grazing and scrub removal.

Dune grassland

Shipman Head and Downs

Shipman Head and Hell Bay

sand as well as a great rarity, the tiny *Viola kitaibeliana* Dwarf Pansy. On Heathy Hill nearby is another rarity, *Ornithopus pinnatus* Orange Bird's-foot. On some stretches of short turf on the island there can be a range of less-common clovers. Of particular note is the red-flowered form of White Clover *Trifolium repens* var.

Gweal Hill

The arable flora here is somewhat different from that on the other islands. This is believed to be due to the change from bulb growing to soft fruit and vegetables, with a different regime and cultivation times. Another puzzle is the presence here of *Fumaria officinalis* Common Fumitory that is found in some of the arable fields on Bryher, although not a rare plant nationally, the species does not apparently occur on any of the other islands.

Bryher fields

Bryher Pool is a saline lagoon; the only true such habitat in the Isles of Scilly. Only *Potamogeton pectinatus* Fennel Pondweed and *Ruppia maritima* Beaked Tasselweed have been found here. The salinity of the pool varies, but after some high tides the water is virtually as salty as sea water. As *Ruppia* is reputed to be more tolerant of high salinity than *Potamogeton pectinatus* this may be why only *Ruppia* is sometimes present in the Pool. A narrow strip of saltmarsh vegetation which includes *Puccinellia maritima* Common Saltmarsh-grass, *Juncus gerardii* Saltmarsh Rush, *Spergularia marina* Lesser Sea-spurrey and *Glaux maritima* Sea-milkwort grow along the sides of the leat and around the edge of the pool.

St Martin's

St Martin's is a long, narrow island with a west-east axis, unlike all the other 'off islands' which are aligned north-south. The island covers 223 ha. Nearby is **White Island** (15 ha) accessible at low tide over a boulder ridge. White Island is uninhabited and is mostly heathland and short coastal grassland. The most notable feature is a deep cleft called Chad Girt that cuts deeply into the rocky cliffs on the seaward side of the island. The geology of White Island is important due to glacial deposits there.

Causeway to White Island

St Martin's has superb white sand beaches along most of the southern and northern shores. The coasts and extremities of the island are open heath and rocky promontories. At the east end of the island the heathland rises up towards the Daymark, a red and white striped tower built in 1683, and the steep cliffs of St Martin's Head (47 m). From here on a clear day it is possible to see Land's End on the Cornish mainland. Three small hamlets, Lower, Middle and Higher Towns are linked

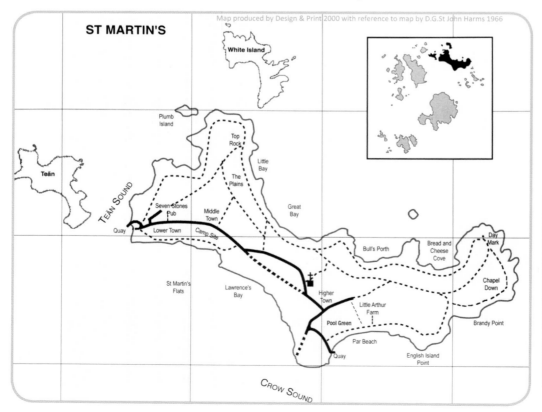

Chapel Downs daymark

Chad Girt on White Island

by the long concrete road that stretches from Lower Town quay in the west to Higher Town in the east and then down the hill to the Higher Town quay. Much of St Martin's is covered in blown sand which in places can be up to two metres deep (Mulville, 2007). The farmland is now mainly concentrated in the centre of the island although there are many abandoned fields invaded by bracken, and walls that are falling down. There are bulbfields, including some with very important weed floras, other arable fields and some under grass. Some of the fields are on almost pure sand, these tend to have a weed flora with heathy or dune elements.

St Martin's field pattern

The very sandy soils are also responsible for some of the botanical 'hotspots' on St Martin's. The cricket field is typical of others on Scilly with a close-mown turf which includes *Chamaemelum nobile* Chamomile and several different clovers with *Trifolium suffocatum* Suffocated Clover growing along the sandy paths nearby. In the corner of the cricket field is a small pool which some years is covered in a dense mat of *Ranunculus baudotii* Brackish Water-crowfoot. Usually *Potamogetum pectinatus* is present and in 2012 *P. crispus* Curled Pondweed, a new record for the islands, was found there. Later that year the pool dried out so it is not yet known whether the *P. crispus* has survived. The pool was invaded by *Azolla filiculoides* Water Fern in 2016.

St Martin's Pool

On the north slope of St Martin's just above Great Bay and the dunes is a flat expanse of short grassland and dune heath over deep sand known as The Plains. Plants recorded here have included *Ophioglossum azoricum*, *Ornithopus pinnatus* and in one place *Pilosella officinarum* Mouse-ear Hawkweed at one of its few Scillonian sites where there may be a patch of more calcareous shell sand.

The Plains

As well as *Carpobrotus edulis* Hottentot-fig, another species of *Carpobrotus*, *C. acinaciformis* Sally-my-handsome can be found on St Martin's. There are large

Great Bay

stands of the plant at Lawrence's Bay and Great Bay. Another alien invader that is becoming a serious problem is *Phormium cookianum* Lesser New Zealand Flax, with many hundreds, perhaps thousands of plants spreading throughout the dunes and heathland along the north side of the island. So far, attempts to control the *Phormium* plants have been unsuccessful; their tough, fibrous leaves seem impervious to chainsaws or herbicides.

St Agnes and Gugh

St Agnes is a small island (106 ha) linked to the even smaller (37 ha) adjoining island of **Gugh** by a narrow bar (a tombolo - a spit that connects an island to the mainland).

St Agnes, meadow and gladiolus

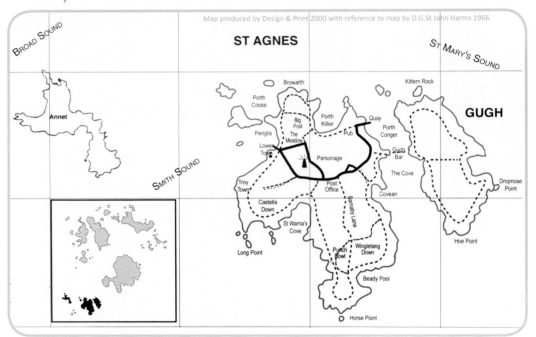

Bulbfield

Gugh Bar

There are three hamlets on St Agnes; Higher, Middle and Lower Towns, and several working farms in the middle of the island. Wingletang Down is an area of heathland in the south and there are smaller heathy areas around the coast. In the north of the island is the dune grassland meadow (also the cricket field), and Big Pool. Despite its small size the island has a varied coast, from exposed rocky bays to more sheltered sandy porths.

Some of the bulbfields on St Agnes are notable for their species-rich arable weed communities. Visitors often used to stop to admire the colourful display of *Glebionis segetum* Corn Marigold, fumitories and *Gladiolus communis* ssp. *byzantinus* Eastern Gladiolus, known on Scilly as 'Whistling Jacks', as they walked up from the quay, probably without realising they are weeds! Other bulbfields on the island also have a rich weed flora, but each farm has a slightly different composition.

St Agnes has its own botanical treasures. Perhaps the most notable is *Ophioglossum lusitanicum* Least Adder's-tongue. This tiny fern is only found on Wingletang and

Wingletang

nowhere else either on Scilly or mainland Britain. *O. azoricum* also occurs on St Agnes and at one time so did *O. vulgatum* Common Adder's-tongue – although the last has not been seen for some years.

Other botanical highlights on St Agnes include the fine colonies of *Asplenium obovatum* ssp. *lanceolatum* Lanceolate Spleenwort on walls especially in Lower Town Lane, and *Asplenium adiantum-nigrum* Black Speenwort is also found on some walls. The meadow (used as the cricket field) is another fine example of mown coastal grassland dominated by *Chamaemelum nobile*, with several different clovers and in some years the tiny orchid, *Spiranthes spiralis* on the dune bank. Paths that cross the field can be very rutted and wet in winter, and tiny *Isolepis* spp. club-rushes and *Eleocharis*

Gugh

St Agnes near Nag's Head

Periglis Bay

St Agnes Leat and Pool

spp. spike-rushes are frequently found there. Big Pool is usually fresh-water but may be slightly brackish at times. It is linked by a leat to the sea and when it overflows onto the meadow in winter, it can be drained into the sea at Periglis bay. The low dune here supports a few typical coastal species such as *Honckenya peploides* Sea Sandwort and *Malva arborea* Tree-mallow as well as a number of garden escapes from tipped garden refuse.

Gugh consists of two hills with a low 'neck' area between them. The hills are covered by heathland and rocky carns with abandoned farmland and two houses in the centre of the island. Where Gugh Bar links the island to St Agnes, there is a low dune and on the opposite side of the island is the small bay, Dropnose Porth. The farm on Gugh has been abandoned, but there are still two houses. There are also an extraordinary number of archaeological sites including several entrance graves, many barrows, a standing stone and field walls among the rocks and heath vegetation. *Ornithopus pinnatus* and *Scrophularia scorodonia* occur on Gugh and it is also one of the few places where *Thymus polytrichus* Wild Thyme can be found in any amount.

THE UNINHABITED ISLANDS

Following the divisions used by Lousley (1971), this account divides the uninhabited islands into several main groups: the Western Rocks, the Norrards (or Northern) Rocks, the Eastern Isles and those near St Helen's. The larger islands such as Annet and Samson are described separately. Over the period covered by this Flora most of the uninhabited islands large enough to support vegetation have been visited at least once. Some smaller islands have no higher plants, just seasonal pools with green algae to lure you into a landing!

Western Rocks and Melledgan

These islands are a series of sharp granite crags emerging from submerged reefs; they are among the most exposed islands in the archipelago. Most of them are bare rocks that are washed over by the sea and cannot support any higher plants. In gales the seas send salt spray right

Western Rocks from St Agnes

across even the larger islands so any plants need to be extremely hardy to survive. On **Rosevear** there are a few tough coastal plants: *Cochlearia officinalis* Common Scurvy-grass, *Atriplex* spp. oraches, *Beta vulgaris* ssp. *maritima* Sea Beet, *Rumex crispus* ssp. *littoreus* the coastal form of Curled Dock, *Spergularia rupicola* Rock Sea-spurrey and most obviously a large stand of *Malva arborea* Tree-mallow. During the building of the Bishop Rock lighthouse (1851-1858) the workmen lived on Rosevear for a few months each summer. It is said they grew a few vegetables there but this seems unlikely due to the lack of shelter.

Beta vulgaris ssp. *maritima*

The neighbouring island of **Gorregan** is the tallest of the group but only *Spergularia rupicola* and *Cochlearia officinalis* have been recorded there, growing in rock crevices.

Bishop Rock lighthouse

Between the main group of the Western Rocks and Annet is another small rocky island, **Melledgan**. This island is a difficult landing, there always seems to be a heavy swell running. It is little more than a heap of rocks but does have a patch of vegetation in the middle where a few plants, *Atriplex prostrata* Spear-leaved Orache, *Cochlearia officinalis* and *Malva arborea* maintain a precarious foothold.

Spergularia rupicola

Malva arborea

Annet

Annet lies about half-way between the Western Rocks and St Agnes. Seen from St Mary's it has a very distinctive outline with a line of pointed rocks – the Haycocks – extending out from Annet Head. Annet is a 'closed' island to protect the important seabird communities that nest there. The island is low-lying but rises to high, rocky carns at the extremities. One of these promontories is Annet Head, the other Carn Irish. In May the island appears to glow pink due to the flowering *Armeria maritima* Thrift that dominates the island. Annet is one of the larger islands at 22.4 ha, but even so salt spray can fetch right across the island in storms. During

Thrift on Annet

winter 2013/14 the waves washed across the island severely damaging the distinctive Thrift tussocks. Unlike Samson and the Eastern Isles, Annet may not have had a permanent resident human population but the remains of a hut, midden and other sites means it was inhabited at some time. From animal remains in the midden it is likely the island was used as grazing for cattle and sheep and as a seal-hunting station.

Annet neck

Tree-mallow stand on Annet

Across the middle of the island is a narrow, sandy neck where plants such as *Tripleurospermum maritimum* Sea Mayweed flourish among the piles of rubbish washed up by the sea. One half of the island is dominated by unusual and distinctive vegetation composed of giant Thrift tussocks. The other end of the island is apparently grassy with *Holcus lanatus* Yorkshire-fog, and unusually tall, grassy-looking *Carex arenaria* Sand Sedge. Around the island are large stretches of boulder beaches that at one time supported small colonies of *Rumex rupestris* Shore Dock. During storms in about 1982 most of the plants were swept away, leaving just one colony in a freshwater seepage higher up the shore. This seepage also contained a small population of *Samolus valerandi* Brookweed and some *Rumex crispus* ssp. *littoreus*. In 2014 the *Rumex rupestris* plants were washed away from here. It will be several years before we know if the loss is permanent.

Samson

Samson is the largest of the uninhabited islands at 36.33 ha. Samson is now a popular place to visit, easily reached by boat from Bryher or by a tripper boat for a few hours. At the lowest spring tides it is even possible to walk across the sand flats to there from Tresco or Bryher.

View from North Hill

Annet

There are no trees or bushes on Annet, but their place is to some extent taken by the metre-high, shrubby plants of *Malva arborea* that grow around the island. These are used by migratory birds for shelter. Their dead stems persist for some time and frequently gulls nest among them.

Until the mid 1800s the island was home to several families though now there are only the ruins of their houses and some field walls. After the island was depopulated on the orders of Augustus Smith in 1853 attempts were made to graze cattle and deer on the island. This was unsuccessful, but the deer-park wall remains today. Other signs of the people who lived on Samson are plants that still grow there. Two *Sambucus nigra* Elder bushes and a *Tamarix gallica* Tamarisk tree still grow beside one of the ruined cottages. Patches of *Primula vulgaris* Primrose that originated from their gardens are common around the buildings. On South Hill a circle of *Iris pseudacorus* Yellow Iris may mark

Sambucus nigra on Samson

where someone once had a small pond and in a few places under the bracken cultivated bluebells can be found.

Samson consists of two low hills with a sandy 'neck' between them. At the northern end of the island there are dunes and the landing beach. There are barrows and archaeological remains on North Hill and elsewhere on the island. As is common on Scilly the tops of the hills are heathy, with dense bracken with brambles lower down. On Samson there is more heathland on North Hill with only a smaller patch on South Hill. Fresh water is scarce on the island, a problem that would have been taxing for the former inhabitants. A small, stone-lined well (sometimes with interesting ferns growing nearby) on the northern slope of South Hill was once the main water supply. The only other fresh water is a pool called Southward Well which lies close to the shore and is muddy and slightly brackish at times. There are one or perhaps two other tiny wells but these have been lost under tall vegetation or filled with sand.

Anacamptis pyramidalis

Samson neck

Samson is home to a few unusual plants; there is large colony of *Scutellaria galericulata* Skullcap around Southward Pool and along the eastern shoreline. For many years this was the only site in the Isles of Scilly for the species until a few plants were found on Tresco by Will Wagstaff in 1999. *Anacamptis pyramidalis* Pyramidal Orchid was first reported from the dunes at the north end of the island in 1997 and has since spread, with more than a hundred flowering spikes by 2011. It is possible the first plant was found as long ago as 1986 but was originally misidentified. Samson is also one of the islands where *Rumex rupestris* was found until recently. Although the last plants were lost in 2014 it is possible some may reappear from buried seed. *Rumex rupestris* did at one time grow on the 'neck' of the island until the area became invaded by tall vegetation. It is believed the sea has inundated the neck at times in the distant past and that when the area recovered in about the 1780s, it formed a 'fine meadow' grassland (Woodley, 1822). In the early 1980s the neck was sandy with scattered dune plants and grasses, but since then it has all grown over and become dense and brambly. The sea came over part of the area in 2014 but only made minor changes.

Samson looking north from South Hill

Samson Well

Just off Samson are two small islands that have some vegetation on them. **White Island** lies to the west of Samson (there is also a White Island off St Martin's) and it has just common resilient coastal species present: *Tripleurospermum maritimum, Atriplex prostrata, Cochlearia officinalis, Armeria maritima* and *Malva arborea*. In summer the island is a breeding place for gulls and other seabirds.

Samson looking north with Puffin Island and the landing beach

The other island, off the north end of Samson is **Puffin Island**, also a nesting place for gulls, but despite the name, no puffins! It is more grassy that White Island and has slightly more species including *Silene uniflora* Sea Campion and *Rumex crispus* Curled Dock. An unusual occurrence was finding *Fumaria capreolata* White Ramping-fumitory there in 1984 growing in a rock crevice, presumably carried to the island by birds.

Puffin Island

White Island off Samson

Norrards or Northern Rocks

This group of islands and exposed rocks lie in the west of the archipelago and are so open to the elements that only a few hardy plants can survive there. Just six species of plants have been recorded on the island of **Illiswilgig**: *Malva arborea, Cochlearia officinalis, Sedum anglicum* English Stonecrop, *Spergularia rupicola, Beta vulgaris* ssp. *maritima* Sea Beet and *Armeria maritima*. The same six species have also been recorded from **Mincarlo.** From **Castle Bryher** there are records of *Spergularia rupicola, Malva arborea, Cochlearia officinalis* and *Beta vulgaris* ssp. *maritima*. Only *Spergularia rupicola* and *Atriplex* spp. had been recorded from **Scilly Rock** by Lousley but I did not find any when I visited in 2002. An *Atriplex* sp., probably *A. prostrata* has been recorded from **Seal Rock.**

Norrards from Bryher

Eastern Isles

The **Eastern Isles** are very different in character from the more exposed islands of the Western Rocks and Norrards. These islands are greener, heavily vegetated and 'softer' in appearance. Great and Little Ganinick, Little and Great Ganilly, Great and Little Innisvouls, Nornour, Menawethan, Ragged Island and Little, Middle and Great Arthur all have good plant lists that reflect their additional habitats and more protected position. The tiny islet of Guther's Island which lies closer to St Martin's is usually included in this group.

Eastern Isles from St Martin's

Unlike the Western and Norrard Rocks it is possible to land on some of the Eastern Isles (although some are closed or seasonally so to protect birds and seals). This can be as part of an organised trip to look at archaeological sites or with an independent boatman. Although there are some paths around the islands, the interiors are dense bracken or very thorny, impenetrable bramble thickets, so most people keep to the paths or the sandy areas. Interestingly there are several woodland species on the Eastern Isles also reflecting a link to former times when the islands were forested.

Great Ganilly from the Arthurs

Most of the Eastern Isles are breeding sites for seabirds whose guano enriches areas around their nests and roosts, resulting in many plants there being much fleshier and succulent than usual. The gulls are also considered responsible for the presence of plants such as *Aeonium* spp., *Drosanthemum floribundum* Pale Dewplant and other 'exotic' species carried there as nest material.

Great Ganinick and Little Ganinick are two small islands to the west of the group. Despite their small size (Great Ganinick is 1.8 ha and Little Ganinick 1.2 ha) both have a good plant list, especially Great Ganinick where the woodland species *Euphorbia amygdaloides* Wood Spurge, *Ruscus aculeatus* Butcher's-broom and *Calamagrostis epigejos* Wood Small-reed are found. A small, stunted oak tree (*Quercus* sp.) which was described by Lousley was still there in 1997 (not seen recently) as was *Fumaria capreolata* White Ramping-fumitory) also found there by Lousley.

Little Ganilly and Little Ganinick

Little Ganilly and Great Ganilly despite their names are not very close to each other. Little Ganilly is 2.7 ha and lies just north of Little Arthur while Great Ganilly is linked to Nornour. According to Thomas (1985) some of the names of the islands have been changed, so what is now Nornour was originally Little Ganilly, and Little Ganilly was one of the Arthurs! Little Ganilly follows the same pattern as most of the other islands; a patch of heath on the top of the hill, bracken and bramble slopes and coastal grassland around the shore. *Hyacinthoides non-scripta* Bluebell is also found on the island.

Great Ganilly with Menawethan in the distance

Great Ganilly at 13.8 ha is the largest island in the group and has a number of unusual species. *Ornithopus pinnatus*, *Thymus polytrichus* Thyme and *Betonica officinalis* Betony have all been recorded growing on the heathland on the top of the southern hill. As with some other islands there is a sandy 'neck' at the lowest part of the island where *Scrophularia scorodonia* Balm-leaved Figwort, *Crambe maritima* Sea Kale and both *Euphorbia portlandica* Portland Spurge and *E. paralias* Sea Spurge can be found.

Great Ganilly grassy "neck" area

North of Great Ganilly is the tiny island of **Nornour** (1.6 ha). It is joined to Great Ganilly by a boulder causeway that is exposed at low water. Nornour is one of the islands regularly visited by people interested in the archaeology. In 1962 storms swept away part of the dunes exposing the remains of a number of small hut circles that had been inhabited during the Bronze and

Iron Ages when the island would have been much larger. When the site was excavated many Roman objects were also found including hundreds of brooches. There was believed to be a small harbour on Nornour which could have been used at the time. Despite attempts to protect the site from the sea it is still being eroded.

Nornour hut circle

There is a small area of heathland on the hill above the archaeological site, and in the disturbed area around the hut circles *Euphorbia portlandica* and *Scrophularia scorodonia* occur. Other species that have been recorded on Nornour include *Ruscus aculeatus* Butcher's Broom, *Euphorbia paralias* Sea Spurge, *Hyacinthoides non-scripta* and *Asplenium marinum* Sea Spleenwort. Although a *Rumex rupestris* site had been known on the island before 1970, the plants may have been washed away by the sea or lost when the area was excavated.

Eastern Isles and St Martin's from the air

South-east of Great Ganilly are the islands of **Great Innisvouls** (1.82 ha) **and Little Innisvouls** (0.98 ha), noted mainly as breeding sites for grey seals and seabirds. Great Innisvouls is basically a small hill which rises to 26 metres and supports a typical vegetation of grasses, bracken and brambles on the slopes, a fringe of *Armeria maritima, Tripleurospermum maritimum* and some *Silene uniflora* around the edge with *Atriplex prostrata, Beta vulgaris* ssp. *maritima* and *Cochlearia officinalis* near the shore. *Hyacinthoides non-scripta, Senecio jacobaea*

Ragwort, *Solanum dulcamara* Bittersweet, *Viola riviniana* Common Dog-violet and other typical herbs have been found on the island. Little Innisvouls has some of the same plants with grassy areas and strandline species. Both islands have *Stellaria media* Common Chickweed present which is associated with nesting gulls.

Menawethan (2.81 ha) is steep-sided and rocky, with extensive areas of vegetation on the faces of the rocky slopes. *Atriplex* spp., *Cochlearia officinalis, Holcus lanatus, Armeria maritima* and *Silene uniflora* are frequent, as are stands of *Malva arborea*. Where a tiny seepage runs from under boulders *Samolus valerandi* has been recorded. A huge expanse of *Carpobrotus edulis* covers most of the shoulder and southern side of the island.

Menawethan with *Carpobrotus*

Between Great Ganilly and the Arthurs is the tiny rocky **Ragged Island**. Most of the usual coastal plants are found there but the occurrence of both *Borago officinalis* Borage and the alien fern *Blechnum cordatum* Chilean Hard Fern there in 1987 is curious. Although it is thought the fern could have arisen from wind-blown spores, it is assumed that the Borage was brought to the island by gulls.

Great Ganilly and Ragged Island

Little Arthur, Middle Arthur and **Great Arthur** (7.8 ha) are joined by boulder beaches to form almost one large, curved island in the middle of the group and to the south of Little Ganilly. At high water the three individual

Great Arthur and Great Ganinick

Little Arthur

islands are revealed - Little Arthur in the north, the narrow strip of land forming Middle Arthur and the larger island of Great Arthur. There are entrance graves and other archaeological features on the islands. All the islands are well-vegetated with a range of habitats: dune, coastal grassland, strandline, bracken and heathland. Little Arthur is one of the few localities on Scilly where *Hypericum pulchrum* Slender St John's-wort is found and *Hyacinthoides non-scripta* has also been recorded there.

Guther's Island

Lying closer to St Martin's than the Eastern Isles is the small **Guther's Island,** just eroded rocks with a small castled shape at one end and an odd, bird-like 'sculpture' at the opposite end. Among the jumble of rocks are a few hardy plants including *Festuca rubra* Red Fescue, *Beta vulgaris* ssp. *maritima*, *Rumex crispus* ssp. *littoreus* and *Malva arborea*.

St Helen's group

There are several islands on the north side of Scilly between Tresco and St Martin's. These include Northwethel, Foreman's Island and tiny Hedge Rock, as as well as St Helen's, Teän and Round Island. Of these some have been inhabited in the past and have more varied habitats and a wider range of plants present. Some of the plants may have been introduced by people who lived and worked on the islands or as a result of their farming activities.

St Helen's (19 ha) is the largest in the group. It has had a long history of human and animal occupation. The buildings and other activities all took place on the flatter ground on the south of the island near the landing. Here the land is more sheltered and the soils deeper and richer – probably partly as a result of previous cultivation. Every year there is a pilgrimage to St Helen's on the 8th August, the saint's day of St Elidius, after whom the island was originally named. In preparation for the event the dense vegetation that grows there is mown to enable people to access the little, ruined church. Also near the landing are the ruins of the former Pest House built in 1764 as a quarantine building to accommodate sick sailors. They would be put ashore here to recuperate (or die), so they did not carry sickness into the islands. Nearby are the Hermitage and the tiny church. At one time there would have been gardens and cultivated areas, but these have all now vanished, although some field walls can still be seen. One important advantage St Helen's had over many of the other islands is that it had a well. In the late 1600s, prior to the building of the Pest House, there had been a family living on the island (Magalotti, 1821). At other times goats, deer and sheep (North, 1850; L'Estrange, 1865) have all been kept on the island.

St Helen's

Carpobrotus on St Helen's

St Helen's and Northwethel

Away from the flat land on the south-east of St Helens the rest of the island is a large hill 42 metres high. The hill has had an unusual history; during the Second World War incendiary bombs set the top of the island on fire, burning right down to the rock. Further fires in 1949 burned for a long time and would have compounded the damage (Lousley, 1971). Since then the vegetation on the top and slopes of the hill has been sparse, eventually becoming colonised by *Carpobrotus edulis*. In 2002 the *Carpobrotus* had been severely knocked back, apparently by cold weather or salt spray but has since recovered and spread back over a large area (D. Mawer pers. comm.).

Northwethel lies just off Tresco, it covers 4.6 ha and is relatively flat. At times the island was used as pasture and may have had a dwelling there (Spencer, 1772). Now much of the island is a dense mat of brambles and bracken, with a sandy area and a seasonal pool near the beach. On one visit the pool had a fringe of *Chenopodium*

Some half dozen or so common plants have been recorded from the even smaller **Hedge Rock** nearby.

Men-a-vaur consists of three tall rock stacks in the north-west of the group. Lousley recorded *Cochlearia officinalis*, *Malva arborea* and *Atriplex* sp., possibly from a boat.

Since the lighthouse went automatic in 1987 **Round Island** no longer has the resident lighthouse keepers and their little garden on top of the rock has been abandoned. All the soil for the garden had been brought to the island

St Helen's and Men-a-vaur

by boat and winched up to the top. In 1987 the vegetable patch was immaculate with not a weed to be seen! Permission to land on Round Island had to be obtained from Trinity House. It was visited by Lousley in 1957 but he only mentioned *Carpobrotus edulis*. There are now

St Helen's

rubrum and *Glaux maritima* with *Potamogeton pectinatus* in the water. Lousley recorded 54 plant species on the island, the latest total is more than 70.

Surprisingly, despite its small size the nearby **Foreman's Island** has a plant list of some two dozen plants including *Asplenium marinum*, *Sonchus* spp. Sowthistle, *Heracleum sphondylium* Hogweed and *Malva arborea*.

St Helen's and Round Island

records of twenty-two species of plants on the island including *Malva arborea* and several common strandline and coastal species. Most of the top of the island is covered in a dense mat of *Carpobrotus* and a smaller succulent *Disphyma crassifolium* Purple Dewplant, the latter forming a slippery surface of 'jellybean-like' leaves. *Drosanthemum floribundum*, *Gazania rigens* Treasureflower, *Aptenia cordifolia* Heart-leaf Ice-plant and *Pittosporum crassifolium* were presumably introduced to the island by the lighthouse-keepers. Some were still seen there by Vicki Heaney in 2015. *Asplenium marinum* was growing in deep crevices on the cliffs.

Although no longer inhabited **Teän** has had people living there intermittently over the centuries. There is a freshwater supply on the island which would have made living there a practical possibility. The presence of

Teän old pump

Round Island

people, their farming practices and their animals would all have had some influence on the vegetation of the island. Some arable farming and keeping sheep and cattle occurred – cattle were recorded there in 1945. Remains of several fields, now reverted to bracken, a cottage and a well are still evident.

Teän and Round Island

Despite its small size, just 16 ha, Teän has a long, sinuous and varied coast with dunes, rocky beaches and sandy bays. Great Hill which dominates the island is 40 metres high and, typically for Scilly, has an area of heathland on its summit. Much of the island, including the former fields, are now covered in bracken. Teän is notable for several very rare plants; *Viola kitaibeliana* Dwarf Pansy, *Ornithopus pinnatus* and a small colony of *Rumex rupestris* (the *Viola* and the *Rumex* could not be found after the storms in 2014). The island is also one of the few places where the declining *Salsola kali* Prickly Saltwort has been recorded recently (2009). Other interesting plants there include *Allium ampeloprasum* var. *babingtonii* Babington's Leek, *Scrophularia scorodonia* Balm-leaved Figwort, *Lotus subbiflorus* Hairy Bird's-foot-trefoil, *Trifolium occidentale* Western Clover and formerly *Polycarpon tetraphyllum* Four-leaved Allseed. Lousley recorded *Trisetum flavescens* Yellow Oat-grass on the island in 1952 but despite searches it has not been found since. It was assumed to have been introduced to St Helen's perhaps with grass seed, or possibly grown to use for weaving straw hats (a minor industry on the islands from 1840-1880).

Great Hill

Teän

The origins of the alien flora

Plant Introductions, Tresco Abbey gardens and the Dorrien Smiths

Prior to the 1800s the landscape of the Isles of Scilly was open and virtually treeless. According to the account by Robert Heath in 1750, there were 'Brambles, Furzes, Broom and Holly' on the islands although 'these never grew above four feet high' (Heath, 1750, republished 1967). George Woodley, who was a SPCK (Society for the Promotion of Christian Knowledge) missionary on St Martin's in 1820, wrote an account of his observations on the local situation at the time (Woodley, 1822). Among his comments on ornamental gardening, cultivation and tree-planting he refers to the flowering of an American Aloe (Agave) in a garden on the Garrison as an example of the mildness of the climate. This was before the arrival of Augustus Smith on Scilly.

Augustus Smith took up residence on Tresco in 1838, four years after becoming Lord Proprietor of the Isles. When he started his garden he had to cope with an exposed and windswept landscape. At first it was necessary to carry out major earth-moving and rock-clearing to set out his garden around the house he built near the ruins of the former Priory. A sheltering wall was built, and many trees including elm, sycamore, oak and poplar planted on the hill behind the house to provide shelter and cover for game. To protect the seedlings he used gorse as a nurse, scattering seed brought from the mainland as he went (King, 1985). This circumstance has been something of a puzzle as gorse is present on the islands in great quantity and appears to have been there before Augustus Smith arrived, why would he need to import seed?

Tresco Priory ruins

Augustus Smith initially set some store by a plant he calls 'Tapac' which had been introduced to the Orkneys from the Falkland Islands and he thought might provide both shelter and fodder on Scilly. This plant was probably *Poa flabellata* Tussac-grass which often reaches more that two metres tall in wet coastal conditions in the Falklands. However after a number of attempts, in 1849 Augustus Smith wrote to Sir William Hooker, who had recommended the plant, to say he had been unable to get it to germinate despite having obtained seed from several sources including from the Governor of the Falklands. A friend of Augustus Smith on Orkney had been more successful, growing luxuriant plants 8 feet high but if Augustus Smith did obtain plants from him they do not appear to have been the success he had hoped for.

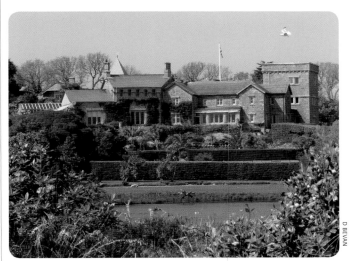
Tresco Abbey

Although the Tussac-grass may have failed, many of the other plants introduced by Augustus Smith thrived. These include many of the trees and shrubs which were planted as shelterbelts around his garden. Those species that survived the fierce storms and salt-laden winds were then adopted for plantings on other islands. It was found that the most successful plants were often those from South Africa, Australia, New Zealand and other places with a Mediterranean type of climate. The mild Scillonian winters, high humidity and light levels enabled plants to thrive that would usually be too tender to grow on the mainland (Nelhams, 2000).

Tresco Abbey gardens

Augustus Smith received cuttings and plants from Kew Gardens and from contacts all over the world including friends on the mainland. Later some of the island sons whom he had forced into education had done well as seamen and brought back plants from their travels.

Tresco Abbey gardens

Within a few years the gardens had expanded as well as Augustus Smith's reputation as a great plantsman.

As Augustus Smith had no legitimate offspring when he died in 1872 he left his estate to his nephew Thomas Algernon Smith Dorrien Smith (later changed to Thomas Algernon Dorrien Smith). Although initially the lease for Scilly had been willed back to the Duchy, they refused to take it back so Thomas inherited the lease as well. It was Thomas who saw the possibilities when a few boxes of flowers sent to Covent Garden got good prices. He then devoted much time and enthusiasm to researching and promoting what was the start of the Scillonian flower industry. Meanwhile the Tresco Gardens continued to flourish as many of Augustus Smith's plantings reached maturity. Some idea of how successful this had been can be seen in the beautiful and very accurate paintings made by Mrs Frances Le Marchant from 1873-1883 (King, 1985). Since then subsequent members of the Dorrien Smith family have continued to manage and improve the gardens and they are now one of the finest 'sub-tropical' gardens in the British Isles. Besides some expansion of the gardens over the years, there has also been planting into the surrounding area of woodland and the dunes at Appletree Banks. Tresco Estate continue their 'exotic' planting around the island which has resulted in many alien plants becoming found away from the Gardens and generally making it difficult to know whether they are established or not. The Hell Bay Hotel on Bryher and other houses and buildings that are part of Tresco Estate have also been subject to similar 'exotic' planting.

The shelterbelts of *Pinus radiata* Monterey Pine, Tamarisk hedges and evergreen shrub 'fences' around the bulbfields are now accepted as typical elements of the Scilly landscape despite their relatively recent introduction. Although some native species such as elm, elder, willows and poplars have been planted on Scilly, none have made as good windbreaks as the fast-growing alien evergreens such as *Pittosporum crassifolium*. Elms grew well in more protected areas such as in the middle of St Mary's, but failed to thrive in exposed places, rarely growing much above the height of the stone hedges.

Monterey Pines at Tresco Abbey gardens

Although Tresco Gardens is still the major source of the alien plants that become established on Scilly such as *Senecio glastifolius* which is now spreading on Tresco and St Mary's they are not the only source. Many Scillonians are keen gardeners and can now order plants over the internet so it is not surprising that many new species of tender plants are continually being introduced to their gardens. In some places you could be excused for thinking all the 'weeds' are garden escapes rather than native species! Therefore, it is expected that the alien flora of the Isles of Scilly will continue to expand and continue to challenge botanists in the future.

Senecio glastifolius at Abbey Gardens

The history of flower farming in the Isles of Scilly

Perhaps the most entertaining account of the start of the bulb industry is in Vyvyan (1953) who tells how in 1865 or 1867 the farmer William Trevillick sent some of the narcissus flowers that grew wild on Scilly to Covent Garden in his wife's hatbox. That first consignment received seven shillings and sixpence and subsequent ones 15 and 30 shillings. Trevillick is then said to have gone around the Garrison with a trowel digging up the narcissus bulbs that grew there and transferring them to his garden. Possibly other farmers soon did the same, growing on and bulking up the bulbs. Narcissus had been in the islands for years, possibly centuries. Some were thought to have been introduced to Scilly by the monks at the Priory on Tresco since at least 1193 (Borlase, 1756), others had arrived with seamen to St Mary's and been given to islanders. There is a story that the wife of the Governor of the Garrison was given some bulbs by a sea captain which she cooked thinking they were onions – when she found them inedible she threw the remaining bulbs into the Garrison moat where they appear to have flourished. Some of the narcissus growing in the islands included the one now known as Soleil d'Or which originated in Morocco and Algiers, and the Scilly White that came from Vigo, Spain.

'Sols' at St Agnes

Thomas Algernon Dorrien Smith may have been involved from the beginning of the flower industry; some accounts say it was actually his idea to send the first flowers to Covent Garden. In his account written to the Royal Horticultural Society (Dorrien Smith, 1890) he gives the credit to Augustus Smith whom he says advised his tenants to send the first bunches to Covent Garden – starting with some he sent in 1863 that received £1 (this is another of the several versions of the story). He was soon growing bulbs too, building on the varieties introduced by Augustus Smith to the Abbey gardens as well as those already growing around the ruins of the Abbey. He had also bought some for £10 off Trevillick to

Narcissus crop

grow in his garden. Writing in 1890 he states that twenty-five years earlier there were 'eight varieties of Narcissus growing in the Isles of Scilly, besides those in the Abbey gardens, some almost wild, some in hedges, and some in the gardens attached to the little farms'. The eight varieties below are from Thomas A. Dorrien Smith's list (Dorrien Smith, 1890).

Narcissus pseudonarcissus 'Telemonius plenus' Campernelli

Tazetta ochroleucus Scilly White

Tazetta aureus Grand Soliel d'Or

Tazetta 'Grand Monarque' (two varieties)

Tazetta x *medioluteus* Primrose Peerless

Narcissus poeticus fl. pl., a double form of Pheasant's-eye

Narcissus poeticus Pheasant's-eye

Narcissus crop

Bulking up the bulbs was slow, so Dorrien Smith went to Holland and the Channel Islands to see how things were done there. He bought 190 varieties of narcissus and brought them back to Tresco. Then from 1881-6 he imported large numbers of bulbs and had ten acres of land laid out to conduct trials.

Soon a number of farmers were successfully growing narcissus as well as their usual potatoes and other crops. It is hard to imagine what a difficult trade this was. Initially the fields were sheltered by stone 'hedges' and reed screens but as the flowers were picked when open the salt winds would often damage the crop. For many years the field-grown flowers became the major industry on Scilly. Protecting the crop from the fierce salt-laden winter winds led to the use of at first Tamarisk hedges to replace the walls and screens; later introduced evergreen shrub hedges were found much more effective. As a result the micro-climate within the small fields where the bulbs are grown can be a degree higher than outside. At first, flowers were packed in heavy wooden boxes and those from the off-islands had to be taken by boat to St Mary's to catch the ferry to the mainland. It was many years before the wooden crates were replaced by specially designed cardboard boxes and the flowers were picked 'in the green'. Now they use plastic crates. Over time many varieties of narcissus were introduced to see how they would do commercially, some were successful, although those that seem to have been most popular and have thrived were often some of the original Scilly varieties such as Soleil d'Or and Scilly White, as well as Pheasant's Eye. It is recorded that Soleil d'Or was by far and away the most popular from 1931-1941 (Pett, 2009), and it is still popular today. When the flower industry was at its zenith the great advantages that Scilly had over flower production elsewhere was that the climate enabled farmers to produce the crop a month earlier than on the mainland.

Bulbfield on St Agnes with tamarisk hedges

The fortunes of the bulb industry fluctuated over time, the First World War depressed markets as many men were away from the islands and buying flowers was not a priority at the time. Flower growing was not considered essential either during World War II as land was given over to the growing of vegetables. Once restrictions on flower growing were lifted in 1946 (Pett, 2009) the islanders could return to growing potatoes and flowers again. Gradually the trade picked up and the ability to send narcissus in cardboard boxes was an improvement though there were still many problems for the growers. By the 1960s some farmers were opening up new fields in more marginal land to increase their crop. Despite the success of the flower industry at this time there were already problems with pests and diseases from using the same fields all the time as well as the difficulties they already had due to weather and transport.

Hedge species

Tamarix gallica Tamarisk was the first hedge species grown to protect the flower fields from the elements but soon it was superseded by more substantial shrub species, mostly from New Zealand, which were salt and wind resistant with glossy leaves. These proved much more successful so they are now the dominant species in the shelterbelts. The following are the most popular hedging used;

> *Pittosporum crassifolium*
>
> *Coprosma repens*
>
> *Euonymus japonicus*
>
> *Hebe* x *franciscana*
>
> *Ligustrum ovalifolium*
>
> *Olearia traversii*

Two of the above species, *Pittosporum crassifolium* and *Coprosma repens* have not only remained as hedging but have spread into the landscape to such an extent that in some places they have become a threat to the native flora and the lichens that grow on the carns (by overshading them).

Bulbfield with evergreen hedges

Changes in flower farming

When we describe flower farming on Scilly we usually mean growing narcissus for cut flowers and bulbs, though over the years many other kinds have been tried out. Some such as *Zantedeschia aethiopica* Altar Lily were grown from at least 1885 and were very popular for a time as can be seen in the Gibson photographs

Zantedeschia aethiopica

Agapanthus field

Nerine field

Freesia, *Convallaria majalis* Lily-of-the-valley, *Viola* spp. violets, *Matthiola incana* 'Brompton Stocks', *Calendula* spp. marigolds, *Myosotis* spp. forget-me-nots, *Erysimum* spp. wallflowers, *Argyranthemum frutescens* Paris Daisy or Marguerite, *Primula* spp. Polyanthus, *Lilium candidum* Madonna Lily and many others.

Recently farmers have turned to farming other flower crops that could grow successfully in Scilly and were more commercial. One of these has been pinks (*Dianthus*) which are grown in the summer. Another plant is *Scilla peruviana* Portuguese Squill, a large, handsome plant from the Western Mediterranean (not Peru) that seems to be catching on. Other flower crops, such as some of the scented-leaved pelargoniums, are being grown for their essential oils to use to perfume products such as soaps and even chocolate!

Pinks

Portugese Squill crop

reproduced in Pett (2009). Although still occasionally sold as cut flowers or used locally in churches the plants are not now grown commercially but can still be found growing in damp corners on St Mary's and Tresco. Some other species such as *Agapanthus praecox* ssp. *orientalis*, *Amaryllis belladonna* Jersey Lily (known on Scilly as Naked Ladies), *Nerine sarniensis* Guernsey Lily, *Sparaxis grandiflora* Plain Harlequin flower, *Ixia* spp. corn-lilies and *Iris* spp. irises are still produced for market, although mostly in small amounts. Other flowers that have been grown at times include *Freesia* x *hybrida*

Some of these other plants that have been tried and later rejected are occasionally now found as weeds in the bulbfields – a notable example being *Gladiolus communis* ssp. *byzantinus* Whistling Jacks. Recently the 'Jacks' have had a slight resurgence with some corms being sold again to visitors who have seen them flowering. Another crop introduction was *Claytonia perfoliata* Springbeauty that may have been grown as a salad and is now a persistent and common weed.

The decline of the bulb and cut-flower industry

The practicalities and cost of transporting cut flowers from the off-islands to St Mary's and then to the mainland has always been a problem. Until about 1950 flowers were packed in heavy wooden boxes. After that cardboard boxes were much easier to handle and they were used until very recently. Light plastic crates have now been introduced. Despite such improvements there is now a serious downturn in the industry. Competition from mainland and foreign markets had been an increasing problem for many years so gradually many of the smaller farmers gave up bulbs and looked for alternative crops.

Modern plastic packing crates

By 2014 there were no longer any full-time flower farmers on Bryher or Tresco and only the largest farms were still viable on St Agnes, St Martin's and St Mary's. Although some farmers kept a few fields in flowers for interest, their income was derived from other sources, mainly holiday lets. So it is difficult to see how the weed communities we have admired are going to survive in any meaningful way. Schemes that pay farmers to keep a few pieces of land in a suitable cultivation regime may keep some of the more interesting species 'ticking-over', as with *Ranunculus marginatus* St Martin's Buttercup on St Martin's, otherwise it is hoped that putting fields down to grass or autumn-sown crops may preserve most of the plants in the seed bank.

Typical bulbfield in summer

Bulbfield plants

A botanist's interest in the bulbfields is often more with the weed species than with the bulbs that are grown, attractive as they are. These weed species fall into three main categories: Native, such as *Spergula arvensis* and *Urtica urens* that were probably present before the first farmers arrived on Scilly; Archaeophytes (plants

Field dominated by *Spergula arvensis*

introduced before AD 1500) which form the bulk of the common weed species; and Neophytes (plants introduced after AD 1500) the most recent arrivals. Neophytes may have arrived with bulbs or other plants very recently and have found the bulbfield regime suited them. Examples are the pernicious *Oxalis pes-caprae* Bermuda-buttercup and *Allium triquetum* Three-cornered Garlic. The former may have been introduced from South Africa either by accident or as a potential garden plant as it is very pretty, and *Allium triquetum* may have been an accidental introduction with other bulbs from the Mediterranean. These two are impossible to control, and are the dominant weeds in the bulbfields and common along roadsides. Some suites of plants associated with disturbed ground have also colonised cultivated ground and are now often more common there than in their original habitats. A good example is *Polycarpon tetraphyllum* which is now much more frequent in the bulbfields and urban areas than in the dunes from where it may have originated. Others have colonised from adjacent habitats, such as the rare *Ornithopus pinnatus*, typically a heathland species, also found in sandy fields on St Martin's and Gugh.

Those weed species which have adopted the bulbfields are often winter annuals whose season fits in well with the narcissus crop or they are species that can germinate at any time of year. The latter can also be found among arable or other flower crops and are not confined to bulbfields. The more typical bulbfield weeds grow among the crop or along the field margins and are ploughed or sprayed out at the end of the season when they will have shed their seeds. It is important for their survival that they are able to germinate in late summer/autumn, complete their life cycle and be over by spring/early summer when the narcissus season finishes.

Rare or threatened arable plants

The Scilly bulbfields are a haven for a number of rare and threatened species that are no longer common on the mainland, including those assessed as Extinct (EX), Critically Endangered (CR), Endangered (EN), Vulnerable (VU) and Near Threatened (NT) in the red lists of vascular plants for England (Stroh, et al., 2014) and for Great Britain (Cheffings & Farrell (eds), 2005), including *Briza minor* Lesser Quaking-grass, *Glebionis segetum* and *Spergula arvensis*. Plants which are generally more common on Scilly than elsewhere in the British Isles include *Polycarpon tetraphyllum*, *Ranunculus marginatus* and *R. muricatus* Rough-fruited Buttercup. The following table shows the range of common, rare and threatened species found in bulbfields with a note on frequency in bulbfields. Species that are Nationally Rare are found in fifteen or fewer 10 km x 10 km squares in Great Britain. Species that are Nationally Scarce are found in sixteen to a hundred 10 km x 10 km squares in Great Britain.

Fumaria occidentalis

Silene gallica

Species name	Red List for England	Red List for Great Britain	Nationally Rare or Scarce	Bulbfield status D = Dominant A = Abundant F = Frequent O = Occasional R = Rare L = Locally EX = Extinct
Aira caryophyllea				F (ssp. *multiculmis* is mainly a bulbfield weed)
Allium ampeloprasum var. *babingtonii*			Scarce	F
Allium roseum				LF
Allium triquetrum				A-LD
Anagallis arvensis ssp. *arvensis* var. *caerulea*				O
Anchusa arvensis				O
Anisantha diandra				F
Anisantha madritensis				O
Anisantha rigida				O
Anthriscus caucalis				R
Aphanes arvensis				F
Aphanes australis				F
Arabidopsis thaliana				LF
Arum italicum ssp. *neglectum*		NT		F
Briza maxima				O
Briza minor			Scarce	A
Calendula arvensis				R
Ceratochloa cathartica				O
Chenopodium album				F
Chenopodium murale	EN	VU		O-LF
Claytonia perfoliata				LF
Convolvulus arvensis				F
Conyza bonariensis				O
Conyza sumatrensis				O
Cotula australis				O
Crassula decumbens				R
Echinochloa crus-galli				R
Erodium moschatum				F-A
Euphorbia helioscopa				O
Filago vulgaris	NT	NT		F
Fumaria bastardii				F
Fumaria capreolata				F
Fumaria muralis ssp. *boroei*				F

Species name	Red List for England	Red List for Great Britain	Nationally Rare or Scarce	Bulbfield status D = Dominant A = Abundant F = Frequent O = Occasional R = Rare L = Locally EX = Extinct
Fumaria occidentalis			Scarce	LF
Fumaria officinalis				LO
Fumaria purpurea	VU		Scarce	EX
Galinsoga quadriradiata				R
Geranium dissectum				F
Geranium molle				F
Gladiolus communis ssp. *byzantinus*				LA
Glebionis segetum	VU	VU		F-LD
Gnaphalium luteoalbum				R
Gnaphalium uliginosum				F
Helminthotheca echioides				O
Hypericum humifusum				LF
Juncus bufonius				A
Kickxia elatine				O
Kickxia spuria				R
Lamium amplexicaule				R
Lamium hybridum				O
Lepidium didymum				F
Lepidium coronopus				F
Lotus subbiflorus			Scarce	LO
Malva neglecta				O
Malva pseudolavatera			Rare	F
Malva pusilla				R
Malva sylvestris				F
Matricaria chamomilla				LO
Matricaria discoidea				LF
Medicago polymorpha			Scarce	O
Mercurialis annua				F
Misopates orontium	VU	VU		O
Myosotis discolor				F
Nicandra physalodes				O
Ornithopus pinnatus			Rare	R
Oxalis pes-caprae				A
Papaver dubium				F
Papaver rhoeas				F
Parentucellia viscosa				O
Polycarpon tetraphyllum			Rare	F
Ranunculus marginatus				R
Ranunculus muricatus				O-LF
Ranunculus parviflorus				O-LF
Ranunculus sardous				LF
Raphanus raphanistrum				O
Rumex acetosella				F
Scandix pecten-veneris	EN	CR		R
Scrophularia scorodonia			Scarce	O
Setaria pumila				R
Sherardia arvensis				O
Silene gallica	EN	EN	Scarce	F
Sinapis arvensis				O
Solanum lacinatum				R
Solanum physalifolium				R
Solanum sarachoides				R
Solanum nigrum				F
Spergula arvensis	VU	VU		F-LA
Spergularia rubra				O

Species name	Red List for England	Red List for Great Britain	Nationally Rare or Scarce	Bulbfield status D = Dominant A = Abundant F = Frequent O = Occasional R = Rare L = Locally EX = Extinct
Stachys arvensis	NT	NT		LF
Thlaspi arvense				O
Torilis nodosa				R
Trifolium dubium				F
Trifolium glomeratum			Scarce	O
Trifolium micranthum				O
Trifolium suffocatum			Scarce	R
Urtica urens				F
Valerianella locusta				F
Veronica agrestis				F
Veronica arvensis				F
Veronica persica				F
Veronica polita				O
Vicia bithynica		VU	Scarce	R
Vicia hirsuta				F
Vicia sativa				F
Viola arvensis				O
Vulpia bromoides				F
Vulpia myurus				R

Bulbfield plant communities

As is mentioned in the section on plant habitats, the accounts of the NVC Open Habitat communities are a little difficult to relate exactly to the bulbfield weed flora today due to some of the changes in flower farming. But two communities OV2 *Briza minor – Silene gallica* and OV6 *Cerastium glomeratum – Fumaria muralis* ssp. *boraei* are closely associated with Scilly, being either unique to the islands or only otherwise found in the southwest of Cornwall and nowhere else. It is interesting to realise these communities have only arisen within the past hundred years or so.

Silene gallica

When the downturn in the flower farming became a threat to the survival of the unique bulbfield plant communities it became enough of an issue to concern both DEFRA and English Nature both of whom commissioned reports on the subject. In 2010 a special new option to Higher Level Stewardship was introduced that paid farmers to manage areas for rare arable weeds. A number of Scillonian farmers have entered into the scheme and eventually it will be seen whether this has been successful in retaining some of these rare plants and unique plant communities.

Comparisons with the flora of Cornwall

Although there are many similarities with the flora of Cornwall and the Isles of Scilly, when arriving on Scilly it soon becomes obvious that there are also many differences. Some species, such as *Ornithopus pinnatus*

Ophioglossum lusitanicum and *Ornithopus pinnatus*

Orange Bird's-foot, *Ophioglossum lusitanicum* Least Adderstongue and *Viola kitaibeliana* Dwarf Pansy, are not found in Cornwall or on mainland Britain. Others such as *Silene gallica* Small-flowered Catchfly, *Polycarpon tetraphyllum* Four-leaved Allseed, *Malva pseudolavatera* Small-flowered Tree-mallow and *Poa infirma* Early Meadow-grass are found in Cornwall and elsewhere. Exceptionally all three taxa of *Ophioglossum* ferns have been recorded on just one island on Scilly (St Agnes).

Other common mainland plants are absent or very rare on Scilly, for example *Tussilago farfara* Colt's-foot (recently extinct), *Betonica officinalis* Betony and even *Ranunculus acris* Meadow Buttercup. Until recently *Pilosella offinarum* Mouse-ear-hawkweed was only known from one place on St Martin's until it was found in a heathy field on St Mary's. As is typical of islands the flora is impoverished, with far fewer species than might be expected in a similar sized area on mainland Cornwall where there would be a larger range of habitats. However as Lousley commented 'what they lack in floristic richness is more than made up by the interest, and often abundance, of the plants which do occur'.

Pilosella officinarum

Botanical links

The isolation of the Isles of Scilly and the small range of habitats are one of the main reasons for the limited number of plant species on Scilly. Some of the plants that are present on the islands have clear affinities with the Mediterranean and the Lusitanian element of the flora of the Iberian peninsula. Many of these species could have spread up the Atlantic coasts to colonise Scilly and the Channel Islands, but some may have come by boats directly to Scilly from the Mediterranean.

These plants are often winter annuals so they are well-suited to the mild winters in the Isles of Scilly. Many found ideal conditions as arable weeds in the bulbfields alongside the winter-flowering narcissus.

Two main groups of taxa overlap in Cornwall and the Isles of Scilly: those of the Mediterranean-Atlantic and those with an Atlantic distribution. There is something of a concentration of these species on Scilly (Preston and Arnold, 2006).

Examples of Mediterranean-Atlantic plants native in Scilly

Arum italicum ssp. *neglectum*
Asplenium obovatum ssp. *lanceolatum*
Beta vulgaris ssp. *maritima*
Calystegia soldanella
Catapodium marinum
Crithmum maritimum
Euphorbia paralias
Fumaria bastardii
Glaucium flavum
Isolepis cernua
Linum bienne
Malva pseudolavatera
Ophioglossum lusitanicum
Parentucellia viscosa
Poa infirma
Polycarpon tetraphyllum
Polygonum maritimum
Raphanus raphanistrum ssp. *maritimus*
Rubia peregrina
Torilis nodosa
Trifolium glomeratum
Trifolium suffocatum
Umbilicus rupestris

Examples of Atlantic species native in Scilly

Anagallis tenella
Atriplex laciniata
Carex binervis
Carex laevigata
Cochlearia danica
Conopodium majus
Dactylorhiza praetermissa
Erica cinerea
Euphorbia portlandica
Euphrasia tetraquetra
Fumaria muralis ssp. *boroei*
Hyacinthoides non-scripta
Mysotis secunda
Puccinellia maritima
Rumex rupestris
Scilla verna
Sedum anglicum
Sibthorpia europaea
Spergularia rupicola
Trifolium occidentale
Ulex europaeus
Ulex gallii

Examples of Mediterranean species alien in Scilly

Acanthus mollis
Allium roseum
Allium triquetrum
Anisantha diandra
Anisantha madritensis
Anisantha rigida
Briza maxima
Calystegia silvatica
Centranthus ruber
Crepis vesicaria
Gladiolus communis ssp. *byzantinus*
Lobularia maritima
Petasites fragans
Soleirolia soleirolii
Spergularia bocconei
Vinca major

Allium roseum

Garden escapes

What immediately strikes the naturalist arriving on Scilly is the extraordinary profusion of alien plants. Often these are escapes from cultivation; typically they are plants from warmer temperate climates including the

Canary Islands, South Africa, Chile, New Zealand and Australia. Unlike elsewhere in Britain, the mildness of the Scillonian climate enables such frost sensitive species to flourish. Although most of the alien species that have become established on Scilly have originated from Tresco Abbey Gardens, some were already present before the Gardens were established. In the days of sail Scilly was an important port. Merchant ships that called into the Isles would have brought in bulbs and plant material. Other plants have arrived accidentally with goods from abroad. Many island men became mariners and some brought back 'exotic' specimens as souvenirs from their travels.

Round Island established aliens

Bird-sown and wind-borne plants

The recent arrival of several water plants, either species new to Scilly, or new to one of the islands where they had been unknown previously, can often be directly attributed to waterfowl. Examples of the former are *Bidens tripartita* Trifid Bur-marigold first found on Tresco in 1994 which has become established, and *Ranunculus sceleratus* Celery-leaved Buttercup on St Agnes in 1987 and 1998 where the plants did not persist. Then we have the appearances of *Ranunculus flammula* Lesser Spearwort by the pool on St Agnes in 2000, a new island for the species, and in 1999 *Scutellaria galericulata* Skullcap was found on Tresco, presumably spread from the colony on Samson.

Pittosporum on carn

Some garden escapes are found on uninhabited islands as the result of gull activity – although most do not survive for long in the exposed conditions. Others, such as *Carpobrotus edulis* Hottentot-fig are more resilient and soon become established and spread further. Some other succulent plants and, unusually, some natives such as *Fumaria capreolata* White Ramping-fumitory have at times been found on uninhabited islands. Work by Gillham showed gulls are frequent agents for this kind of dispersal (Gillham, 1956 & 1970). It is not unusual to see gulls pick up and carry away all manner of inedible things, not always for nesting material. Dr Jack Oliver mentions seeing a Herring Gull drop a viable fragment of *Fascicularia bicolor*. So it seems likely this is a common way roots and plant material get translocated.

Fumaria capreolata flower

Coprosma on Samson

Thrushes and blackbirds are believed to be responsible for spreading *Pittosporum* and *Coprosma* seeds when they feed on the sticky fruit. It is also possible they and other passerines spread seeds of brambles and other plants in their droppings.

A few plants may also be wind-borne. There are numerous examples of wind-borne invertebrates that have reached Scilly which shows how easy this could be. The arrival of *Anacamptis pyramidalis* Pyramidal Orchid on Samson is believed to have been from wind-borne seed as was a similar arrival of *Gymnadenia borealis* Fragrant Orchid in 1971. Fern spores are very light and also can travel great distances, which may explain the alien fern *Blechnum cordatum* on Ragged Island, in the Eastern Isles.

Fascicularia bicolor

Coprosma repens

Plant habitats

Although there have been a number of habitat surveys on Scilly, some aimed at National Vegetation Classification (NVC), most have been variations on Phase 1 mapping. No complete vegetation classification of the islands has yet been made although the IoSWT have Phase 1 maps of their holdings on the islands. Phase 1 surveys by CeC (the Cornwall Wildlife Trust's consultancy) as part of their habitat mapping for the IoSWT Management Plan (2010-2015) have been consulted. Because the habitats on Scilly were not easy to classify using the NVC, in the case of dune habitats Dargie (1990a; 1990b; 1990c) went a step further and compiled his own very interesting and detailed classification, although this does not seem to have been adopted generally. As the Isles of Scilly lie on the edge of the geographical range covered by the NVC, it is not surprising they provide a relatively 'poor-fit' to some of the published communities. Some plant communities on Scilly often lack the 'constant' species and typical components of the national classification. This leads to botanists having to use some inspired guesswork to assign vegetation to the published communities!

As is common to small islands, besides their small area, they also lack many habitats or only have small samples, such as the tiny strip of salt marsh on Bryher. This also leads to a paucity of species both overall and within those habitats present.

The following is an attempt to describe the main habitats found on the Isles of Scilly with some reference to the NVC (Rodwell (ed.) 1991 - 2000). Many comparable communities can be recognised but are often not that close to the published accounts, so although the NVC community classifications are included, these are just a guide and should not be taken too critically.

Heathland

This is one of the most important habitats on Scilly and one of the most revered as an emblematic island landscape. Historically heaths would have been exploited by the islanders in many ways. 'Turf' was cut for thatching, building cottages and for burning on their fires. Bracken and gorse were also cut as firing, bedding and other uses. Farm stock would have grazed the heathland so this was always a very much managed and valuable resource.

Today most remaining heathland on the Isles of Scilly is on the coast, often reaching right to the edge of the cliff. In places these areas can be quite extensive, for example on headlands such as Castle Down, Tresco and Chapel Down, St Martin's. Small patches of heath are found on the tops of hills both on the inhabited and uninhabited islands. It is presumed most inland heaths have been ploughed and cultivated so that only fragments can now be found. It is not unusual to find heathland elements persisting in arable fields. Low-lying sandy areas have developed into dune heath or dune grassland.

Chapel Down

Chapel Down waved heath

Salt damaged heath

Waved heath extreme type

Lichen heath, Wingletang

Besides the changes due to management, the heathlands on Scilly are continually subject to the influence of the sea and salt-laden winds. This affects not just the coastal edges, the raking winds and storms have a pruning affect on heath right across the islands. The beautiful 'waved heath' that is such a feature on the islands is the result of wind erosion, where the shrubs are rolled over away from the wind direction to form the characteristic ripples. In the most extreme waved heath, the way the stems and roots of the heather are exposed within ribbons of bare ground make the similarity to waves even more apt. Another feature of some of the Scilly heathlands are areas of quite beautiful 'lichen heath' where the density of lichens, often *Cladonia* species, result in areas where the dwarf shrubs are almost totally obscured.

Most of the *Calluna/Erica* heathland in the Isles of Scilly needs little or no management. But there are areas where gorse, bracken and bramble have invaded the heath and are shading or out-competing the dwarf shrub element and rarer plants. Attempts to control this by some kind of active management has been tried by both farmers and by the Wildlife Trust with mixed results. The most extreme management tools are herbicides and burning. Burning has often been a popular option with some farmers but without careful planning (size of area, fire breaks, protection of rare plants and lichens) the results have sometimes been unsuccessful or worse. Without follow-up grazing all that results is an proliferation of gorse seedlings. Using herbicide is sometimes suggested to control bracken but rarely recommended due to the difficulty of protecting other ferns such as *Ophioglossum*. Cutting and bracken-rolling have been used where there is no grazing and if done regularly the results can be excellent – as on the north side of St Martin's where many heathland plants have benefitted. Grazing too has effectively restored other overgrown areas to beautiful heathland.

St Martin's gorse cutting

Bryher H7 heath (Tresco, with Cromwell's Castle in background)

In places a dense bracken-bramble scrub is found on the fringes of the heathland, this and gorse that has become invasive have overwhelmed some areas of heath where there has been little or no management.

In the Isles of Scilly heathlands appear to fall into three or four communities. The small range of species, especially compared with the Lizard heaths in Cornwall, is soon obvious at close examination. Frequently there is an intimate mosaic of the different types of heathland and grassland within a larger area of heath.

H7 *Calluna vulgaris – Scilla verna* heath

This is considered to be the most common and most maritime of the heathland communities that occur in the Isles of Scilly. *Scilla verna* is usually one of the constants in this community, but on Scilly it is a very local plant. Two sub-communities can be recognised; the H7a *Armeria maritima* type that grows nearer the sea and the H7e *Calluna vulgaris* type slightly further inland. The following constant species are present on Scilly; *Calluna vulgaris, Erica cinerea, Holcus lanatus, Hypochaeris radicata, Lotus corniculatus, Potentilla erecta,* and on Bryher and a discrete area of St Mary's, *Scilla verna*. In places some of the H7 can be 'waved heath'. The H7a *Armeria maritima* sub-community is more species-rich and supports a number of rarities including *Spiranthes spiralis, Trifolium occidentale* and sometimes in more sandy places *Euphorbia portlandica*.

Gugh

H8 *Calluna vulgaris – Ulex gallii* heath

This community is seen at its best along the eastern side of St Mary's between Salakee Down and Porth Hellick where in late summer/autumn it is marked by the golden-yellow flowers of *Ulex gallii* flowering alongside the purples of *Erica cinerea* and *Calluna vulgaris*. Also found in this community are *Potentilla erecta, Galium saxatile, Rumex acetosella*, and *Polygala* spp. Frequently the *Danthonia decumbens* sub-community (H8b) can also be recognised growing on paths and gaps between the dwarf shrubs. Some surveyors distinguish H7 from H8 by the presence or absence of *Ulex gallii* based on Malloch's assertion (Rodwell *et al.*, 1991) that it only grows on places sheltered from the sea. However on Scilly this does not seem always to hold true.

H8 St Mary's

H10 *Calluna vulgaris – Erica cinerea* heath

Although listed as present on Scilly this is a more northern community and has not been confirmed for Scilly.

H11 *Calluna vulgaris – Carex arenaria* heath

This is a common plant community on Scilly where it forms a link between dune communities and other heathlands. It is also a stand alone vegetation type that makes up 'dune heath' on many of the sandier areas on Scilly, even relatively inland. Besides the strong presence of *Carex arenaria*, often as tall, grass-like plants, there are often several species of lichens. *Trifolium suffocatum* and *Euphorbia portlandica* are frequently found in this community.

vegetation grows in the crevices between rocks and forms patches of *Crithmum maritimum, Festuca rubra, Spergularia rupicola* and other typical plants. In places *Plantago coronopus, Cochlearia* spp. and *Daucus carota* ssp. *gummifera* are also represented. *Asplenium marinum* is at home here, but in deep clefts under rocks.

H11 Dune heath, Tresco

MC1 Garrison, Scilly

H11 Dune heath, St Martin's

Coastal grassland and cliffs

Although the cliffs on Scilly are not very high they support a range of distinctive types of vegetation and colourful plants. Combined with the extraordinary range of lichens that cover the rocks they are one of the most attractive landscapes in the islands. Some of the cliffs are just a low bank or rocky edge but most of the cliff edges are vegetated right down to the edge of HWM. Along the cliff or coastal edge is usually a fringe of short grassland which extends inland to merge with heath or dune communities.

Maritime cliff communities

MC1 *Crithmum maritimum – Spergularia rupicola* maritime rock-crevice community

This is a distinct and common community around the cliffs, rocky coast and even on some man-made structures, for example on the Garrison walls on St Mary's, and Cromwell's Castle on Tresco. Usually the

MC1 Shipman Head

MC5 *Armeria maritima – Cerastium diffusum* maritime therophyte community

Another well represented community on Scilly. Here the plants grow on thin soils of the cliffs, on carns, shallow soils over granite and in similar places on the tops of walls near the coast. Many of the plants are winter annuals (ones that germinate in autumn, grow through winter and complete their life-cycle in spring so are able to withstand summer droughts). This community may also be found in a mosaic with MC1 and other coastal communities. The grasses found here include fine-leaved fescues *Festuca rubra* and *F. ovina*, (and occasionally *Koeleria macrantha* in places on Bryher). This is often the community where rarer *Trifolium* spp. such as *T. occidentale* are found, also *Poa infirma*, *Aira praecox*, *Armeria maritima*, *Plantago coronopus*, *Sedum anglicum*, *Catapodium marinum*, and in a few places the very rare lichen *Heterodermia leucomela*. The sub-communty most frequently seen is close to the MC5b *Anthyllis vulneraria* sub-community (although *Anthyllis* is not found on Scilly), or MC5c *Aira praecox* sub-community. Examples include short, species-rich turf with *Anagallis arvensis*, *Galium verum*, *Centaurium erythraea*, *Euphorbia portlandica*. In places *Ornithopus perspusillus*, *O. pinnatus*, several lichens, *Trifolium* spp., *Spiranthes spiralis* and *Ophioglossum azoricum* occur. On Heathy Hill, Bryher there is a particularly rich area of the community which includes *Ophioglossum azoricum*, *Ornithopus pinnatus*, *O. perpusillus*, *Radiola linoides*, *Scilla verna*, *Aira caryophyllea*, *Trifolium occidentale*, *Lotus subbiflorus*, *Trifolium micranthum* and *T. ornithopodioides*.

St. Mary's MC5

St. Mary's MC5

Ornithopus pinnatus and *O. perspusillus* on Heathy Hill

MC6 *Atriplex prostrata* – *Beta vulgaris* ssp. *maritima* sea-bird community

This community is best seen on some of the smaller islands where seabirds nest, although patches also occur along cliffs where gulls stand. Due to enrichment from the seabird guano the plants can often be big and fleshy; typically they include *Festuca rubra*, *Dactylis glomerata*, *Cochlearia officinalis*, *Rumex crispus* ssp. *littoreus*, *Tripleurospermum maritimum* and *Atriplex* spp. On some of the smaller islands *Malva arborea* can be a significant part of the vegetation with gulls frequently nesting among the stands of old, dead stems.

MC5

Annet

Bird nest vegetation

MC7 *Stellaria media – Rumex acetosa* sea-bird cliff community

Another community influenced by the presence of seabirds. The species list is similar to above but the significant constant is a fleshy form of *Stellaria media*. Possibly MC7 occurs in a mosaic with MC6, otherwise there does not seem much difference between them.

MC8 *Festuca rubra – Armeria maritima* maritime grassland

This is one of the most characteristic and widespread plant communities found around the coasts on the Isles of Scilly. *Daucus carota* ssp. *gummifer* and *Festuca rubra* are abundant, *Trifolium occidentale* is seen in spring and in a few places *Scilla verna* may also occur. One of the characteristics of this community are the dense, tall 'mattresses' of *Festuca rubra* on some cliffs and some of the smaller uninhabited islands. The relatively species-poor MC8e *Plantago coronopus* sub-community is found as a closely grazed sward (although in places it appears to be maintained by wind-pruning) and on pathways dominated by *Festuca rubra* (although not as mattresses), *Armeria maritima*, *Plantago coronopus*, *Agrostis stolonifera* and *Daucus carota*. Another sub-community is the *Holcus lanatus* sub-community (MC8d) which also includes those grasses, plus *Rumex acetosa*, *Hypochaeris radicata* and *Achillea millefolium*. The *Crithmum maritimum* sub-community (MC8b) is transitional between MC1a and MC8 and can occur where *Crithmum maritimum* grows on the cliffs.

Bryher MC8 on path

An extreme form of what is apparently the *Armeria maritima*-dominated sub-community (MC8g) is found on the uninhabited island of Annet where huge *Armeria maritima* tussocks, often a metre across and almost as high, form large areas with just *Sedum anglicum* and occasional *Festuca rubra* between the tussocks. It has been suggested this type of vegetation could be associated with former puffin and shearwater burrows. Certainly at one time almost the whole of Annet was a vast puffin colony.

Armeria maritima, Annet

MC9 *Festuca rubra – Holcus lanatus* maritime grassland

This grassland is frequent on slightly more sheltered coasts than MC8. The species composition is fairly species-rich, dominated by grassy tussocks of *Holcus lanatus*, *Festuca rubra*, *Agrostis stolonifera* and *Dactylis glomerata* with *Plantago lanceolata* and *Armeria maritima*.

MC9 coastal grassland

MC9 cliff edge with *Plantago maritima*

In places *Galium verum*, *Achillea millefolium*, *Potentilla erecta* and *Trifolium occidentale* can occur in places where the sward is short. The *Dactylis glomerata* sub-community (MC9b) is the most common on Scilly but the *Plantago maritima* sub-community (MC9a) also occurs in a few places along the coast on St Mary's.

MC10 *Festuca rubra – Plantago* spp. maritime grassland

On the Isles of Scilly it is the *Armeria maritima* sub-community (MC10a) that is likely to be encountered. *Plantago coronopus*, *Festuca rubra*, *Plantago lanceolata* and *Agrostis stolonifera* are found here, sometimes with the rarities *Trifolium occidentale* and where found, *Scilla verna*. Often this grassland is found in close proximity to maritime heath.

MC10 grassland on White Island

Chamaemelum nobile grassland

A very distinctive grassland dominated by *Chamaemelum nobile* and with several clovers including *Trifolium subterraneum*, *T. repens* (including var. *townsendii*), *T. occidentale*, *T. ornithopodioides* and *T. micranthum*. It is found on close-mown 'lawns', cricket fields, mown areas around scheduled monuments and as short turf on some cliffs and heathlands. Tentatively it has been included with MC10 as it contains many of the constants but this is not a very satisfactory conclusion.

Chamomile 'lawn' St Agnes

MC12 *Festuca rubra – Hyacinthoides non-scripta* maritime bluebell community

This community is found on many coastal areas around the islands. On the cliff slopes *Hyacinthoides non-scripta*, *Festuca rubra*, *Holcus lanatus* and other grasses, and *Rumex acetosa* are dominant. *Ficaria verna* and in some places *Hedera helix* ssp. *hibernica* are found in the understorey. *Pteridium aquilinum* can also appear here.

Bluebell slope

Coastal pools

There are just a few pools on Scilly. Some may be natural but most of the smaller ones seem to have been modified for stock. Larger pools such as Big Pool on St Agnes have in the past been used for drinking water. Great Pool on Tresco has not been studied in detail but does appear to have the same aquatic flora.

Most of the pools, although freshwater, have variable salinity due to their proximity to the sea.

A21 *Ranunculus baudotii* community

Found in freshwater and slightly brackish pools, although often now without any *Ranunculus baudotii* which has become a very rare plant on Scilly. The other constants, *Potamogeton pectinatus* and *Ruppia maritima* are usually the only other species present.

St Martin's Pool

Salt marsh

The only area of salt marsh vegetation that occurs on Scilly is around the Pool and leat on Bryher. It is just a narrow strip of vegetation a few metres wide so probably not very typical. The following are the possible communities present, but none very distinctive.

SM15 *Juncus maritimus – Triglochin maritima* salt-marsh community

This forms taller stands with tussocks of *Juncus maritimus*, *Agrostis capillaris* and *Festuca rubra*. *Triglochin maritima* is not present on Scilly.

Bryher leat and saltmarsh

SM16 *Festuca rubra* salt-marsh community

Slightly shorter vegetation especially around the pool edges with *Festuca rubra*, *Juncus gerardii* and *Agrostis stolonifera* as the most obvious constituents. *Glaux maritima* and *Atriplex* spp. including *Atriplex glabriuscula* also occur here.

Bryher leat

SM23 *Spergularia marina – Puccinellia distans* salt-marsh community

This community seems sometimes to merge with SM16. It forms a narrow band of shorter turf along the leat (sometimes on raised places in the leat) and around the Pool on Bryher.

Sand dunes

The dune communities in the Isles of Scilly are the least well-matched to the NVC. Dargie (1990a; 1990b; 1990c) came up with his own vegetation classification which is very detailed and describes all the variations of community that he found on Scilly. However, for the sake of consistency Dargie's scheme has not been adopted here. One of the problems with the dune series on Scilly is that the sequence is usually very truncated and lacks several of the expected stages in the range from strandline and foredune through to fixed dune and beyond. In some places such as Rushy Bay on Bryher the dune habitats are composed of a mosaic of tiny patches of several communities (Coleman and O'Reilly, 2004). The winter storms and strong winds can modify the dunes on an almost annual basis. Often sand is blown inland and large drifts can be deposited over shorter vegetation totally obscuring the plants.

Dune sequence SD1-SD6.

Most of the dune and other coastal sites in Scilly are constantly subject to inundation or erosion by the sea. So areas can change markedly from year to year both physically and in species composition. The 2014 storms in many places completely remodelled the beaches; on some all vegetation was swept away, as was shingle and rocks to expose bedrock. In other places deep sand was deposited – as on the *Viola kitaibeliana* sites on Teän, or boulders as on Rushy Bay, Bryher.

Boulders deposited by storm, Bryher

SD1 *Rumex crispus – Glaucium flavum* shingle community

Rumex crispus in the coastal form ssp. *littoreus* is a constant here with *Glaucium flavum, Beta vulgaris* ssp. *maritima, Atriplex prostrata* and on some beaches *Crambe maritima*. This vegetation is found on Wingletang Bay, St Agnes and on some of the Eastern Isles but is vulnerable to storms so sometimes most of the plants get washed away. *Crambe* and deep-rooted plants usually only lose their top-growth and grow back from their roots, others return from seed that has been buried or washed back on shore.

SD1, Teän

SD2 *Honckenya peploides – Cakile maritima* strandline community

This is a typical strandline community with masses of *Honckenya peploides* although *Cakile maritima* numbers can fluctuate markedly from year to year. *Rumex crispus* ssp. *littoreus* can also occur here.

SD2 type St Martin's sandy beach

SD4 *Elymus farctus* ssp. *boreali-atlanticus* foredune community

This community occurs as a narrow fringe in front of the SD6 dune or sometimes on its own with no Marram present. The colonising vegetation on the disturbed, sandy 'neck' area of Samson appears to be close to this.

SD6 *Ammophila arenaria* mobile dune community

The typical sand dune community dominated by *Ammophila arenaria* Marram. Sometimes there may be little else but Marram but a few other plants can be found: *Euphorbia portlandica, E. paralias, Scrophularia scorodonia, Allium ampeloprasum* var. *babingtonii* and *Calystegia soldanella*. On occasion more grassland species or strandline plants occur – for example in the dunes on Lower Town Bay, St Martin's the following occur: *Galium verum, Achillea millefolium, Elytrigia atherica, Atriplex glabriuscula* and *Cirsium arvense*.

SD6 dune

SD7 *Ammophila arenaria – Festuca rubra* semi-fixed dune community

A stage on from SD6 is this community which has more *Festuca rubra* and *Euphorbia portlandica* and possibily *E. paralias,* although Marram still features.

SD7 dune vegetation

SD4 Samson neck

SD8 *Festuca rubra – Galium verum* fixed dune grassland

Here the dune has developed a closed turf community over the sand with the grasses *Festuca rubra* and *Holcus lanatus*, plus *Plantago lanceolata*, *Lotus corniculatus*, and sometimes *Galium verum* and *Trifolium occidentale*. On Gugh *Thymus polytrichus* appears here at one of its few Scilly locations. The SD8c *Tortula ruralis* ssp. *ruraliformis* sub-community is more species-rich with bryophytes including *Tortula*, *Hypnum lacunosum* and also flowering plants such as *Spiranthes spiralis*, *Sedum anglicum*, *Chamaemelum nobile* and *Rumex acetosella*.

SD8 Gugh

SD8 Gugh

SD12 *Carex arenaria – Festuca ovina – Agrostis capillaris* dune grassland

The dune grassland above Great Bay, St Martin's may be this community.

The Plains St Martins

SD19 *Phleum arenarium – Arenaria serpyllifolia* dune annual community

This is a very interesting community. It occurs on areas of pure sand and forms a very close turf of tiny, often ephemeral annuals and tiny rosette-forming perennials. In many years the annuals dry up in summer and in especially droughty years even the perennials can shrivel. Although *Phleum arenarium* does not occur on Scilly, *Arenaria serpyllifolia* is found at Rushy Bay, Bryher.

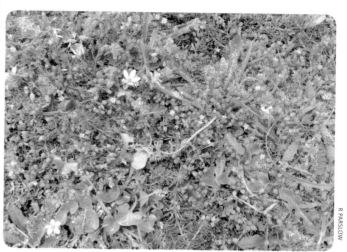

SD19 with *Arenaria serpyllifolia*

Grassland, meadows and pastures

In the late 1950s there were still some working horses on St Agnes and small meadows that were cut for hay. Now on Scilly any remaining grasslands are mainly pasture and often have been resown; there are only a few left that now have anything approaching a typical hay-meadow flora. Other interesting grasslands can be found on the coast and in some wet meadows. There are a couple of apparent hay meadows with *Leucanthemum vulgare* and other meadow species, there was one in Barnaby Lane, St Agnes, but these will have been sown with a wild flower seed mixture. *Leucanthemum vulgare* is an unusual plant on Scilly and is only found naturally in a few small remnant grassland areas such as the edge of the Airfield and near Harry's Walls.

U1 *Festuca ovina – Agrostis capillaris – Rumex acetosella* dune grassland

This is the typical grassland found on the islands that frequently occurs in a mosaic with heathland. Usually found in somewhat more inland sites the constant species are *Festuca rubra* (rather than *F. ovina* which is uncommon on Scilly), *Agrostis capillaris* and *Rumex acetosella*. The *Hypochaeris* sub-community (U1f) is typical, with *Aira praecox*, *Hypochaeris radicata*, *Plantago coronopus* and *P. lanceolata*. This sward is usually closely maintained by rabbits and where the surface is broken, plants such as *Sedum anglicum* may occur.

U1 grassland site

Solidago virgaurea heath St Martin's

U4 *Festuca ovina – Agrostis capillaris – Galium saxatile* grassland

Another common type of grassland dominated by *Festuca rubra, Agrostis capillaris, Anthoxanthum odoratum*, with *Galium saxatile* and *Potentilla erecta* found in between the patches of dwarf shrubs on some heathland areas. *Danthonia decumbens* can also occur, often being abundant along pathways. Once again *Festuca rubra* usually replaces *F. ovina*.

U4 Wingletang St Agnes

U20 *Pteridium aquilinum – Galium saxatile* community

This is the bracken dominated community, where the bracken is growing within a heathy or acid grassland sward. In Scilly it now covers large areas and in places can be so dense as to be a monoculture. Where the structure is less dense a few forbs may be found, although in Scilly *Galium saxatile* is relatively uncommon. Here bramble is rare or at low density. Where the bracken has been cut or rolled, for example on the north side of St Martin's, there has been a good response from the underlying heathland species such as *Solidago virgaurea*.

MG5 *Cynosurus cristatus – Centaurea nigra* grassland

Something akin to this grassland occurs on Scilly but infrequently, it may have been more common when there were hay meadows. There is an example on St Martin's that is close to the *Danthonia decumbens* sub-community (MG5c). Here there were a number of grasses, *Danthonia decumbens*, both *Festuca rubra* and *F. ovina, Holcus lanatus, Agrostis stolonifera, Agrostis capillaris, Cynosurus cristatus* as well as *Centaurea nigra, Trifolium repens* (including var. *townsendii*), *Crepis capillaris, Rumex acetosella, Achillea millefolium, Rumex pulcher* and *Conopodium majus*.

MG5 with *Conopodium majus*, St Martin's

U20 bracken slopes St Martins

MG5 *with Hyacinthoides non-scripta*, St Martin's

MG6 *Lolium perenne – Cynosurus cristatus* grassland

This occurs in a few rather poor fields that may have been agriculturally improved, or have reverted from bulbfields, for example those below Samson Hill, Bryher. In some cases some of the bulbfield plants may persist at low densities.

MG6 Bryher former fields

MG10 *Holcus lanatus – Juncus effusus* rush-pasture

Found on more or less permanently moist soils with *Agrostis stolonifera*, *Holcus lanatus*, *Juncus effusus* and *Ranunculus repens*. On St Mary's there are examples of the typical sub-community (MG10a) with *Trifolium repens*, *Rumex acetosa* and *Rumex* spp. The *Iris pseudacorus* sub-community (MG10c) with *Iris* and *Lotus pedunculatus* is also found on St Mary's, especially on fields near Porthloo. Sometimes the community may occur in a mosaic with MG13. In the past the fields may have been more species-rich, but have deteriorated due to lowering of the water table. A suite of fields near Lower Moors with particularly good examples of this community was lost when it was used to store building materials during the construction of the new school.

MG11 *Festuca rubra – Agrostis stolonifera – Potentilla anserina* grassland

A wet grassland community that can be found in both freshwater and brackish situations. The species-poor *Lolium perenne* sub-community (MG11a) is found near the pool on Bryher and is typically dominated by the grasses *Festuca rubra*, *Holcus lanatus* and *Agrostis stolonifera* with some *Anthoxanthum odoratum* and *Lolium perenne*. Only a few herbs are typically found here; *Potentilla anserina*, *Ranunculus repens* and *Rumex crispus* (with *R.crispus* ssp. *littoreus* on coastal sites). The *Honckenya peploides* sub-community (MG11c) can be found in seepages and around freshwater pools near the coast. It is found in small amounts in standing water where *Potentilla anserina* may be prominent.

MG11 Little Pool, St Agnes

MG13 *Agrostis stolonifera – Alopecurus geniculatus* grassland

This wet grassland community is also on Scilly although it often occurs without *Alopecurus geniculatus* which is uncommon on the islands. Several species of rush can occur here, also *Ranunculus flammula* in some places. There are examples near the cricket field pool, St Martin's and in the Porthloo fields.

MG10 iris field

MG10 rush-pasture, now lost

MG13 *Ranunculus flammula* at Big Pool, St Agnes

Arable weeds and drawdown habitats

There are two main types of arable land on Scilly – bulbfields and other cultivated land. Currently there are considerable changes happening to farmland on Scilly as more farmers change from bulb and flower growing to alternative crops, or put the fields down to grass, game crops or quinoa. Unfortunately this is resulting in the loss of many fields with the unique arable plant communities that are only known from the Isles of Scilly and southwest Cornwall. Some of the species found in the bulbfields are ones with a Lusitanian or Mediterranean distribution. Species composition can often vary from season to season and sometimes more than one community may appear to be present in a field. The following divisions are not always easy to distinguish which may be due to the limited amount of survey material on which the descriptions were based or possibly because there have been changes in cultivation methods since Silverside wrote his thesis on which these communities were based (Silverside, 1977).

Abandoned narcissus field, Gugh

OV2 *Briza minor – Silene gallica* community

Only found on the Isles of Scilly, this is one of the most species-rich bulbfield communities. Besides *Silene gallica* and *Briza minor* there is a long list of associated species including common plants such as *Rumex acetosella*, *Anagallis arvensis* and *Trifolium dubium*, as well as others which are rare or scarce: *Ranunculus muricatus*, *R. parviflorus*, *Polycarpon tetraphyllum* and *Anisantha diandra*. Although the NVC account includes *Scrophularia scorodonia* and *Trifolium suffocatum* these would now be unusual here.

OV2 species-rich bulbfield

OV4 *Chrysanthemum segetum – Spergula arvensis* community

A frequent community on Scilly, often occurring as patches within the fields with another community but sometimes alone. Where the field is dominated by either *Glebionis (Chysanthemum) segetum* or *Spergula arvensis* the visual effect can be very dramatic. A number of other annuals occur in this community but at low densities.

OV4 with *Glebionis segetum* and *Spergula arvensis*

OV4 with dominant *Spergula arvensis*

OV6 *Cerastium glomeratum – Fumaria muralis* ssp. *boraei* community

This vegetation community only occurs on Scilly and West Cornwall. It is quite distinctive due to the large amount of *Fumaria* spp. fumitories found here along with *Cerastium glomeratum*. In the Isles of Scilly the typical fumitories are *Fumaria muralis* ssp. *boraei* and *Fumaria bastardii*. On St Mary's *Fumaria capreolata* and occasionally *F. occidentalis* may also occur here.

OV6 with *Fumaria* spp.

OV13 *Stellaria media – Capsella bursa-pastoris* community

This community seems to be a stage on from the previous communities as it appears to occur where there has been a change in the cultivation regime including the time of cultivation and use of fertilisers. There are far more weedy species present and a decrease of species such as fumitories although they are sometimes present as in the *Fumaria officinalis – Euphorbia helioscopia* sub-community (OV13c) which has been identified on Bryher. The constant species are *Capsella bursa-pastoris, Chenopodium album, Senecio vulgaris, Polygonum aviculare* and *Stellaria media*. This community also may be found on farmland on Peninnis, St Mary's and other fields growing potatoes or root crops. Additionally these fields are often cultivated in spring rather than in late summer/autumn as are bulbfields. The use of fertiliser can lead to fields with a much ranker weed flora of *Sonchus* spp., *Urtica dioica* and *Rumex* spp. and are generally less interesting botanically.

OV13 with *Euphorbia helioscopia*

OV35 *Lythrum portula – Ranunculus flammula* community

An interesting assemblage found around the muddy edge of Abbey Pool, Tresco and a few other drawdown areas around freshwater pools. *Hydrocotyle vulgaris, Anagallis tenella, Littorella uniflora, Lythrum portula* and *Samolus valerandi* appear on the exposed mud in dry summers. A very tiny, procumbent form of *Ranunculus flammula* is found on the mud around Abbey Pool; it is quite unlike the same species elsewhere on the islands which are more typical, tall, upright plants.

OV35 Higher Moors

Drawdown, Porth Hellick pool

Reedbeds and swamps

There are typical reed beds on Scilly but also the more interesting or confusing swamp areas on the St Mary's Moors where *Juncus maritimus* is such a feature.

S4 *Phragmites australis* swamp and reed-beds

This is the typical reed bed community found around Great Pool, Tresco and Higher and Lower Moors and Porthloo duck pond on St Mary's. In the St Mary's Moors sites there are areas with *Juncus maritimus* either co-dominant or part of the understorey. This may possibly have resulted if the habitat was formerly a *Juncus* community when the area was more brackish, that later became invaded by *Phragmites*? The *Juncus maritimus* on St Mary's is attributed to the rare variety *atlanticus* with a shorter lower bract and lax habit which is only known from Scilly. The rest of the community includes a range of wetland plants including *Galium palustre* and *Hydrocotyle vulgaris*.

S4 Higher Moors in summer

S4 *Phragmites/Juncus maritimus* swamp, Lower Moors

S21 *Scirpus maritimus* swamp

This community is found around brackish and freshwater pools such as Big Pool, St Agnes. It is dominated by *Bolboschoenus maritimus* with *Juncus gerardii* and *Agrostis stolonifera* sometimes present. The

S21 edge of Big Pool, St Agnes

Potentilla anserina sub-community (S21d) is found around St Agnes pools as well as the Water Meadow Pool on Bryher. Besides *Bolboschoenus maritimus* are usually *Agrostis stolonifera*, *Potentilla anserina* and *Chamaemelum nobile*. Sometimes *Parentucellia viscosa* is common on the Bryher site.

S21d with *Ranunculus flammula* and *Potentilla anserina*

Woodland and scrub

Scilly has no ancient woodland, only planted copses, or willow carr or scrub. All the trees have probably been planted relatively recently except for the older elms and willows.

W1 *Salix cinerea – Galium palustre* woodland

There are several areas of *Salix* carr on St Mary's and Tresco. Most are dominated by *Salix cinerea* and its hybrids. Generally there are few other trees present, other than some suckering elm in places. Usually these areas are wet and have a ground flora of *Oenanthe crocata* and *Apium nodiflorum*.

W1 *Salix* carr

W23 *Ulex europaeus – Rubus fruticosus* scrub

This scrub community is very common on the islands. Dominated by *Ulex europaeus* it forms dense thickets especially on some hillsides such as on Bryher and St Martin's. In places there are grassy areas between the gorse bushes, often where there has been some clearance. This suggests the W23a *Anthoxanthum odoratum* sub-community, though other sub-communities may be present. W23c *Teucrium scorodonia* sub-community is probably the most common with dense *Ulex europaeus* and *Rubus fruticosus* agg., but not much can grow underneath the shrubs although some rank grasses, *Holcus lanatus* and *Dactylis glomerata*, and tall plants including *Teucrium scorodonia*, *Digitalis purpurea* and *Silene dioica* may just squeeze in.

W23 on heathland

W25 *Pteridium aquilinum* – *Rubus fruticosus* underscrub

It is not clear whether this community is represented on Scilly, though something very similar, a bracken-bramble community, is frequent but without woodland components. It is not easy to decide whether this community, with its *Hyacinthoides non-scripta* sub-community (W25a), is the best fit, although something like this occurs of some of the Eastern Isles. The *Teucrium scorodonia* sub-community (W25b) is also possible where there is a somewhat more heathy feel.

Bracken-bramble scrub is an important habitat on the uninhabited islands, on some of which it is almost impossible to force a way through the dense, bramble thickets. In some places *Scrophularia scorodonia* is found here. On the inhabited islands it usually reflects lack of management.

Ulmus woodland (no NVC)

Ulmus x *hollandica* copses, for example those in Holy Vale and elsewhere on St Mary's, consist only of elm trees with a ground flora of *Athyrium filix-femina*, *Heracleum sphondylium*, *Arum italicum* ssp. *neglectum* and *Hyacinthoides non-scripta*. The trees were originally planted, presumably as shelterbelts and are now mature, tall specimens.

Great Ganilly

Ulmus woodland - Doiley Wood

Habitat maps of the islands

The schematic maps of all islands with vegetation cover on the following pages have been produced by David Mawer based on aerial imagery flown by Plymouth Coastal Observatory in 2008, 2010 and 2014.

Key for maps on following pages

heathland and maritime grassland

heather dominated heathland

dense bracken/bramble/gorse and non native plants

farmland

grassland

broadleaf woodland with willow carr and major elm hedges

woodland - conifer plantations

marsh

rocky foreshore

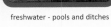
sand

dune

Tresco Abbey garden

freshwater - pools and ditches

building

built up area

Habitat maps of the islands

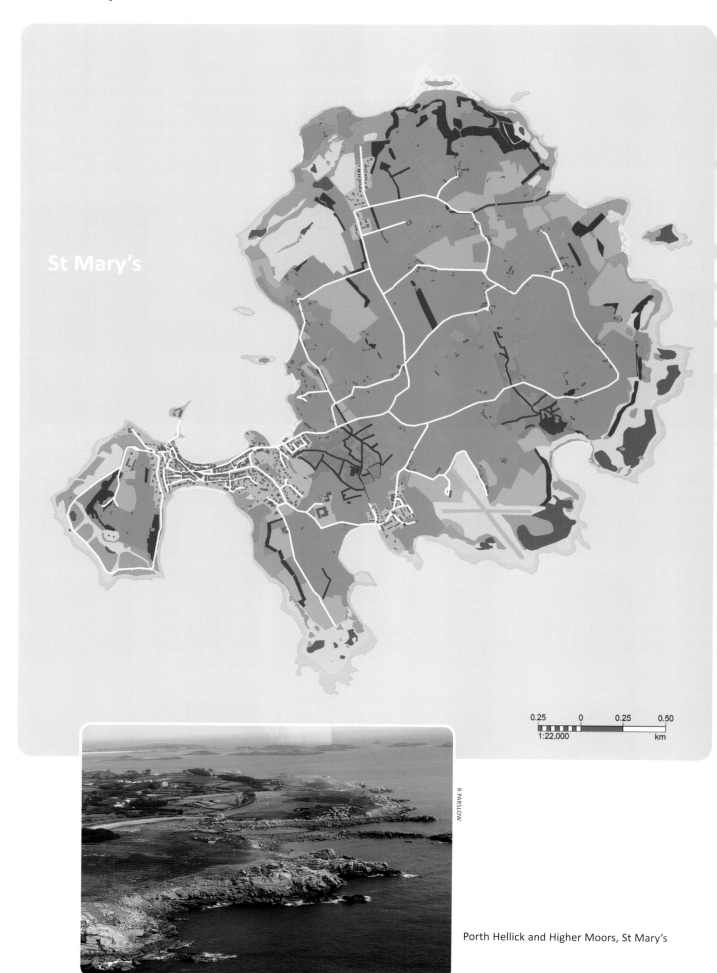

St Mary's

Porth Hellick and Higher Moors, St Mary's

Plant habitats

key

- heathland and maritime grassland
- heather dominated heathland
- dense bracken/bramble/gorse and non native plants
- farmland
- grassland
- broadleaf woodland with willow carr and major elm hedges
- woodland - conifer plantations
- marsh
- rocky foreshore
- sand
- dune
- Tresco Abbey Garden
- freshwater - pools and ditches
- building
- built up area

Bryher and Tresco

the Isles of Scilly

Habitat maps of the islands

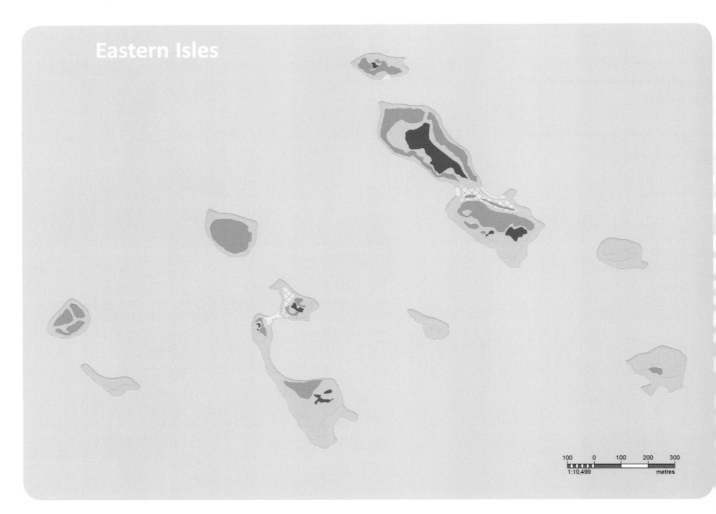

key
- heathland and maritime grassland
- heather dominated heathland
- dense bracken/bramble/gorse and non native plants
- farmland
- grassland
- broadleaf woodland with willow carr and major elm hedges
- woodland - conifer plantations
- marsh
- rocky foreshore
- sand
- dune
- Tresco Abbey Garden
- freshwater - pools and ditches
- building
- built up area

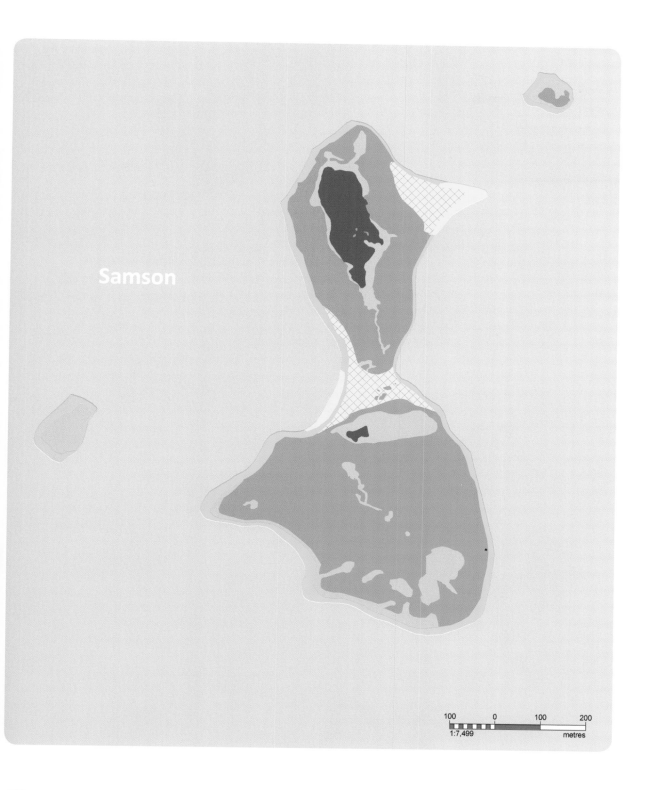

Samson

Samson, Norrards and Bryher

Habitat maps of the islands

key

- heathland and maritime grassland
- heather dominated heathland
- dense bracken/bramble/gorse and non native plants
- farmland
- grassland
- broadleaf woodland with willow carr and major elm hedges
- woodland - conifer plantations
- marsh
- rocky foreshore
- sand
- dune
- Tresco Abbey Garden
- freshwater - pools and ditches
- building
- built up area

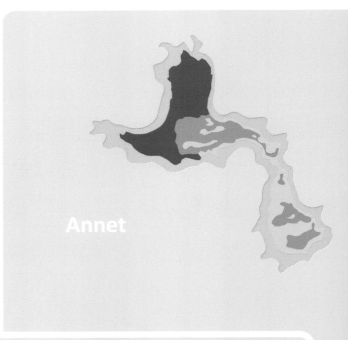

Annet

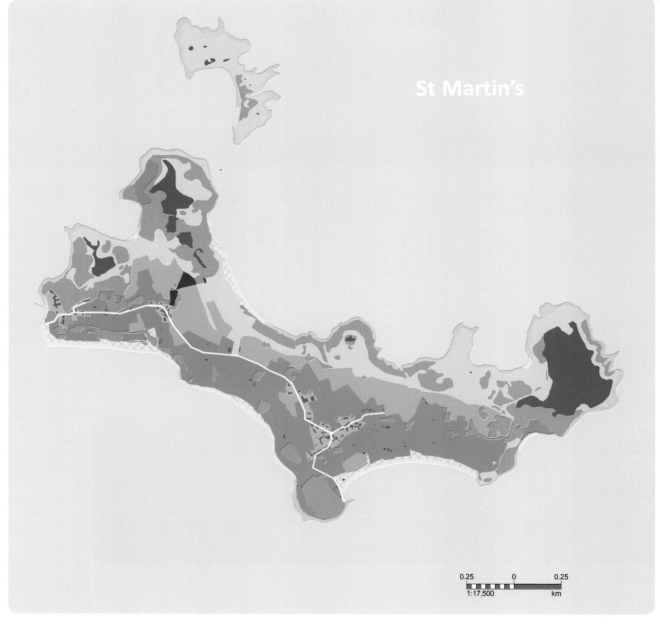

St Martin's

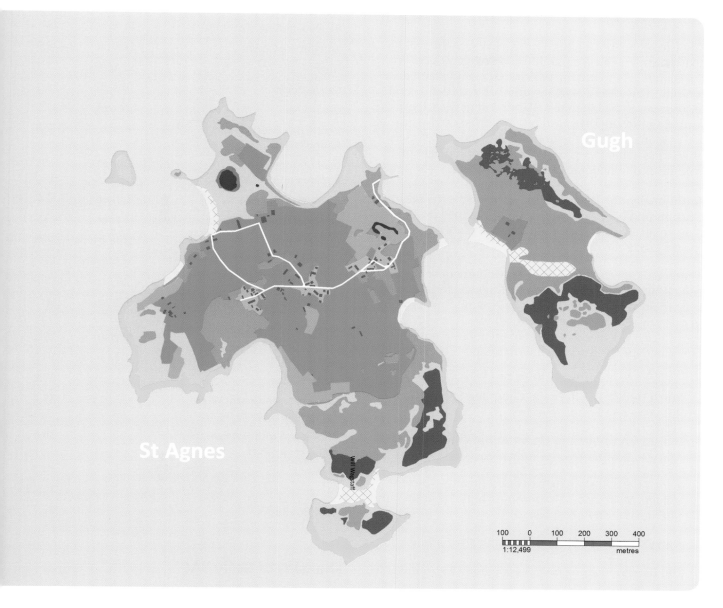

St Martin's and Eastern Isles

Nature conservation in the Isles of Scilly

The Isles of Scilly Wildlife Trust (IoSWT) is the main local charity responsible for wildlife and historical sites on the islands. They care for 1958 ha, 60% of the land area of Scilly. This includes all the uninhabited islands and large areas of the inhabited islands (other than Tresco) including most of the coastline. The Trust carries out active management to maintain the wonderful variety of wildlife habitats, to conserve them for the benefit of wildlife and to maintain access for people.

Additionally, the Isles of Scilly is the smallest Area of Outstanding Natural Beauty (AONB) in the UK. Despite their small size they have a diversity of habitats and species that belie their small scale.

There are 238 Scheduled Ancient Monuments on Scilly, eight of which are maintained by English Heritage who keep them very tidy and with close-mown surroundings. In most cases these have developed into interesting and species-rich grassland, several of the monuments have rare plant species associated with them.

Porth Hellick Down Burial Chamber, St Mary's A Bronze Age Entrance grave

Innisidgen Lower and Upper Burial Chambers, St Mary's Bronze Age communal burial chambers

Bant's Carn Burial Chamber and Halangy Ancient Village, St Mary's A Bronze Age burial chamber and nearby an Iron Age village

Harry's Walls, St Mary's An unfinished artillery fort

Old Blockhouse, Tresco A 16th century gun tower

Garrison Walls, St Mary's 16th-18th century fortifications

King Charles's Castle, Tresco Mid-16th century coastal artillery fort and later a Royalist Civil War garrison

Cromwell Castle, Tresco Cromwellian Civil War fortification

Site and species protection

There are 25 Sites of Special Scientific Interest (SSSI) on the Isles of Scilly of which only three are notified solely for their geological interest. All the others have biological, or both biological and geological interest. Most of these sites are leased to the IoSWT who manage them for nature conservation. Not all the important plants are in SSSI and there are no designations covering arable farmland such as bulbfields.

Biological SSSI table

SSSI name	Island	Hectares
Big Pool and Browarth Point	St Agnes	25
Wingletang Down	St Agnes	28.9
Gugh	St Agnes	37.7
Castle Down	Tresco	58.1
Great Pool	Tresco	17.5
Pentle Bay, Merrick and Round Islands	Tresco	42.8
Pool of Bryher and Popplestone Bank	Bryher	5.9
Rushy Bay and Heathy Hill	Bryher	12.2
Shipman Head and Shipman Down	Bryher	40.7
Chapel Down	St Martin's	34.9
Plains and Great Bay	St Martin's	15
White Island	St Martin's	16.6
Higher Moors and Porth Hellick Pool	St Mary's	16.2
Peninnis Head	St Mary's	16.1
Lower Moors	St Mary's	10.2
Eastern Isles		83.8
Samson (with Green, White, Puffin and Stony Islands)		38.7
St Helen's (with Northwethel and Men-a-vaur)		26.6
Teän		18.8
Annet		119.5
Western Rocks		62.7

The Isles of Scilly SAC

Special Areas of Conservation (SACs) are strictly protected sites designated under the EC Habitats Directive. Article 3 of the Habitats Directive requires the establishment of a European network of important high-quality conservation sites that will make a significant contribution to conserving the 189 habitat types and 788 species identified in Annexes I and II of the Directive (as amended). The presence of Shore Dock *Rumex rupestris* on Scilly was one of the reasons for the designation.

There are historical records of Shore Dock from seven of the larger islands, as well as from several small rocky outcrops. More recent survey of the Eastern Isles suggest that it had already gone from all except just four islands; Tresco, Annet, Samson and Teän. The last plant was lost from Tresco in 2012 and in 2014 surveys to check the remaining populations failed to find any plants. This was after the devastating storms early in 2014. Whether this is the last of the species on Scilly or whether it will return (there is apparently plenty of suitable habitat) remains to be seen.

Threatened and rare species

Assessments of 'threat' to native and archaeophyte species in England is summarised in *A Vascular Plant Red List for England* (Stroh, *et al.*, 2014), and in Great Britain (England, Wales and Scotland) in *The Vascular Plant Red Data List for Great Britain* (Cheffings & Farrell (eds.), 2005). Species were assessed using IUCN criteria and were categorised as not threatened (Least Concern), threatened (Critically Endangered, Endangered or Vulnerable) and 'Near Threatened'. Some species are now considered Extinct. Species may also be assessed as 'Data Deficient'. Species that are Nationally Rare are found in fifteen or fewer 10 km x 10 km squares (hectads) in Great Britain, and species that are Nationally Scarce are found in sixteen to a hundred 10 km x 10 km squares in Great Britain.

The following table includes the native and archaeophyte species in the Isles of Scilly that are threatened, near threatened, rare or scarce and are of 'conservation concern'. Though native or archaeophyte in Great Britain, the following threatened, near threatened, rare or scarce species have been recorded in the Isles of Scilly but as deliberate introductions, casuals or garden escapes - *Agrostemma githago*, *Anthemis arvensis*, *Bromus secalinus*, *Buxus sempervirens*, *Centaurea cyanus*, *Cynodon dactylon*, *Cyperus longus*, *Erica vagans*, *Gastridium ventricosum*, *Gnaphalium luteoalbum*, *Gnaphalium sylvaticum*, *Hippophae rhamnoides*, *Lythrum hyssopifolia*, *Mentha suaveolens*, *Menyanthes trifoliata*, *Ranunculus arvensis* and *Viola tricolor*.

Hyacinthoides non-scripta, *Mentha pulegium*, *Ophioglossum lusitanicum*, *Polygonum maritimum*, and *Rumex rupestris* are on Schedule 8 of the Wildlife and Countryside Act 1981. The Act makes it an offence (subject to exceptions) to pick, uproot, trade in, or possess (for the purposes of trade) any wild plant listed in Schedule 8, and prohibits the unauthorised intentional uprooting of such plants.

Achillea maritima, *Chamaemelum nobile*, *Fumaria purpurea*, *Mentha pulegium*, *Oenanthe fistulosa*, *Ranunculus tripartitus*, *Rumex rupestris*, *Salsola kali* ssp. *kali*, *Scandix pecten-veneris* and *Silene gallica* are all listed as 'Priority Species' on the UK List of Priority Species and Habitats published in 2007. All the above except *Achillea maritima* are listed in Section 41 (England) of the Natural Environment and Rural Communities Act 2006 and are species "of principal importance for the purpose of conserving biodiversity".

Rumex rupestris is listed in the Habitats and Species Directive, in Annex II - non-priority species (animal and plant species of Community interest (i.e. endangered, vulnerable, rare or endemic in the European Community) whose conservation requires the designation of special areas of conservation); and Annex IV (animal and plant species of Community interest (i.e. endangered, vulnerable, rare or endemic in the European Community) in need of strict protection; they are protected from killing, disturbance or the destruction of them or their habitat).

Ruscus aculeatus is listed in the Habitats and Species Directive, in Annex V (animal and plant species of Community interest whose taking in the wild and exploitation may be subject to management measures).

Rumex rupestris is listed on Schedule 5 (European protected species of plants) of The Conservation of Habitats and Species Regulations 2010.

Rumex rupestris and *Zostera marina* are listed in Appendix 1 (Special protection ('appropriate and necessary legislative and administrative measures') for the plant taxa listed, including prohibition of deliberate picking, collecting, cutting, uprooting and, as appropriate, possession or sale) of the Bern Convention.

More information on conservation statuses can be seen on the Joint Nature Conservation Committee (JNCC) website - www.jncc.defra.gov.uk/page-3408

Oenanthe fistulosa Tubular water-dropwort

Species name	Red List for England	Red List for Great Britain	Nationally Rare or Scarce	Notes on the presence in the Isles of Scilly
Achillea maritima	Extinct	Extinct		Last seen in 1936
Adiantum capillus-veneris			Scarce	Extinct as native in 1967 but present since 1999 as garden escape
Allium ampeloprasum			Scarce	Both vars. *ampeloprasum* and *babingtonii* present
Anthemis cotula	Vulnerable	Vulnerable		Last seen before 1971
Apium inundatum	Vulnerable			
Artemisia maritima	Near Threatened			
Arum italicum ssp. *neglectum*		Near Threatened	Scarce	
Asplenium obovatum ssp. *lanceolatum*	Near Threatened	Near Threatened	Scarce	
Baldellia ranunculoides	Vulnerable	Near Threatened		Last confirmed record 1988
Botrychium lunaria	Vulnerable			
Briza minor			Scarce	
Bromus secalinus	Near Threatened	Vulnerable		
Calluna vulgaris	Near Threatened			
Calystegia sepium ssp. *roseata*			Scarce	
Calystegia soldanella	Vulnerable			
Carex echinata	Near Threatened			Last seen before 1971
Centunculus minimus	Endangered	Near Threatened		
Chamaemelum nobile	Vulnerable	Vulnerable		
Chenopodium bonus-henricus	Vulnerable	Vulnerable		Last seen in 1967
Chenopodium murale	Endangered	Vulnerable		
Cichorium intybus	Vulnerable			
Cuscuta epithymum	Vulnerable	Vulnerable		
Daucus carota ssp. *gummifer*			Scarce	
Erica cinerea	Near Threatened			
Eriophorum angustifolium	Vulnerable			Last seen in 1998
Erodium lebelii			Scarce	
Eryngium maritimum	Near Threatened			
Erysimum cheiranthoides	Near Threatened			
Euphorbia peplis	Extinct	Extinct		Last seen in 1920 or 1936
Euphrasia micrantha	Endangered	Data Deficient		Last confirmed record 1939
Euphrasia nemorosa	Near Threatened			
Euphrasia tetraquetra	Near Threatened	Data Deficient		
Filago vulgaris	Near Threatened	Near Threatened		
Fumaria occidentalis			Scarce	
Fumaria purpurea	Vulnerable		Scarce	Last seen in 1988
Glaucium flavum	Near Threatened			
Glebionis segetum	Vulnerable	Vulnerable		
Hydrocotyle vulgaris	Near Threatened			
Hyoscyamus niger	Vulnerable	Vulnerable		
Hypericum elodes	Near Threatened			
Inula helenium	Near Threatened			Last seen in 1876
Jasione montana	Vulnerable			
Lathyrus japonicus			Scarce	
Lotus subbiflorus			Scarce	
Malva pseudolavatera			Rare	
Medicago polymorpha			Scarce	
Mentha pulegium	Critically Endangered	Endangered	Scarce	Last seen in 1915
Misopates orontium	Vulnerable	Vulnerable		
Oenanthe fistulosa	Vulnerable	Vulnerable		
Oenanthe lachenalii	Near Threatened			Last seen in 1923
Ophioglossum azoricum			Scarce	
Ophioglossum lusitanicum	Vulnerable	Vulnerable	Rare	
Ornithopus pinnatus			Rare	

Species name	Red List for England	Red List for Great Britain	Nationally Rare or Scarce	Notes on the presence in the Isles of Scilly
Pedicularis sylvatica	Vulnerable			
Poa infirma			Scarce	
Polycarpon tetraphyllum			Rare	
Polygala serpyllifolia	Near Threatened			
Polygonum maritimum	Endangered	Vulnerable	Rare	
Potentilla erecta	Near Threatened			
Radiola linoides	Vulnerable	Near Threatened		
Ranunculus flammula	Vulnerable			
Ranunculus tripartitus	Endangered	Endangered	Scarce	Last seen in 1887
Rumex acetosa ssp. *hibernicus*	Data Deficient	Data Deficient	Rare	
Rumex rupestris	Vulnerable	Endangered	Scarce	
Ruppia maritima	Near Threatened			
Salsola kali ssp. *kali*		Vulnerable		
Salvia verbenaca	Near Threatened			Last seen in 1966
Scandix pecten-veneris	Endangered	Critically Endangered		
Scrophularia scorodonia			Scarce	
Sibthorpia europaea			Scarce	
Silene flos-cuculi	Near Threatened			
Silene gallica	Endangered	Endangered	Scarce	
Silene noctiflora	Vulnerable	Vulnerable		
Solidago virgaurea	Near Threatened			
Spergula arvensis	Vulnerable	Vulnerable		
Spiranthes spiralis	Near Threatened	Near Threatened		
Stachys arvensis	Near Threatened	Near Threatened		
Trifolium fragiferum	Vulnerable			Last seen in 1909
Trifolium glomeratum			Scarce	
Trifolium occidentale			Scarce	
Trifolium suffocatum			Scarce	
Valeriana dioica	Near Threatened			
Valeriana officinalis	Near Threatened			
Veronica officinalis	Near Threatened			
Vicia bithynica		Vulnerable	Scarce	
Viola kitaibeliana	Near Threatened	Near Threatened	Rare	
Wahlenbergia hederacea	Near Threatened	Near Threatened		Last seen in 1963
Zostera marina	Vulnerable	Near Threatened		

Rumex rupestris on Samson

Visiting the Isles of Scilly

Visitors at Tresco Abbey Gardens

Tourism is now the mainstay of the Scillonian economy. Of the some 90,000-100,000 visitors to Scilly every year, most come to enjoy the beautiful scenery and pleasant climate. Many take an interest in the wildlife when visiting the islands even if it is not their primary reason for holidaying there. Others will be birdwatchers or have an interest in wild flowers or other wildlife. More serious visitors may come to study the flora or fauna or as volunteers to work for the Wildlife Trust.

Although botanists usually visit Scilly in spring, especially in May which is probably the best month to see most of the special plants, the islands have plenty to offer at any month. Winter visits are essential to see *Ophioglossum lusitanicum* when it is at its best, although some years it may appear as early as October at one site. Some of the vernal species appear very early on Scilly, so visits in March can often be rewarding. In September *Spiranthes spiralis* is in flower as well as the heathland at its purple best. The only disadvantage of visiting Scilly out of season is the lack of transport between the islands before Easter and after October. Currently the RMV *Scillonian* does not regularly sail in winter (although there may be a limited service in future) leaving the more expensive option of flying the only way to get to the islands.

St Mary's is the best starting point for a first visit as it allows easy access to the other inhabited islands and to special boat trips. Staying on one of the 'off islands' is ideal if you want more time to explore that island, but it gives you less options and flexibility for getting around to other islands. The following are some of the best botanical 'hotspots' to look out for.

St Mary's

1   **Porth Minick** Coastal plants and House Holly-fern
2   **Porth Hellick Down** Burial Chamber. Mown heath around the monument
3   **Halangy Down Ancient village** Heath species and Scilly Pigmyweed
4   **Peninnis Head** Heathland species
5   **Innisidgen** The two burial chambers have species-rich mown heath and grassland surrounds
6   **Salakee Down** Waved heath and wet craters
7   **Lower Moors and Shooter's Pool** Wetland plants
8   **Higher Moors and Porthellick** Wetland plants and Greater Tussock-sedge
9   **The Garrison** Coast and heathland species on wall tops, Autumn Lady's-tresses, Smith's Pepperwort, Black and Sea Spleenwort
10 **Harry's Walls** Autumn Lady's-tresses
11 **Tolls Island, Watermill Cove and Helvear Hill** Coastal plants including Spring Squill

Tresco

12 **The Abbey Gardens** are always a 'must' on Tresco, not only for the cultivated plants but for some of the weeds there such as Orange Bird's-foot and Hairy Bird's-foot-trefoil!
13 **Old Blockhouse** Orange Bird's-foot
14 **Appletree Banks** Heath and dune species, also established aliens
15 **Pentle Bay** Dunes and coast, also established aliens
16 **Abbey Pool** Wetland plants
17 **Great Pool** Wetland plants and Tubular Water-dropwort
18 **Abbey Wood and Monument Hill** Ferns, woodland species
19 **Castle Down** Waved heath

St Martin's

St Martin's has the most beautiful white sand beaches at Great Bay and Higher Town Bay.

20 **The Plains** Heathland species and dunes
21 **Chapel Down** Heathland, coastal species including Goldenrod and Pignut
22 **Cricket field and pool** Chamomile-rich grassland, pool with Brackish Water-crowfoot
23 **White Island** Coastal species and heathland

Bryher

24 **Shipman Head and Badplace Hill** Heathland and coastal
25 **Gweal Hill** Heathland and coastal plants
26 **Bryher Pool** Saltmarsh species
27 **Heathy Hill** Heathland with Orange Bird's-foot
28 **Rushy Bay** Coastal and dune species, Dwarf Pansy

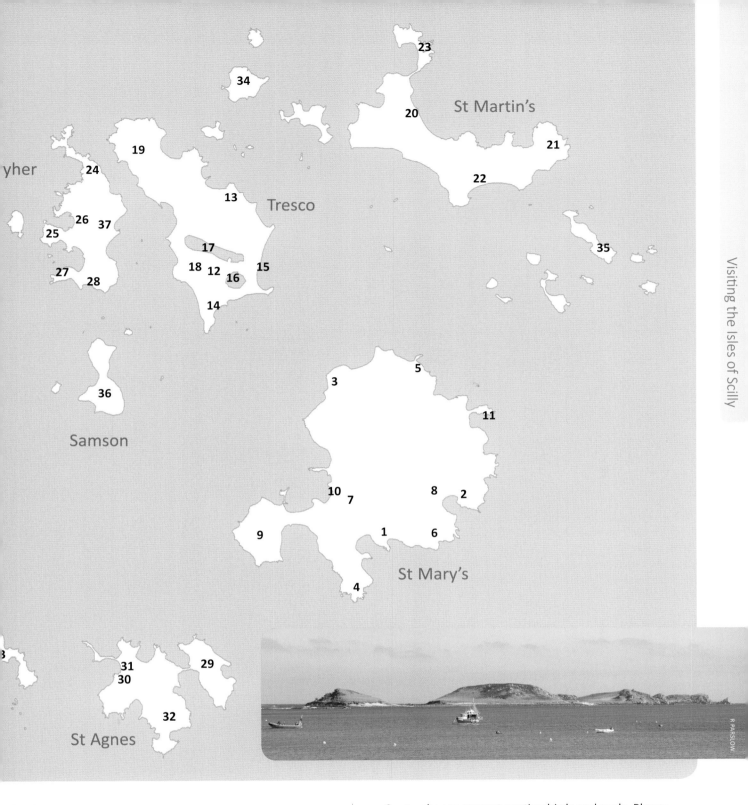

St Agnes

29 **Gugh** Heathland and coastal, Orange Bird's-foot

30 **Lower Town Lane** Lanceolate Spleenwort

31 **The Meadow and Big Pool** Dune grassland, coastal and wetland plants

32 **Wingletang Down** Heathland and coastal, *Ophioglossum lusitanicum* and *Ophioglossum azoricum*

Samson and other uninhabited islands

33 **Annet** is a closed island that can only be visited by permit for research purposes. Some other islands are either totally closed or have some areas closed from April to September to protect nesting birds and seals. Please check the IoSWT maps or web site for details

It is sometimes possible to be landed on other uninhabited islands such as **34 St Helen's** or **35 Great Ganilly** from one of the independent boats

36 **Samson** is easier to get to; tripper boats to **37 Bryher** will sometimes make a detour to drop people off there for a few hours. Besides looking at the ruined village and heathland on the hill tops, Samson is the only site for Pyramidal Orchids and the easiest place to see Skullcap

Finding and photographing plants at Peninnis Head, St Mary's.

SPECIES ACCOUNTS

Glebionis segetum Corn Marigold
Galium luteoalbum Jersey Cudweed
Gnaphalium uliginosum Marsh Cudweed
Helminthotheca echioides Bristly Oxtongue
Hypericum humifusum Trailing St John's-wort
Juncus bufonius Toad Rush
Kickxia elatine Sharp-leaved Fluellen
Kickxia spuria Round-leaved Fluellen
Lamium amplexicaule Henbit Dead-nettle
Lamium hybridum Cut-leaved Dead-nettle
Lepidium didymum Lesser Swine-cress
Lepidium squamatus Swine-cress
Lotus subbiflorus Hairy Bird's-foot-trefoil
Malva neglecta Dwarf mallow
Malva pseudolavatera Smaller Tree-mallow
Malva pusilla Small Mallow
Malva sylvestris Common Mallow
Matricaria discoidea Pineappleweed
Matricaria recutita Scented Mayweed
Medicago polymorpha Toothed medick
Mercurialis annua Annual Mercury
Misopates orontium Weasel's Snout
Myosotis discolor Changing Forget-me-not
Nicandra physalodes Apple-of-Peru

the Isles of Scilly

SPECIES ACCOUNTS

In almost all the following accounts we have used the *New Flora of the British Isles* (Stace, 2010) with the exception of grasses where we have used *Grasses of the British Isles* (Cope and Gray, 2009) but have retained *Anisantha* for the group of alien grasses that seem very distinct from other *Bromus* ssp. On a few occasions we have used the *Flora of Great Britain and Ireland* (Sell and Murrell, 1996 - 2014) where they have described varieties that are found in Scilly but are too local to be included in Stace (2010), for example *Trifolium repens* var. *townsendii*. Throughout the accounts reference is made to Lousley's *The Flora of the Isles of Scilly* (1971) for historical context.

PTERIDOPHYTES	**Ferns and Fern Allies**	**90**
LYCOPHYTES	*Clubmosses*	*90*
EUSPORANGIATE FERNS	*Adder's-tongues and Moonworts*	*90*
CALAMOPHYTES	*Horsetails*	*93*
LEPTOSPORANGIATE FERNS	*True Ferns*	*94*
GYMNOSPERMS	**Conifers**	**108**
ANGIOSPERMS	**Flowering Plants**	**109**
PRE-DICOTS	*Primitive Angiosperms*	*109*
EU-DICOTS	*True Dicotyledons*	*110*
MONOCOTS	*Monocotyledons*	*427*

How to use the species accounts pages

LEPTOSPORANGIATE FERNS ← Major group
DRYOPTERIDACEAE Buckler-fern family ← Family
Cyrtomium falcatum (L.f) C. Presl ← Species
House Holly-fern ← Common name
St Mary's ← Where it is found

DRYOPTERIDACEAE Buckler-fern family
Dryopteris filix-mas (L.) Schott ← Photo of species
Male-fern
Found on all the inhabited islands, also Samson and Teän

Flora Stone
Fernleigh Frond

Neophyte. This well-known alien Asian fern has become established in several places on the island; among rocks on the shore at Porth Minick, in a copse by the Lifeboat Station and in hedgebanks elsewhere.

Native. Although frequently found on most of the inhabited islands it was only located on Bryher in 2009. Hedgebanks, ditches and in wetlands including Higher and Lower Moors.

← Photographer
← Information
← Main section group (Pteridophytes, Gymnosperms or Angiosperms) and their common name

PTERIDOPHYTES FERN AND FERN ALLIES

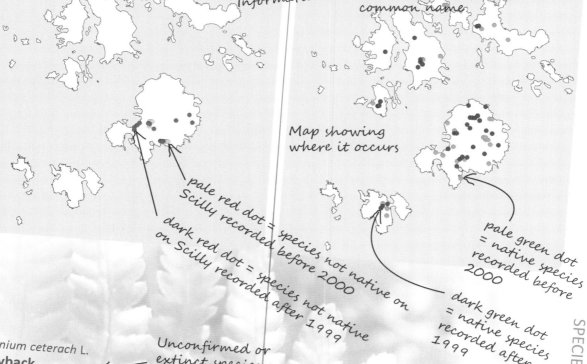

Map showing where it occurs

pale red dot = species not native on Scilly recorded before 2000
dark red dot = species not native on Scilly recorded after 1999

pale green dot = native species recorded before 2000
dark green dot = native species recorded after 1999

Asplenium ceterach L.
Rustyback
Extinct

← Unconfirmed or extinct species are shown in purple

Native. Last recorded in 1992 from a barn wall on the track to Chapel Down, St Martin's; but lost when the barn was converted to a holiday cottage.

531

SPECIES ACCOUNTS

LYCOPHYTES

SELAGINELLACEAE Lesser Clubmoss family
Selaginella kraussiana (Kunze) A.Braun

Mossy or Krauss's Clubmoss

Tresco

Neophyte. This native of southern Africa has long been established as a weed in Tresco Abbey Gardens, where Lousley first recorded it in the lower part of the Garden in 1954. It grows in flowerbeds, lawns and even on the bark of some tree ferns, especially in shady areas. The first record outside the Gardens was from near Abbey Pool in 2016.

EUSPORANGIATE FERNS

OPHIOGLOSSACEAE Adder's-tongue family
Ophioglossum vulgatum L.

Adder's-tongue

St Agnes

Native. Although this fern has been recorded from several places in Scilly in the past, it is now thought that the only reliable records are those from a site near Browarth on St Agnes (SV878086). Where it has been possible to check the records from other sites they have subsequently been found to be *O. azoricum*. At the St Agnes site a large colony of the fern grew in damp, sandy ground under bracken and was still there in the mid 1990s but has not been seen recently. The area had become overgrown with bramble and although this was cut back the fern did not reappear. More recently there has been dumping and other damage to the site.

OPHIOGLOSSACEAE Adder's-tongue family
Ophioglossum azoricum C.Presl.

Small or Lesser Adder's-tongue

St Agnes, Bryher, Tresco, St Mary's and St Martin's

Native. First found near Bar Point, St Mary's by Millett in 1852 (Marquand, 1893). On the inhabited islands the fern grows in a number of discrete sites although it is nowhere as common as on St Agnes. Although the fern was found on the small Toll's Island off St Mary's in 1984, it has not been seen there since. There are no records from any other uninhabited islands, but this may be due to the general inaccessibility of these islands at an appropriate time of year. Also it has been found that the *Ophioglossum* ferns can aestivate for several years before making a reappearance so this can mask their distribution. Even so the fate of many of the sites for this fern is precarious. Some of its former heathland sites have become overgrown by bracken and gorse, others have been severely burned over. In the past at least one grassland site has been ploughed up and once, many years ago, the herbicide *Asulox* was used to control bracken also decimated the *Ophioglossum* growing in the same area.

On Wingletang Down, St Agnes, *O. azoricum* grows in several of the same sites as *O. lusitanicum*, often around the same granite pavements, but with *O. azoricum* in the deeper soils. Where the fern grows in wet places, such as the flooded craters on Salakee Down, St Mary's, the fronds are often taller than usual. Generally, *Ophioglossum azoricum* is widespread in a range of habitats including heathland, maritime grassland, bracken fields in short turf, around granite rocks, wet hollows and in a freshwater seepage on Wingletang Down, St Agnes.

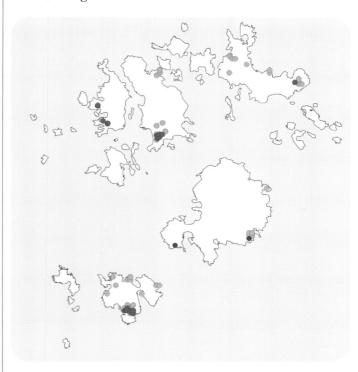

EUSPORANGIATE FERNS

OPHIOGLOSSACEAE Adder's-tongue family

Ophioglossum lusitanicum L.

Least Adder's-tongue

St Agnes

Native. Known from Guernsey for almost a century before it was found on St Agnes in 1950 by John Raven (Raven, 1950). It appears he had found the fern on a previous visit but wanted to confirm the identification. Despite searching the rest of the island, this colony of more than one hundred fronds was the only one he found. This is interesting in that we now know of a number of small sites on Wingletang Down so did Raven overlook these, or has the fern spread since his visit?

Sterile fronds

This very rare native fern reaches its northern outpost in Scilly; it has only been found on Wingletang Down, St Agnes where it grows in areas of very shallow turf over granite rocks. The whole plant is tiny, barely 1.5 cm tall with narrow, linear to somewhat boat-shaped sterile fronds and a fertile frond with between 3-6 sporangia either side. As the sterile fronds usually grow flat on the ground in very short turf or on mossy edges aound flat rocks they are difficult to find. Usually the sporophytes (the aerial fronds) appear in late autumn, although some colonies have appeared as early as September in some years. Spores are shed from December to January and the fronds die down by spring. The dying fronds turn yellowish which helps to distinguish them from the emerging tips of *O. azoricum* fronds that appear just as *O. lusitanicum* is going over. Although several new (or previously undiscovered) colonies have been found on Wingletang, several others have recently become submerged under taller vegetation and apparently lost. It is not always easy to distinguish the sterile fronds of this species from similar fronds of *O. azoricum*. Additionally it is also possible to mistake the leaves of small *Plantago lanceolata* or undivided leaves of some *P. coronopus* plants for the fern!

Most of the sites for *O. lusitanicum* on St Agnes have been mapped and photographed over the years although changes in vegetation cover and management can make it difficult to relocate the colonies every year. As the best time to map the plants is in mid-winter revisiting the sites is a time-consuming and not very easy process. Recently the use of GPS readings is making mapping the ferns more accurate.

It has been a puzzle as to why *O. lusitanicum* is only found on the one small area of St Agnes and apparently nowhere else on the islands. Botanists have looked for it in suitable habitat elsewhere on the islands including on Gugh, but without success. A possible clue is suggested by Page (1988) when he comments on the colonies of *Ophioglossum* ferns being restricted to open patches of turf on cliff-top sites that had not been formerly wooded. There is some evidence that St Agnes and some parts of the westernmost islands may have been outside the woodland that once covered Scilly. Additionally, St Agnes was cut off from the main archipelago by a deep channel very early in the separation of the islands.

As already mentioned, the *Ophioglossum* ferns can remain underground for years, where their roots can survive on nutrients derived from their close association with mycorrhizal fungi, only producing the sporophyte generation when conditions are suitable. This probably explains the reappearance of the ferns after long periods submerged under dense gorse and other tall vegetation when the area is cleared.

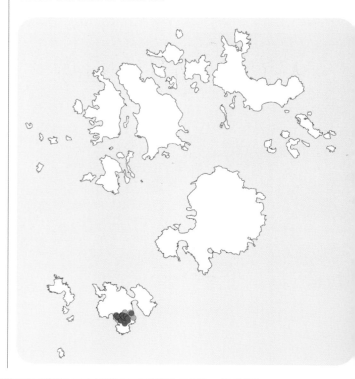

OPHIOGLOSSACEAE Adder's-tongue family
Botrychium lunaria (L.) Sw.
Moonwort
Tresco

Native. This fern was previously only known from Bar Point, St Mary's where it was last seen in 1980. The dune area where it grew has been greatly disturbed since then and attempts to refind the species have proved unsucessful. Then in 2007 a frond was recorded in dunes in the south of Tresco (approximately at SV893135).

CALAMOPHYTES

EQUISETACEAE
Equisetum arvense L.
Field Horsetail
St Agnes, St Martin's and St Mary's

A native species that is probably a casual in Scilly. There have been just a few scattered records from St Agnes, St Martin's and St Mary's; most recently in 1999 when plants were found in a potato field on St Mary's and a bulbfield on St Agnes. A record for *E.* x *litorale* from near Higher Town, St Martin's in 1995 was unconfirmed.

LEPTOSPORANGIATE FERNS

OSMUNDACEAE Royal Fern family
Osmunda regalis L.

Royal Fern

St Mary's, Tresco and Annet

Native. A species usually associated with wetland habitats. On Annet a plant was found in 2008 in a small damp overhang below a cliff at the north of the island. On Lower and Higher Moors, St Mary's the fern has responded to the positive management and has become much more common. Some of the plants in the Moors have grown into very large, handsome specimens. Found in a number of different wetland habitats; around pools, under willow carr, in marshes and along ditch sides, small plants are also sometimes found in water-filled pits on heathland and in freshwater seepages at the base of cliffs.

SALVINIACEAE Water Fern family
Azolla filiculoides Lam.

Water Fern

St Mary's, St Martin's and Tresco

Neophyte. Originally from tropical America this tiny alien fern used to be popular with aquarists and pond-keepers. But in the wild it is a pernicious and invasive species of freshwater sites often completely covering the water surface. In the past (1980s and 1990s) strenuous, but ultimately unsuccessful attempts were made, mechanically and with herbicides to eradicate it from the pools at Newford. Eventually it apparently spontaneously disappeared from there. Since then it has at times invaded Porthellick Pool, ditches on Lower Moors, and both the Abbey and Great Pool, Tresco and the pool on St Martin's in 2016. It has continued to fluctuate in abundance although at times it completely vanishes. As tiny fronds are capable of soon multiplying at great speed and that these are also easily carried by water birds from place to place, it seems unlikely the fern will now be ever completely lost from the islands.

DICKSONIACEAE Tree-fern family
Dicksonia antarctica L'Hér.
Australian Tree-fern
Tresco and St Mary's

DENNSTAEDTIACEAE Bracken family
Pteridium aquilinum (L.) Kuhn.
Bracken
Inhabited and the larger uninhabited islands

Neophyte. Known as planted specimens or associated with gardens. In 2004 a few sporelings were found in woodland on Tresco outside the Gardens. See below.

Native. Historically bracken was an important and valued plant on Scilly. It was cut and utilised in many ways from animal bedding and thatching to fuel. On some islands the bracken fields were divided into lots called 'splats' that were owned by specific islanders.

Now changes in farming have caused many areas and fields to be abandoned so the bracken is no longer kept in check and former heathland and coastal grasslands have been overwhelmed by the fern. Now bracken-rolling and other techniques to control it are being used in some places and this is enabling the heathland plants that had been suppressed under the canopy to reappear. The use of herbicides to control bracken has usually been avoided due to the proximity of other fern species.

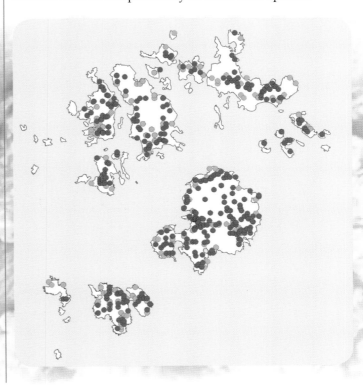

LEPTOSPORANGIATE FERNS

PTERIDACEAE Ribbon Fern family
Adiantum capillus-veneris L.
Maidenhair Fern
St Mary's and Tresco

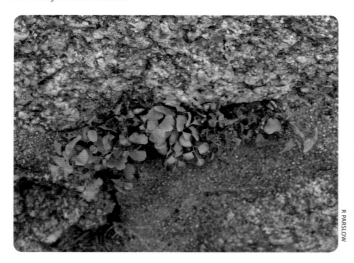

Extinct as a native in Scilly. The last reference to this fern as native was one Lousley refers to growing in a damp hollow on a sea cliff at Tregear Porth in 1959 that was lost in a cliff fall in 1967. All the plants now found have originated from gardens or pot plants. It has been recorded from the Abbey Gardens, Tresco where it was introduced as part of the fern collection. For some years, since at least 2001, a colony was known from inside a building in Well Cross Lane, Hugh Town, from where it has since spread to a nearby wall and along the lane. Plants were also found inside a building at Borough Farm, St Mary's but would have been lost in the renovations (pers. comm. M. and A. Gurr, 2012).

PTERIDACEAE Ribbon Fern family
Pteris cretica L.
Ribbon Fern
Tresco

Neophyte. This alien fern is becoming naturalised in woodland outside the Abbey Gardens.

ASPLENIACEAE Spleenwort family
Asplenium scolopendrium L.
Hart's-tongue
St Agnes, Tresco, St Martin's and St Mary's, also Teän.

Native. May now be more common on St Mary's than when Lousley published his account. Appears to be absent from Bryher. Grows on walls and banks in shady places.

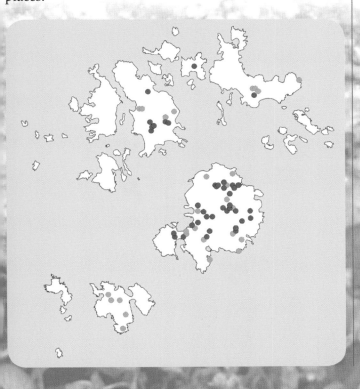

ASPLENIACEAE Spleenwort family
Asplenium adiantum-nigrum L.
Black Spleenwort
Inhabited islands, Samson, St Helen's and Great Ganilly.

Native. In Scilly this is one of the most frequent and typical ferns that grow along the roadside walls and on buildings. It is also found in more natural habitats growing in rock crevices, especially on uninhabited islands. Plants growing on the exposed, seaward side of the Garrison walls, St Mary's appear to be very similar to the unusual form found by Chris Page on the Lizard with very blunt tips to the fronds.

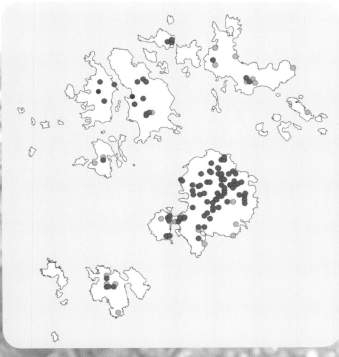

LEPTOSPORANGIATE FERNS

ASPLENIACEAE Spleenwort family
Asplenium x *sarniense* Sleep
(*A. adiantum-nigrum* x *A. obovatum*)
Guernsey Fern
St Mary's

Native. Only one example of this rare hybrid between *A. adiantum-nigrum* and *A. obovatum* ssp. *lanceolatum* has ever been known from Scilly. A plant on a wall on St Mary's, was confirmed by experts from the Natural History Museum in 2004. Although the hybrid is known from the Channel Islands this was the first record from Great Britain. Sadly the condition of plant started to deteriorate from about 2007 and it had eventually disappeared completely by 2009.

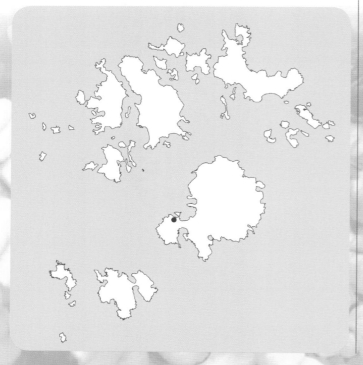

ASPLENIACEAE Spleenwort family
Asplenium obovatum ssp. *lanceolatum*
(Fiori) P.Silva
Lanceolate Spleenwort
St Agnes, St Mary's, Bryher, Tresco and St Helen's

Native. Occurs most frequently on St Agnes and St Mary's, with just a few records from Tresco, Bryher and St Helen's. On St Helen's the fern is found in the Oratory where it was first recorded in 1983 and was formerly recorded from Samson where it was last seen in 1983. It mostly grows on granite walls but occasionally on rocks. On roadside walls it is frequently found alongside *A. adiantum-nigrum* but despite much searching there has only been one record of the hybrid.

ASPLENIACEAE Spleenwort family
Asplenium marinum L.
Sea Spleenwort
All inhabited and most uninhabited islands

Native. Predominately coastal, although not found on the soft coasts within the ring of the archipelago. Small plants are frequently found inland, on rocks and walls, such as on buildings in Hugh Town. Away from habitations it is found more naturally in caves and rock overhangs where it can grow very large, for example at Carn Adnis, St Agnes one year some fronds were as much as 70 cm in length.

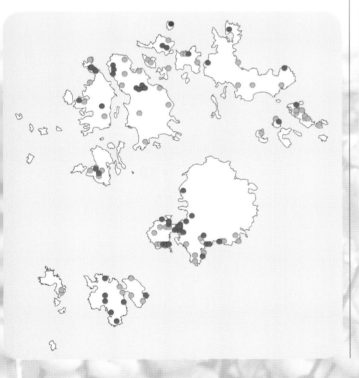

ASPLENIACEAE Spleenwort family
Asplenium trichomanes ssp. *quadrivalens* D.E.Mey
Maidenhair Spleenwort
St Agnes, St Mary's, Tresco and St Martin's

Native. Less common than formerly, post 2000 records are of a few plants on walls at Porthloo, St Mary's and Higher Town, St Martin's. The fine display of the fern on a wall of a house at Higher Town, St Martin's was lost when the wall was repointed some time after 2000. Most other records are associated with old greenhouses and other buildings.

LEPTOSPORANGIATE FERNS

ASPLENIACEAE Spleenwort family
Asplenium ruta-muraria L.
Wall-rue
St Mary's and St Martin's

Native. Known from just two places; a garden wall at Porthloo, St Mary's and on a derelict building and nearby wall at Lower Town, St Martin's. Both sites are vulnerable and the plants on the building on St Martin's are likely to be lost when it is renovated. Fortunately a few ferns were found on a new wall nearby in 2011. The plants at Porthloo appear to be slightly more fleshy than usual, possibly due to their proximity to the sea.

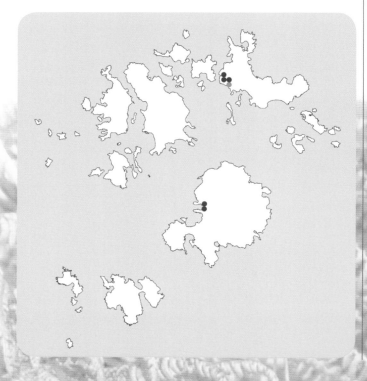

Asplenium ceterach L.
Rustyback
Extinct

Native. Last recorded in 1992 from a barn wall on the track to Chapel Down, St Martin's; but lost when the barn was converted to a holiday cottage.

WOODSIACEAE Lady-fern family
Athyrium filix-femina (L.) Roth
Lady-fern
St Mary's and Tresco

Native. Widespread and locally frequent in wetland habitats, in Higher and Lower Moors, along Holy Vale, at Watermill and Salakee on St Mary's, woodland and pool-sides on Tresco. Pool edges, ditches and under willow carr.

BLECHNACEAE Hard-fern family
Woodwardia radicans (L.) Sm.
Chain Fern
Tresco

BLECHNACEAE Hard-fern family
Blechnum cordatum (Desv.) Hieron
Greater Hard-fern
Tresco and St Mary's

Neophyte. Originally from S. America. Found in woodland outside the Gardens on Tresco and has been long established in a ditch at Salakee, St Mary's (where it was first recorded in 1957). It was found on Ragged Island in the Eastern Isles in 1987, but apparently did not persist.

Neophyte. Originally from S.W. Europe. Recorded on Tresco where it appears to be naturalising in woodland outside the Gardens.

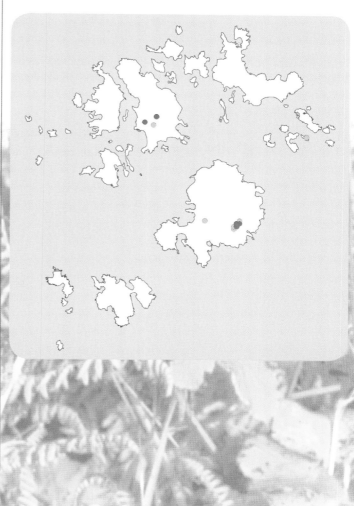

LEPTOSPORANGIATE FERNS

BLECHNACEAE Hard-fern family
Blechnum spicant (L.) Roth
Hard-fern
St Mary's and Tresco

Native. Although well-known in Cornwall this species is suprisingly rare on the Isles of Scilly. Lousley considered it extinct, but it has occurred in a few places near the Abbey on Tresco (Peter Clough pers. comm.). Until it was found on Tresco (SV892145) in 2015 the most recent records were from Holy Vale, St Mary's in 1992 and near Pentle Bay, Tresco in 1982. Ditches and hedgebanks.

DRYOPTERIDACEAE Buckler-fern family
Polystichum setiferum (Forssk.) T. Moore ex Woyn
Soft Shield-fern
St Agnes, St Mary's, Tresco and St Martin's, also Samson, Teän and Northwethel

Native. Scattered records and apparently uncommon, but often makes a temporary reappearance after taller vegetation is cut back. Shady places under trees, hedgebanks and wet flushes.

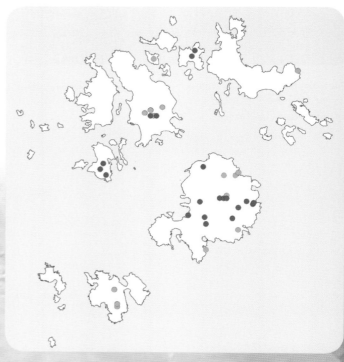

DRYOPTERIDACEAE Buckler-fern family
Cyrtomium falcatum (L.f) C. Presl
House Holly-fern
St Mary's

Neophyte. This well-known alien Asian fern has become established in several places on the island; among rocks on the shore at Porth Minick, in a copse by the Lifeboat Station and in hedgebanks elsewhere.

DRYOPTERIDACEAE Buckler-fern family
Dryopteris filix-mas (L.) Schott
Male-fern
Found on all the inhabited islands, also Samson and St Helen's

Native. Although frequently found on most of the inhabited islands it was only located on Bryher in 2009. Hedgebanks, ditches and in wetlands including Higher and Lower Moors, St Mary's.

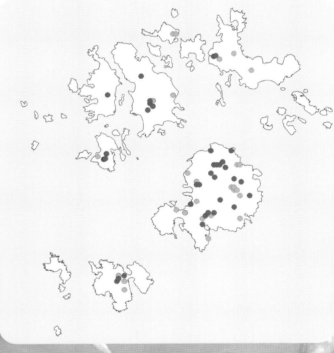

LEPTOSPORANGIATE FERNS

DRYOPTERIDACEAE Buckler-fern family
Dryopteris affinis agg.
Golden-scaled Male-fern
St Mary's and Tresco

Native. *Dryopteris affinis* ssp. *affinis* and *D. borreri* are almost certainly under-recorded. Both were recorded by members of BPS in woodland on Abbey Hill in June 2004. Found on hedgebanks and in wooded areas.

DRYOPTERIDACEAE Buckler-fern family
Dryopteris carthusiana (Vill.) H.P. Fuchs
Narrow Buckler-fern
St Mary's

Native. The only recent record is from Holy Vale in 1993.

DRYOPTERIDACEAE Buckler-fern family
Dryopteris dilatata (Hoffm.) A. Gray
Broad Buckler-fern
On all the inhabited islands, also Teän and Samson

Native. Most common on St Mary's, with a scatter of records from the other islands. Hedgebanks, damp areas on heathland and under trees.

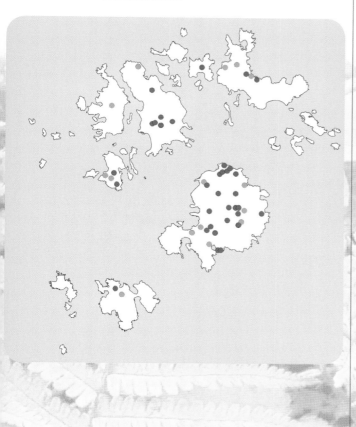

Dryopteris aemula (Aiton) Kuntze
Hay-scented Buckler-fern
Extinct

Although Lousley accepted records for this species, last recorded on St Mary's in 1893, it was not seen by him.

POLYPODIACEAE Polypody family
Polypodium agg. L.
Polypodies
All inhabited islands, most uninhabited islands

Native. Most records of polypody from Scilly do not distinguish between the taxa. The ferns are very commonly found on walls, rocks and as epiphytes on trees.

LEPTOSPORANGIATE FERNS

POLYPODIACEAE Polypody family
Polypodium vulgare L.
Polypody
All inhabited islands, Teän, Samson and Great Ganilly

Native. Records for this species in Scilly have only recently been separated from those of the aggregate. Earlier records for the species may be suspect.

POLYPODIACEAE Polypody family
Polypodium interjectum Shivas
Intermediate Polypody
All inhabited islands, Samson and Teän

Native. This is the common polypody in Scilly - at least on the inhabited islands. Records for this species in Scilly have only recently been separated from *P. vulgare*.

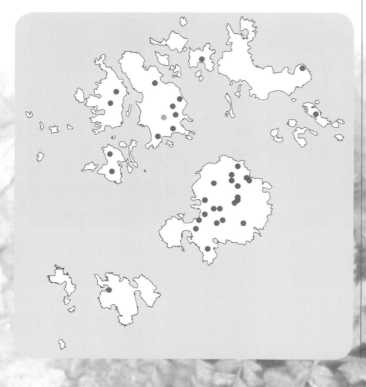
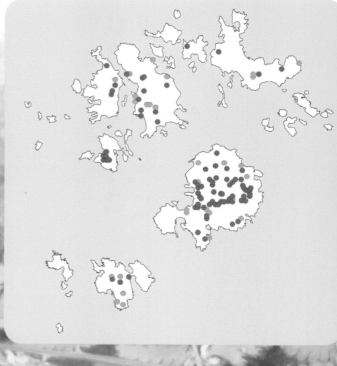

POLYPODIACEAE Polypody family
Polypodium x *mantoniae* Rothm. and U. Schneid.
(*P. vulgare* x *P. interjectum*)
Hybrid Polypody
St Mary's

Native. Only one record of this hybrid, on a roadside at SV91591220 in 2003 (determined by Rose Murphy).

POLYPODIACEAE Polypody family
Phymatosorus diversifolius (Willd.) Pic. Serm.
Kangaroo Fern
Tresco and St Mary's

Neophyte. This alien fern is native to New Zealand and Australia. On Scilly can be found in damp places in woodland outside Tresco Abbey Gardens and in Rocky Hill Lane, St Mary's.

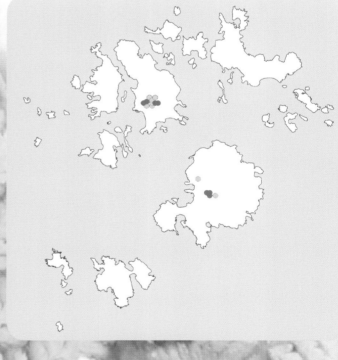

GYMNOSPERMS CONIFERS

PINACEAE – Pine family
Pinus contorta Douglas ex Loudon
Lodgepole Pine
Inhabited islands

Neophyte. This pine originated from North America but is now commonly planted. Used as a windbreak on the inhabited islands, it does not have the same visual impact as *Pinus radiata* but was popular for a time.

PINACEAE – Pine family
Pinus radiata D. Don
Monterey Pine
Inhabited islands

Neophyte. One of the trees that grow on the coast of California that was introduced to Scilly by Augustus Smith. Planted on the inhabited islands as windbreaks, these pines have become a feature in the Scillonian landscape since they were discovered to be tolerant of salt-laden winds. Unfortunately when mature they can topple easily due to their shallow root-plate and went out of favour for a time. Recently some new planting has taken place. Self-seeded specimens are rare as the seeds rarely germinate on Scilly, but occasionally do occur, for example at Lunnon Farm, St Mary's. Older, mature pines support interesting lichens, and bats are known to use deeper cracks for roosting.

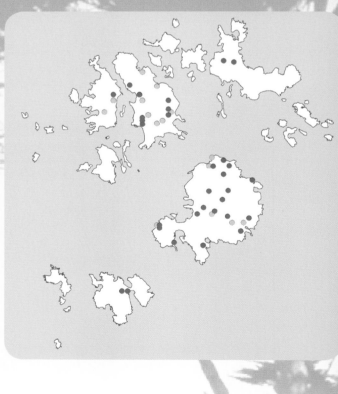

Pinus pinaster Aiton
Maritime Pine
St Mary's and St Agnes?

A record from St Mary's and one from St Agnes need confirmation.

CUPRESSACEAE – Juniper family
Cupressus macrocarpa Hartw. Ex Gordon
Monterey Cypress
Inhabited islands

Neophyte. Also originally a species of the Californian coast. Frequently planted on the inhabited islands. As with *Pinus radiata* this species was one of the trees that was discovered to withstand salt winds. The trees can grow large, but tend to lose the foliage from the lower branches as they mature. They appear to be popular with goldcrests and firecrests that can be found in their cover. No self-seeded plants have been recorded.

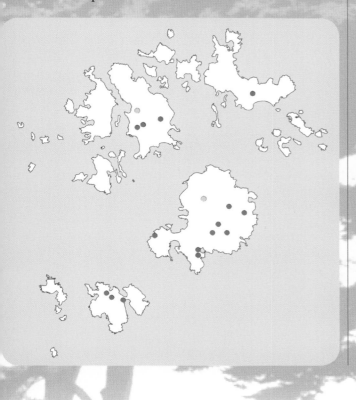

PRE-DICOTS Primitive Angiosperms
NYMPHAEACEAE – Water-lily family
Nymphaea alba L.
White Water-lily
Bryher

Native, but introduced to Scilly. Planted in the Little Pool, Bryher before 2002. May still be there although attempts have been made to remove it to restore the native vegetation. More recently the pool has become invaded by *Typha latifolia*.

EU-DICOTS True Dicotyledons

PAPAVERACEAE – Poppy family
Papaver pseudoorientale (Fedde) Medw.
Oriental Poppy
Tresco and St Mary's

Neophyte. Garden escape found in dunes on Tresco and at Old Town Bay, St Mary's.

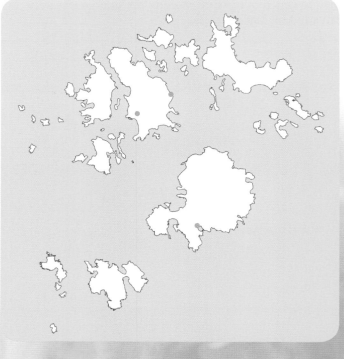

PAPAVERACEAE – Poppy family
Papaver somniferum L.
Opium Poppy
St Agnes, Tresco, St Martin's and St Mary's

Archaeophyte. There are scattered records from waste or disturbed ground and from where garden rubbish has been dumped. The flowers can be single or double and in hues from lilac to pink or red.

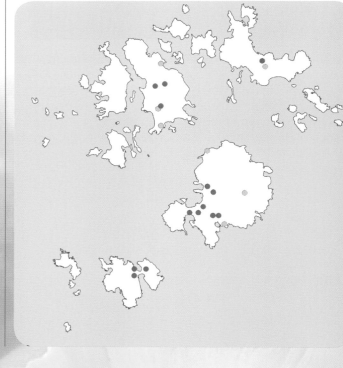

PAPAVERACEAE – Poppy family
Papaver rhoeas L.
Common Poppy
Inhabited islands

Archaeophyte. Found on all the inhabited islands but uncommon on St Agnes, on arable and disturbed land. Some plants on St Mary's and St Martin's have very striking black-spotted centres and have been identified as 'ladybird' or 'Flanders' poppies, sometimes separated as *P. rhoeas* var. *commutatum*. The poppies occasionally seen in abundance in bulbfields on St Martin's (SV919161) were first recorded by Lousley in 1975 as *Papaver arenarium* M.Bieb. (see photo below). Cultivated and disturbed ground.

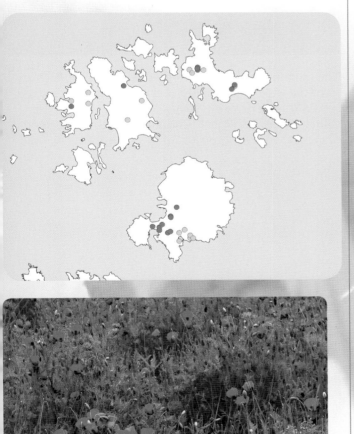

PAPAVERACEAE – Poppy family
Papaver dubium L.
Long-headed Poppy
Inhabited islands

Archaeophyte. This is the common poppy in Scilly; found on all the inhabited islands although less frequent on Tresco. It is a more vermillion or pinkish colour than the scarlet of *P. rhoeas* and tends to drop its petals almost as soon as they open, revealing the tall capsule. It is not unusual to find both this and *P. rhoeas* growing in the same field. Arable and disturbed land.

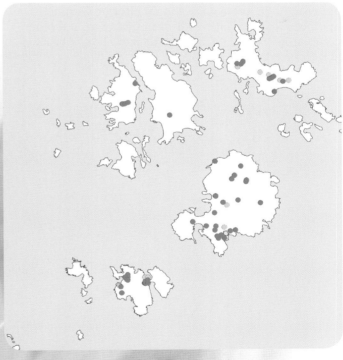

EU-DICOTS True Dicotyledons

PAPAVERACEAE – Poppy family
Glaucium flavum Crantz
Yellow-horned Poppy
Inhabited islands, Teän and Great Ganilly

Native. As with all strandline species the population fluctuates from season to season. The poppy occurs at the top of the shore on both sandy and shingle beaches on the inhabited islands, also on Teän and Great Ganilly. It is very vulnerable to extreme high tides when it can be swept away. Lousley recorded it from Samson, but there are no recent records. There are references to the islanders growing the poppy in their gardens, apparently the roots were sought after by Cornish apothecaries for treating 'disordered lungs' and other complaints (Borlase, 1756). Sandy and shingle beaches and dune edges.

PAPAVERACEAE – Poppy family
Fumaria L.
Fumitories

The identification of fumitories in Scilly has benefitted from the expertise and advice of Rosaline Murphy and her BSBI handbook on *Fumitories of Britain and Ireland* (Murphy, 2009) in identification of this confusing and often misrecorded group.

PAPAVERACEAE – Poppy family
Fumaria capreolata ssp. *babingtonii*
(Pugsley) P.D.Sell
White Ramping-fumitory
Inhabited islands

Native. Although recorded from all the inhabited islands this species is most frequent on St Agnes and St Mary's. Records from Puffin Island off Samson in 1984, and Great Ganinick in the Eastern Isles in 1990 are presumed to have been due to accidental introductions by gulls. A handsome plant that sometimes gets misidentified as *F. occidentalis* (especially as the flowers often flush pink after fertilisation) but can be distinguished from that species by the flowers becoming recurved in fruit. Mainly found in hedgebanks, disturbed ground and arable fields.

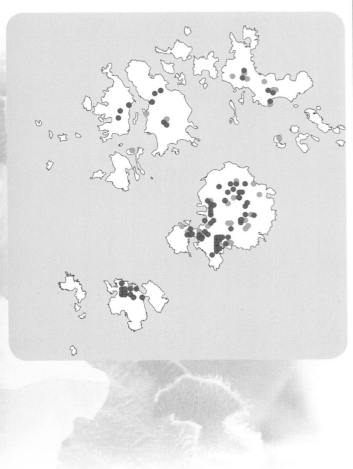

PAPAVERACEAE – Poppy family
Fumaria occidentalis Pugsley
Western Ramping-fumitory
St Mary's and St Martin's

Native. Virtually confined to St Mary's until 1988 when a plant was found on a hedgebank at Higher Town, St Martin's. It is suspected that seeds had been accidentally introduced from St Mary's. The population on St Martin's has persisted and in 1997 plants were also found on the local rubbish dump. By 2004 they were found to have spread to Churchtown Farm, where there were still a few plants in 2011. On St Mary's the population has slightly extended its range since the map in Lousley's Flora was published. This species is considered quite vulnerable, populations are subject to local fluctuations and exterminations, especially when hedgebanks are treated with herbicides. This is more so as it is more frequently a scrambler over walls and hedges rather than as an arable weed, although it does sometimes grow in bulbfields among the crop. Some records for this species are suspected of being errors for *F. capreolata* which can be equally showy. Cultivated ground and hedgebanks.

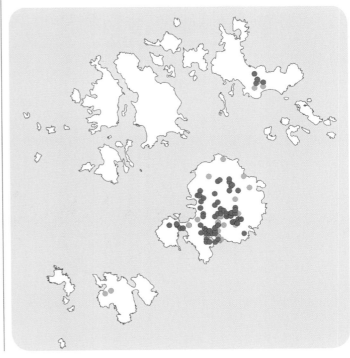

EU-DICOTS True Dicotyledons

PAPAVERACEAE – Poppy family
Fumaria bastardii Boreau
Tall Ramping-fumitory
Inhabited islands

Native. Frequent on all the inhabited islands, but less commonly on Tresco. Often found in the same fields as *F. muralis* ssp. *boroei* and in Bryher occurs in mixed populations with *F. muralis* ssp. *boroei* and *F. officinalis*. The variety *hibernica* was been recorded from Old Town Churchyard, St Mary's in 2006 and var. *gussonei* from both Tresco and St Agnes in 1982. It can sometimes be picked out from *F. muralis* ssp. *boroei* by the slightly different pink of the flowers – but this is not a reliable character. Arable fields and disturbed ground.

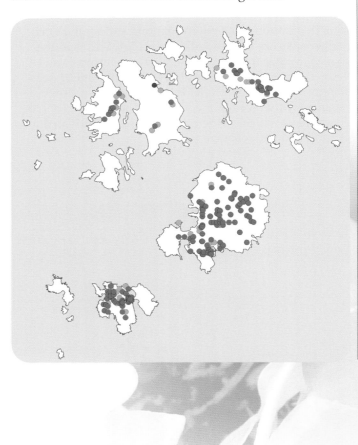

PAPAVERACEAE – Poppy family
Fumaria muralis ssp. *boroei* (Jord.)
Common Ramping-fumitory
Inhabited islands

Native. Probably the most widespread and most variable fumitory found in the Isles of Scilly. Frequently found in mixed populations with other species. The variety *major* has been recorded from several places on St Mary's and Bryher, although it is deeper pink in colour, it is also larger, so measurements of the flower parts are needed for confirmation. The species occurs in bulbfields, other cultivated habitats and disturbed land.

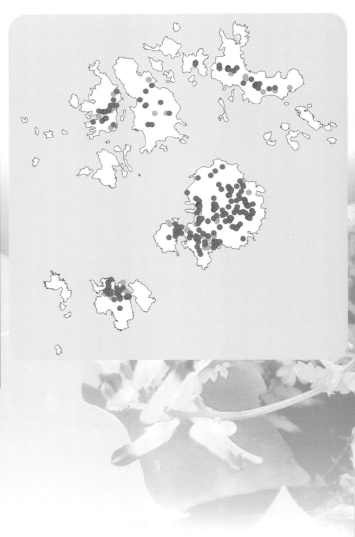

PAPAVERACEAE – Poppy family
Fumaria purpurea Pugsley
Purple Ramping-fumitory
St Mary's

Native. Recorded from bulbfields at Old Town, St Mary's (SV910102 and SV912103) in 1987 but not seen there since 1988. One of the fields was later grassed over. Despite the area being searched in subsequent years this species has not been seen since. Records from St Martin's and elsewhere appear to have been misidentifications for *F. muralis* ssp. *boroei*.

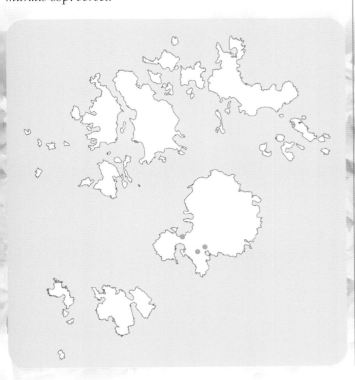

PAPAVERACEAE – Poppy family
Fumaria officinalis L.
Common Fumitory
Bryher (and possibly St Agnes)

Native. In recent years the only reliable records have been from arable fields on Bryher, although there are unconfirmed records from St Mary's in 1988 and from Troy Town, St Agnes in 2005. Lousley recorded *F. officinalis* as frequent from all the inhabited islands, but that is now not the case. It is possible *F. officinalis* may have been more widespread at that time when there were many more bulbfields on the islands. Both ssp. *officinalis* and ssp. *wirtgenii* collected from Bryher recently have been confirmed by Rose Murphy. Arable fields and disturbed land.

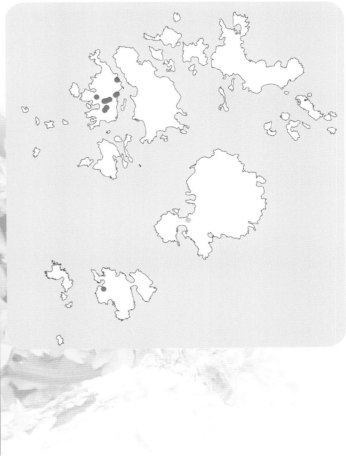

EU-DICOTS True Dicotyledons

RANUNCULACEAE – Buttercup family
Caltha palustris L.
Marsh-marigold
Tresco and St Mary's

A native plant in the British Isles, but not on Scilly, where it has been introduced as a garden plant to pools on Tresco and St Mary's. Some of the Tresco plants appear to be a large-flowered cultivar.

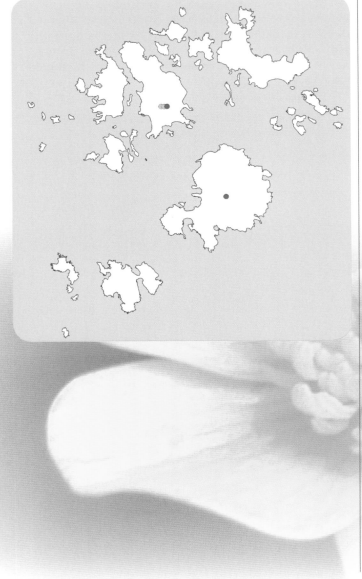

RANUNCULACEAE – Buttercup family
Clematis vitalba L.
Traveller's-joy
St Mary's

Another native species that is almost certainly of garden origin on Scilly. Established in the area between Longstone and the adjacent part of Holy Vale, St Mary's since 1992 and Bryher from a track behind Little Popplestone in 2006.

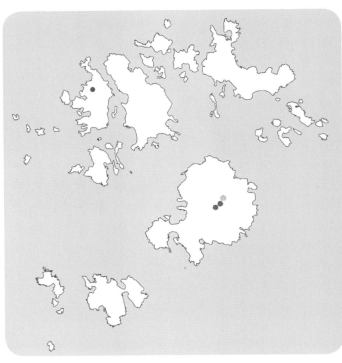

Clematis flammula L.
Virgin's-bower
Tresco

Neophyte. Garden escape near Pentle Bay, Tresco, 2004.

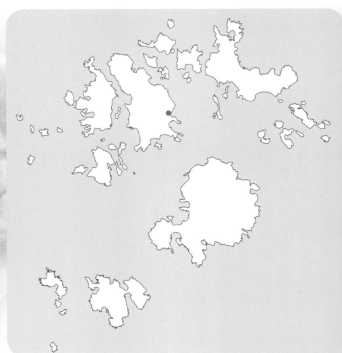

RANUNCULACEAE – Buttercup family
Ranunculus acris L.
Meadow Buttercup
St Agnes, St Martin's, Tresco and St Mary's

Native. An uncommon plant on Scilly, found in just a few scattered grassland sites. On St Mary's this included a lawn at Porthcressa (where it may have been introduced during restoration work) in 1988 which was still there in 2008. This area has been greatly disturbed by building work in 2012 so the buttercup may have been lost. Other recent records are from Higher Moors, St Mary's in 2007; on a wall at Higher Town in 2009 and a field near English Island Point, St Martin's in 1994. On St Agnes it has been found in a field near Troy Town in 2010 and occasionally recorded on The Meadow. On Tresco it has been recorded in the churchyard in 1975 and again in 2002. Most of these records are of just one or a few plants, there appear to be no places where the buttercup is found to be common or in typical grassland situations.

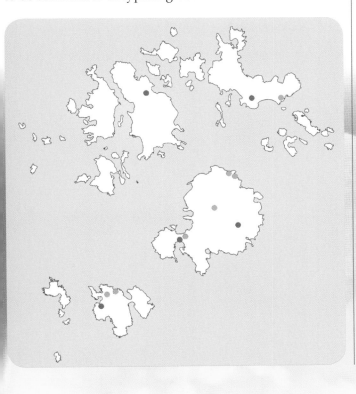

RANUNCULACEAE – Buttercup family
Ranunculus repens L.
Creeping Buttercup
Inhabited islands, also St Helen's, Great Ganilly and Samson

Native. The common buttercup in Scilly. Plants growing along the Holy Vale stream, St Mary's can be unusually large. Found in a range of habitats; roadsides, field boundaries, cultivated ground and wetlands.

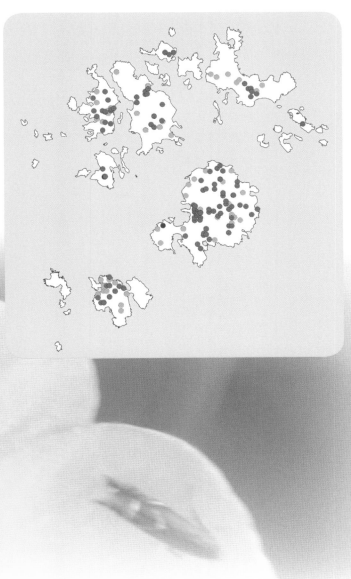

EU-DICOTS True Dicotyledons

RANUNCULACEAE – Buttercup family
Ranunculus bulbosus L.
Bulbous Buttercup
Inhabited islands, also Samson

Native. Nowhere abundant. Lousley recorded the species from Teän. Plants attributable to the variety *dunensis* have been occasionally recorded from coastal sites. Mainly found in grasslands and around the coast.

RANUNCULACEAE – Buttercup family
Ranunculus sardous Crantz
Hairy Buttercup
Inhabited islands, also Samson

Native. Common in the northern half of St Agnes, elsewhere there are scattered records from all the inhabited islands and recently Samson (2008). Formerly found on Teän in 1938 (Dallas, in Lousley, 1971). It can usually be separated from other buttercups such as *R. bulbosus* by the line of tiny tubercles around the edge of the achenes (use a lens). Occurs in two main types of habitat; either in wet grassland, such as around Big Pool, St Agnes or in bulbfields, both within the crop and on the grassy paths and headlands.

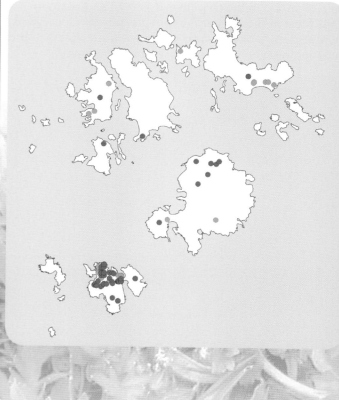

RANUNCULACEAE – Buttercup family
Ranunculus marginatus var. *trachycarpus*
(Fisch. and C.A. Mey)
St Martin's Buttercup
St Martin's

Neophyte. This native of the Eastern Mediterranean was only recognised on the Isles of Scilly in the 1950s when material collected was grown on to obtain fruits for identification (Lousley, 1971). The buttercup was known for many years from a few fields at Higher Town on St Martin's where at one time it was in such abundance as to turn the fields yellow. Later the use of herbicides on the farm almost eliminated it so that by the 1980s the population was down to a few plants. Since 1997 there has been some recovery although the main site was lost when a large packing shed was built in the field. From 2007 it was found in other fields on the farm with several dozen plants growing along the edge of the crop. As *R. sardous* grows in the same fields this has sometimes led to confusion as the two species look somewhat similar when not in fruit. The population now appears to be relatively stable, with plants flowering in May/June along the edges of bulbfields. In 2005 Chris Pogson found a group of about 40 plants growing in a rutted track (SV93111541) between Higher Town and the cricket field. In 2010 there had been a slight increase in the number of plants there. It is frequently sought by botanists, but where it grows on private land they should seek permission before entering the farm. Now plants have been found elsewhere on St Martin's it is possible it is spreading to new sites. Cultivated and disturbed ground.

Ranunculus marginatus fruit

Ranunculus arvensis L.
Corn Buttercup
Extinct

Archaeophyte. Only one casual record, by Lousley from the garden of Star Castle in 1940.

EU-DICOTS True Dicotyledons

RANUNCULACEAE – Buttercup family
Ranunculus muricatus L.
Rough-fruited (or Scilly) Buttercup
Inhabited islands

Neophyte. According to Lousley the first published reference to this species was in 1923. Although recorded from all the inhabited islands this buttercup is now only widespread on St Mary's. At one time considered a pernicious weed of bulb fields it has now declined on all the islands. Lousley failed to find it on St Martin's, but it was found there in the vicinity of the church and adjoining fields between 1986 and 1988, although not since. An easily recognised plant; it has relatively small flowers, but the distinctive fruits have large thorns on the achenes. On St Mary's a large patch of the plant was found growing under a bench on Holgate's Green in 2011. By 2014 the patch had grown and spread over disturbed sand in the corner of the Green. There are also a few records from Bryher, St Agnes and Tresco. Cultivated and disturbed ground.

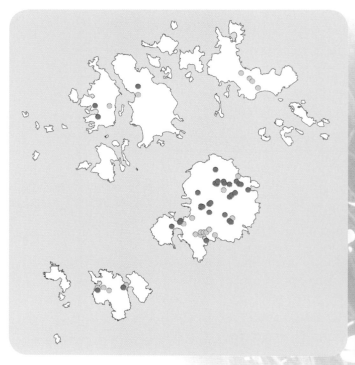

RANUNCULACEAE – Buttercup family
Ranunculus parviflorus L.
Small-flowered Buttercup
Inhabited islands

Native. Although the species has been recorded from all the inhabited islands, it is uncommon on St Agnes (as also noted by Lousley). Now mainly associated with bulbfields and arable fields but occasionally found in disturbed dune or coastal habitats from which it presumably originated. Cultivated and disturbed ground.

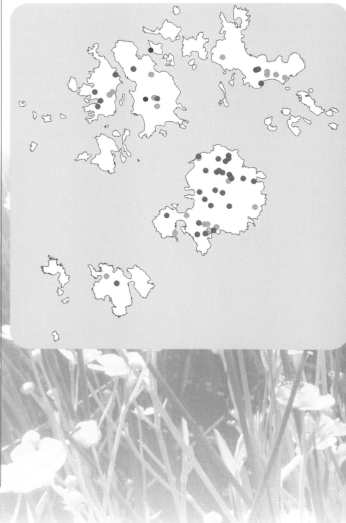

RANUNCULACEAE – Buttercup family
Ranunculus sceleratus L.
Celery-leaved Buttercup
St Agnes and St Mary's

Native. Just three casual records; in two cases they were probably birdsown, growing in wet places on St Agnes, Little Pool (1987) and Teän Plat Point (1998). Also found growing in a flower pot in Hugh Town, St Mary's (2009).

Ranunculus tripartitus DC.
Three-lobed Crowfoot
Extinct

Formerly found in the acid bog in Higher Moors. But long gone, the last record was 1887 (Hanbury in Lousley, 1971).

RANUNCULACEAE – Buttercup family
Ranunculus flammula L.
Lesser Spearwort
Inhabited islands

Native. Found in pools and other wetland sites on the inhabited islands. It has also been found in temporary pools on heathland, for example along tracks on Tresco and St Mary's. In 2000 a few plants were found in the vegetation fringing Big Pool, St Agnes, presumably introduced by waterbirds. The plants on Lower and Higher Moors, St Mary's are robust, upright plants, typical of the species. A tiny creeping, procument form of Lesser Spearwort is found on the drawdown edge of Abbey Pool but has not yet been specifically determined (see right image above). It is very possible this form has been known on the islands since at least 1864 (recorded as *R. reptans* by Townsend and represented in his Herbarium according to Lousley). It was believed to only grow around Abbey Pool until it was recorded by E. Sears in 2009 growing in a temporary pool on a heathland track on Castle Down above Beacon Hill, Tresco and in 2012 by P. Tompsett in a similar place near Charles's Castle. When grown in a garden situation the stems grow longer, but the leaves and flowers still remain tiny. Pools and wetlands.

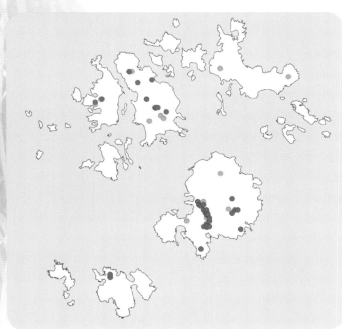

EU-DICOTS True Dicotyledons

RANUNCULACEAE – Buttercup family
Ranunculus hederaceus L.
Ivy-leaved Crowfoot
St Mary's

Native. The only recent records are from Salakee Lane and Watermill Stream in 2004 and Higher Moors in 2002 and 2013 (still there 2015). Never very common, it appears to have been lost from most former sites when these dried out. There had been a large population in a wet field at Salakee where it grew with *Sibthorpia europaea*. When the ditch that drained into the field was re-routed some time before 2004 both species were lost. The Higher Moors plants were found where Jim Askins had cut mist-net lines through the reeds which suggests the plant may still exist somewhere within the reed beds. Found on wet mud or ditch-sides.

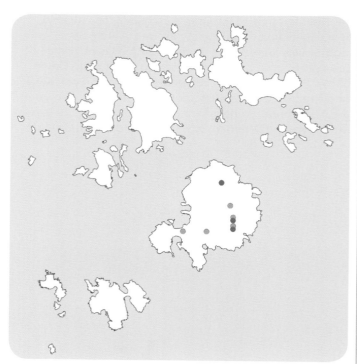

RANUNCULACEAE – Buttercup family
Ranunculus baudotii Godr.
Brackish Water-crowfoot
Bryher, St Martin's and St Mary's

Native. The best population of this attractive buttercup is in the pool on Pool Green, St Martin's. Some years it completely covers the water surface, other times there may be just a few plants. Occasionally the pool overflows and plants spread into the surrounding grassland. Other years, for example after the drought in 2011, the pool dried up completely and only a few plants could be seen on the mud at the bottom. It has also been recorded in 2007 from Little Pool, Bryher and in 2015 from a ditch on St Mary's (SV9123210280) where a large pit had been dug. At one time the species could be found in pools on all the inhabited islands including the small pools on heathlands such as Wingletang and Nag's Head Down, St Agnes. The reason for the decline is not known. Freshwater ditch and pools.

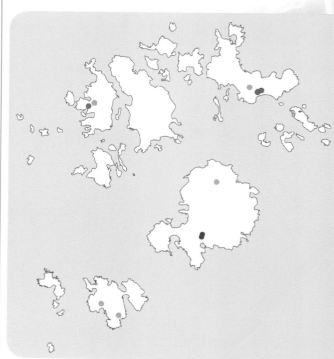

RANUNCULACEAE – Buttercup family
Ficaria verna Huds.
Lesser Celandine
Inhabited islands, also larger uninhabited islands

Native. Apparently all the celandines on Scilly are the ssp. *fertilis* which does not have tubers in the leaf axils. It is widespread on all the inhabited islands as well as on Samson and Annet. There are also records from St Helen's, Great Ganilly, the Arthurs and Teän, where it almost certainly still occurs, although not recorded recently (possibly because no botanists have been there early in the year). The ssp. *ficariiformis* that was found in the copse near the school, St Mary's in 2005 may have been of garden origin, but it has not been seen recently. Open habitats, including heathland, often under bracken, in hedges and in the bulbfields.

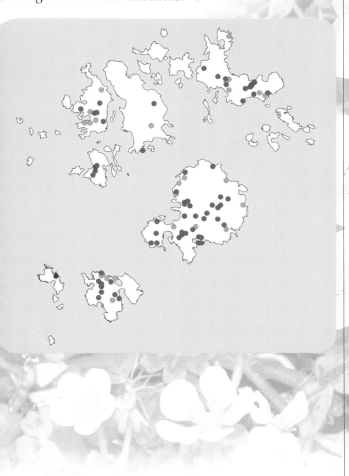

RANUNCULACEAE – Buttercup family

Ranunculus trichophyllus Chaix
Thread-leaved Water-crowfoot
Extinct

Native. Lousley included this species on the basis of a specimen from Tresco. But he did not find the plant himself and there are no other records.

Thalictrum minus L.
Lesser Meadow-rue
Extinct

Formerly found in the acid part of Higher Moors. But long gone, the last record was 1887 (Hanbury in Lousley, 1971).

EU-DICOTS True Dicotyledons

BUXACEAE – Box family
Buxus sempervirens L.
Box
Tresco, St Martin's and St Mary's

Native in the British Isles but a neophyte on Scilly. An occasional garden escape; recorded on Tresco in 1997 (SV894145), 2010 (SV895151) and 2012 (SV895138 and SV896153); St Martin's in 1995 (SV931154) and in Old Town churchyard, St Mary's, 1995 (SV911101). Recorded from several habitats; hedgebanks, woodland and dunes.

GUNNERACEAE – Giant-rhubarb family
Gunnera tinctoria (Molina) Mirb.
Giant-rhubarb
Bryher, Tresco and St Mary's

Neophyte. Occasional garden escape. Plants found in a ditch near Bar Point (SV911126), St Mary's, later colonised an abandoned field near Pendrathen (SV91161267) where 53 plants were found in 2015. DNA results are awaited to confirm species. In 2010 a tiny seedling *Gunnera* sp. was found growing in the bark on top of a fallen tree on Abbey Drive. It did not persist.

Gunnera manicata Linden ex André
Brazilian Giant-rhubarb
Tresco

Neophyte. Planted near Abbey Gardens.

SAXIFRAGACEAE – Saxifrage family
Bergenia crassifolia (L.) Fritsch
Elephant-ears
Tresco

Neophyte. A plant recorded from woodland on Tresco by A. Underhill in 1993 would have been a garden escape.

CRASSULACEAE – Stonecrop family
Crassula decumbens Thunb.
Scilly Pigmyweed
St Mary's

Neophyte. First recorded in 1959 when specimens were identified and an account later published (Lousley, 1960). Only known from the path below Halangy Down Ancient Village and Bant's Carn Farm until recently. The plant is an annual and usually very tiny, although slightly larger plants, a few centimetres tall can occur in cultivated fields. Usually found early in the year, the plants are prostrate and turn bright red as they go over and are easier to spot. Occasionally may produce a second flush later in the year (October 1993 and 2008). Plants have been spreading very slowly along the path towards the golf course. In May 2013 plants were found by Mark Spencer growing in the uneven surface of the pavement at Rocky Hill SV91531115, with about 100 tiny plants over several metres of pavement. This suggests it had been present for some time. In 2015 the plants had increased to several hundreds. Also in May 2015 about 1,000 plants were found growing on the grassy mound of the Bant's Carn chambered tomb. Occurs both in cart tracks, a pavement, short turf and as a weed in cultivated fields.

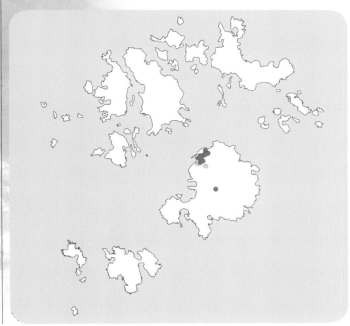

EU-DICOTS True Dicotyledons

CRASSULACEAE – Stonecrop family
Crassula multicava Lemaire
Fairy Crassula or Shrubby Stonecrop
Tresco, St Martin's and St Mary's

Neophyte. Garden escape first recorded from Abbey Hill, Tresco by Julie Clarke in 1996. That and other records from Tresco in 1997 and 1998 were published by Eric Clement in *BSBI News* (Clement, 2003). More recently it has been found again on Tresco near the Monument (SV892142) in 2012; a hedgebank on St Martin's (SV931156) in 2010 and 2011 and at Porthloo, St Mary's (SV909113) in 2012. This species was awarded an Award of Merit by the RHS in 1983 (Lancaster, 1984), recommending it as a plant for a cool greenhouse. On walls and hedgebanks.

CRASSULACEAE – Stonecrop family
Umbilicus rupestris (Salisb.) Dandy
Navelwort
Inhabited islands, larger uninhabited islands

Native. Recorded from Samson, Annet, St Helen's, Northwethel, Round Island, Foreman's Island, Hedge Rock, Teän, and in the Eastern Isles on Little Ganilly, Great Ganilly, Great Innisvouls, Nornour, Menawethan and Great Ganinick. Surprisingly there are no records from the Arthurs, but this may be an oversight. Common on walls, rocks and buildings, occasionally grows on the ground on heathland.

CRASSULACEAE – Stonecrop family
Aeonium arboreum (L.) Webb & Berthel.
Tree Aeonium
Inhabited islands, also Guther's Island.

CRASSULACEAE – Stonecrop family
Aeonium cuneatum Webb & Berthel.
Aeonium
Inhabited islands

Neophyte. Garden escape. As with the next species identifying genuinely naturalised plants can be difficult. Found on all the inhabited islands and was recorded on Ragged Island in 1984 and Guther's Island in 1999, presumably resulting from pieces taken there by gulls (Parslow, 2002b). There is a strong suspicion that this plant is not *A. arboreum* but a similar species *A. balsamiferum* Webb and Berthel. (Mark Spencer, pers. comm.). Walls, rocks hedgebanks.

Neophyte. Garden escape. This plant, originally from Tenerife has been in Tresco Abbey Gardens since at least 1894 (King, 1985). Although Aeoniums are found on all the inhabited islands it is difficult to distinguish naturalised from planted specimens. Also a large number of additional *Aeonium* species and cultivars have been introduced to the Abbey Gardens, Tresco and elsewhere in Scilly as garden plants. A note by Clement in *BSBI News* (2001) gives some of the earliest records and a key to species. What is probably the first record of a plant in the wild was made by Mrs M.C. Foster from dunes on Bryher in June 1978, In October 1992 it was found on an old wall in Hugh Town, St Mary's by R.A. Barrett and in 1996 Alan Underhill found it growing wild on Tresco (Clement, 2001). As pieces of the plant root easily they can spread vegetatively as well as by seed. Plants can be found growing in rock crevices and high up on buildings where they clearly have not been planted. This species has occasionally been recorded from uninhabited islands, for example Little Ganilly in 1983 and Great Ganinick and Samson in 1984 , but the plants do not apparently persist for long (Parslow, 2002b). Rocks, walls, trees and hedgebanks.

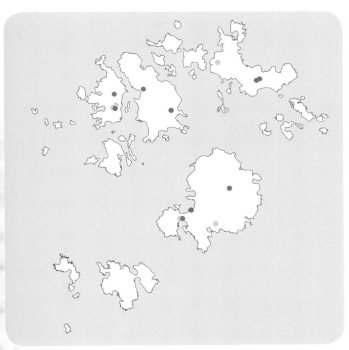

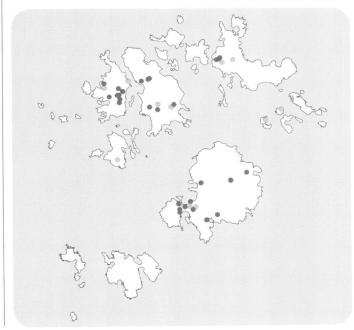

Aeonium spp.
A number of different *Aeonium* spp. are in cultivation on the islands and at least two in addition to those shown here have been found seeding freely and are likely to become naturalised. One, which has smaller rosettes and pink flowers is already growing away from the parent near the quay on St Agnes. Another species, possibly *A. haworthii* Webb & Berthel., is also spreading away from gardens.

EU-DICOTS True Dicotyledons

CRASSULACEAE – Stonecrop family
Sedum kimnachii V.V. Byalt
Mexican Stonecrop
St Mary's

Neophyte. Garden escape. Recorded in Old Town, St Mary's (SV914100) in 1996, but the plant is common in gardens making it difficult to recognise genuinely established plants. Usually recorded as *Sedum confusum* Hemsl. but according to Ray Stephenson (BSBI referee) this is not the correct name.

CRASSULACEAE – Stonecrop family
Sedum acre L.
Biting Stonecrop
Bryher, Tresco and St Mary's, Samson, Teän and Great Ganilly

Native. Much less common than *S. anglicum* although both may occasionally be found together. Apparently absent from St Agnes. Sandy places near the shore.

CRASSULACEAE – Stonecrop family
Sedum album L.
White Stonecrop
St Mary's and St Agnes

Archaeophyte. Occasional garden escape. On the east side of the Garrison, Buzza Hill and near habitations on St Mary's. A record from rocks by St Warna's well, St Agnes in 2001, well away from any gardens, was presumably bird-sown.

CRASSULACEAE – Stonecrop family
Sedum anglicum Huds.
English Stonecrop
Inhabited islands, most uninhabited islands

Native. Recorded from even some of the smaller rocky islets such as Illiswilgig and Mincarlo in the Norrard Rocks; Puffin Island off Samson; Great Ganilly, Little Ganilly, Great Innisvouls, Middle Arthur, Little Arthur, Nornour and Menawethan in the Eastern Isles; Foreman's Island and Pednbrose off Teän.

This species is a feature of the coast and islands such as Annet in midsummer when it is in flower. There seem to be few islands where it has not been recorded. Grows on rocks, walls, bare ground and sandy places on coasts and heathlands.

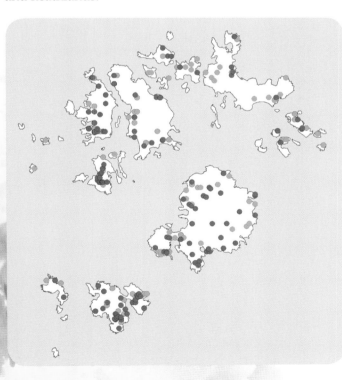

EU-DICOTS True Dicotyledons

CRASSULACEAE – Stonecrop family
Aichryson laxum (Haw.) Bramwell
Tree of Love
Tresco

Neophyte. Garden escape (Parslow, 2002b) common in vicinity of Abbey Gardens, usually on walls and pavements where it self-seeds freely. Is likely to become established elsewhere. Another *Aichryson* tentatively identified as *A. pachycaulon* Bolle by Mark Spencer is also self-seeding on Tresco and could also become established away from the Gardens. Although similar to *A. laxum* which has very hairy leaves, *A. pachycaulon* has glaucus, kite-shaped leaves (sensu Bramwell and Bramwell, 1974).

HALORAGACEAE – Water-milfoil family
Myriophyllum aquaticum (Vell.) Verdc
Parrot's-feather
Tresco and St Mary's

Neophyte. Planted in pools in Tresco Abbey Gardens and a water-filled tank in Holy Vale, St Mary's.

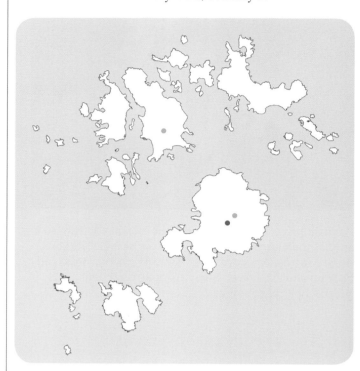

Myriophyllum alterniflorum DC
Alternate Water-milfoil
Extinct

Last mentioned by Lousley, recorded in both Abbey and Great Pools, Tresco in 1936.

FABACEAE – Pea family
Lotus corniculatus L.
Common Bird's-foot-trefoil
Inhabited islands, most uninhabited islands

Native. Very common throughout the islands and even found on the smaller islands with the exception of the more rocky islets of the Norrard Rocks, Western Rocks and Gweal. Individual flowers can be deep orange or even red in colour. At times can occur in such abundance as to colour the ground golden. It is an important food plant for a number of invertebrate species. Occurs in a variety of habitats including heathland, coastal cliffs, dunes, wall-tops, path edges and grasslands.

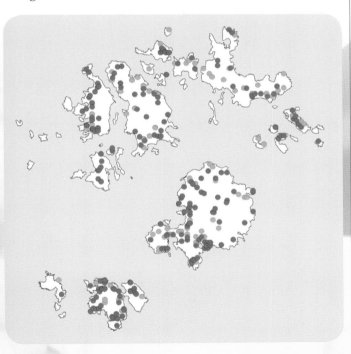

Anthyllis vulneraria L.
Kidney Vetch
Extinct or error

Lousley included references to a record from the Misses Millett (1852) whom he considered reliable, although he rejects a comment by Ralfs (1879) that it was common, as his statements were often 'wildly inaccurate'. A 1963 record from St Helen's was found to be an error.

FABACEAE – Pea family
Lotus pedunculatus Cav.
Greater Bird's-foot-trefoil
Inhabited islands, some larger uninhabited islands

Native. Found on Samson, St Helen's, White Island (St Martin's), Northwethel, Teän, Great and Little Ganilly and Middle Arthur. As well as damp grassland it frequently grows in drier habitats on Scilly than on the mainland. Grasslands, heaths, marshy places and roadsides.

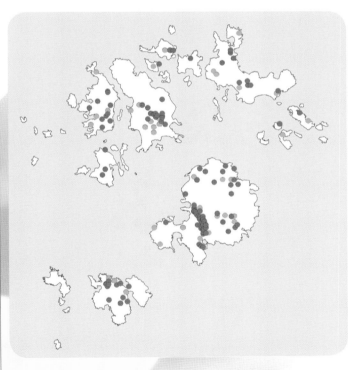

EU-DICOTS True Dicotyledons

FABACEAE – Pea family
Lotus subbiflorus Lag.
Hairy Bird's-foot-trefoil
Inhabited islands. St Helen's, Teän and Samson

Native. Widespread and frequent on the inhabited islands. May still be present on St Helen's where it was last recorded in 1983. Found in two main types of habitat; either in arable fields, where it can be extraordinarily common to the extent of forming dense mats over the ground, or alternatively in heathland and coastal habitats.

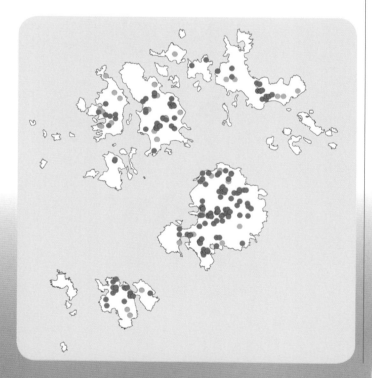

FABACEAE – Pea family
Ornithopus perpusillus L.
Bird's-foot
Inhabited islands, Teän and Samson

Native. Found on scattered sites on the inhabited islands and also on Teän and Samson. Occasionally can be found in sandy arable fields. Frequently found growing in the same areas as the rare *O. pinnatus*. Occurs on sandy or heathy areas, often in short turf over rocks.

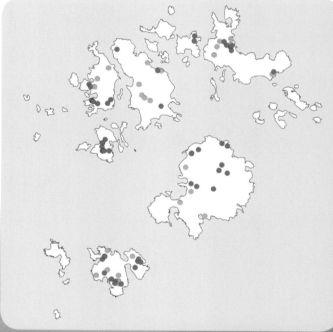

FABACEAE – Pea family
Ornithopus pinnatus (Mill.) Druce
Orange Bird's-foot
St Agnes, Gugh, St Martin's, Tresco, (St Mary's), and Bryher, also Teän, Samson and Great Ganilly

Native. First found by Miss Matilda White on Tresco in 1838, this was new for England (Lousley, 1971). Although recorded as a wild plant from St Mary's in 1877 it later disappeared and Lousley reports being unable to find it on the island. There is now a healthy colony on private land near Old Town, St Mary's that originated from scattered seeds. On the other islands the populations can vary from year to year.
Ornithopus pinnatus often grows in the same places as *O. perpusillus*. It probably needs bare or open ground to germinate and can often appear after heath fires. Populations can fluctuate from year to year; with second or even a third flowering in some years. In December 2008 Will Wagstaff found it still in flower on Samson and in 2012 it was in flower on Tresco at the end of September. Occasionally found as an arable weed, for example in sandy fields behind English Island Point on St Martin's and also still persists where rabbits have been scratching in the abandoned fields on Gugh. Sandy and heathy areas, including short turf over rocks.

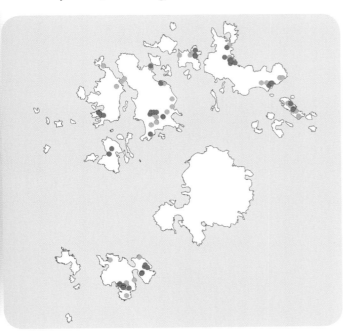

FABACEAE – Pea family
Coronilla valentina L.
Shrubby Scorpion-vetch
Tresco

Dunes where *Coronilla* and other aliens grow, Tresco

Garden escape found growing in dunes on Tresco (SV891140) in 2012.

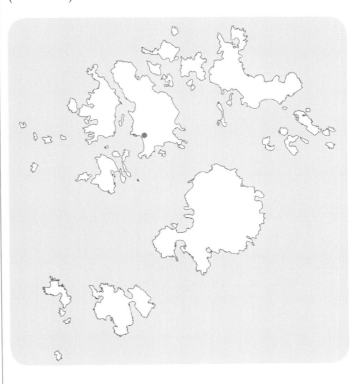

EU-DICOTS True Dicotyledons

FABACEAE – Pea family
Vicia cracca L.
Tufted Vetch
Bryher, Tresco, St Mary's and St Martin's, also Teän

Native. Scattered records from the inhabited islands, but not recorded from St Agnes since Lousley. It was found on Teän in 2007, where it had previously been found on the nearby islet of Old Man in 1952, an unusual record for the site. Grows in grassy areas, disturbed ground and along roadsides.

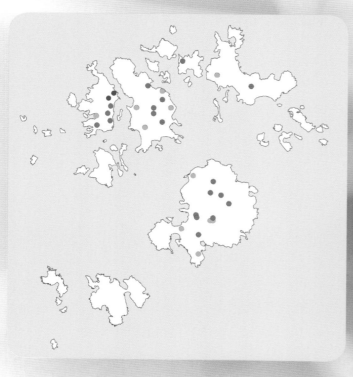

FABACEAE – Pea family
Vicia hirsuta (L.) Gray
Hairy Tare
Inhabited islands

Native. Very common, although probably overlooked as the flowers are diminutive (but delightful in close-up). Found along field edges, waste ground, hedgebanks and especially along roadsides as indicated by the map.

FABACEAE – Pea family
Vicia tetrasperma (L.) Schreb.
Smooth Tare
Bryher, St Agnes, St Mary's, St Martin's and St Helens

Native. Most records are from Bryher and St Mary's, with just a few from St Agnes. The last record from St Martin's was in 1988, and from St Helen's in 1990. May be easily overlooked when not in flower; it is much less common than *V. hirsuta*. Grows in grassy areas, along field edges and roadsides.

FABACEAE – Pea family
Vicia sepium L.
Bush vetch
Inhabited islands

Native. There are records from the inhabited islands, including a small concentration around Lower Town on St Agnes and scattered sites on St Mary's and St Martin's. Lousley does not include *V. sepium* in his Flora, but it seems unlikely that the species has arrived since he wrote it. Found along field edges and roadsides.

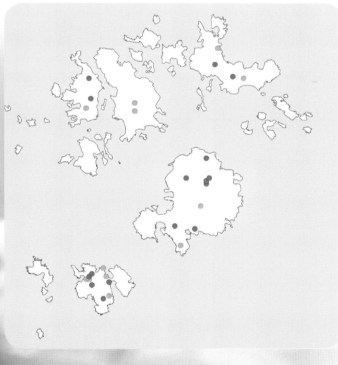

EU-DICOTS True Dicotyledons

FABACEAE – Pea family
Vicia sativa L.
Common Vetch
Inhabited and some uninhabited islands

Native and Archaeophyte. Two of the subspecies of *Vicia sativa* occur in Scilly. Ssp. *nigra* is native and is found on all the inhabited and some larger uninhabited islands including St Helen's, Samson, Teän and Great Ganilly. Ssp. *segetalis* is an archaeophyte that has only been recorded from the inhabited islands, except for a record from Samson (SV87981245) in 2008 – possibly associated with the former cultivations. Records of ssp. *sativa* recorded in the past are now considered to be errors for ssp. *segetalis*. Both subspecies are found on cultivated and disturbed ground, often along roadsides, with ssp. *nigra* more commonly in grassland, heathland, dunes and tracks near the sea.

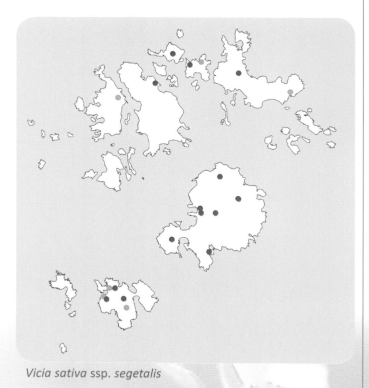

Vicia sativa ssp. *segetalis*

FABACEAE – Pea family

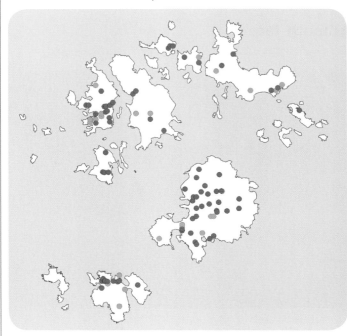

Vicia sativa ssp. *nigra*

Vicia lathyroides L.
Spring Vetch
Bryher and Tresco

Native. A rare plant in Scilly only known recently from below Samson Hill (SV879141 – SV880440) and Rushy Bay SV87571423 Bryher. Has been recorded on Appletree Banks, Tresco (SV895139) in 2002. Last known from School Green, Tresco in 1986, an area that has been much changed recently by building work. As the plant flowers early in the year and may superficially resemble a depauperate *V. sativa* it may be overlooked. Sandy grassland near the coast and on dunes.

FABACEAE – Pea family
Vicia bithynica (L.) L.
Bithynian Vetch
St Martin's and St Mary's

R PARSLOW

Native in the British Isles but possibly a neophyte in Scilly. There are very few records of this lovely vetch, but on St Mary's it seems to have been associated with the area around Trewince (SV913114). The most recent appearances are from the roadside at SV915113 in 2012 and not far away at SV915115 in 2013. In May 2014 and 2015 up to nine plants were found flowering on the hedgebank between SV91501142 and SV91511137. It was frequent in an arable field (SV919164) on St Martin's in 2001, but has not been seen there since. Bulbfields and on hedgebanks.

Vicia faba L.
Broad Bean
St Martin's

Escape from cultivation in a field on the east end of the island in 1994.

FABACEAE – Pea family
Lathyrus japonicus ssp. *maritimus* (L.) P.W. Ball
Sea Pea
St Mary's - formerly St Martin's and Gugh

M F WELLS

Native. The plants that were found on Porth Minick, St Mary's in 2007 were the first since a plant was found on Gugh Bar, St Agnes in 1985. Prior to that there had been a few records from beaches around the islands, but the plants had not persisted. In 2008 the Porth Minick colony had increased to some thirty plants, by 2009 there were approximately 75 plants and they were spreading further up the beach. In 2011 there were nearly a hundred growing on the sand and among the rocks at the top of the beach. By 2012 and 2013 there were approximately 300 plants including many well up the beach that survived the severe storms of early 2014. The colony was still present in 2016.

It is not known whether the plants derived from buried seed or more likely, as drift seeds from North America. Certainly they appear slightly different from those growing in East Anglia and it would seem very unlikely that would be the origin of the seed – the coastal currents being from the west. Sea Pea grows on sandy or shingle beaches above HWM where it is very vulnerable to extreme high tides.

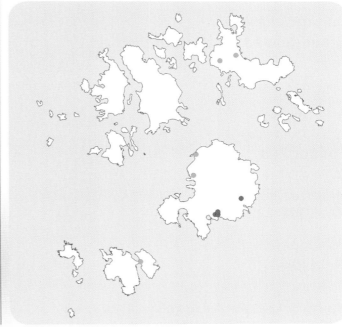

EU-DICOTS True Dicotyledons

FABACEAE – Pea family
Lathyrus pratensis L.
Meadow Vetchling
St Mary's

Native. Always an uncommon plant in Scilly, previously Lousley had recorded it as 'very rare' from St Mary's, St Agnes and Tresco. All the recent records are from St Mary's. The species has been found in a cluster of sites near the Airport, at Pelistry (SV928117), in a hedgebank opposite Newford Duck ponds and at Old Town church area (SV911101). Grows on grassy roadside verges and hedgebanks.

Lathyrus latifolius L.
Broad-leaved Everlasting-pea
Bryher

Neophyte. A garden plant occasionally found away from houses.

Lathyrus annuus L.
Fodder Pea
St Martin's and St Mary's

Neophyte. Escape from cultivation. Last recorded from McFarland's Lane, St Mary's in 1973

Pisum sativum var. *arvense* (L.) Poiret
Field Pea
St Mary's

Neophyte. Last recorded in 1982 with no locality. Presumed escape from cultivation.

FABACEAE – Pea family
Ononis repens L.
Common Restharrow
St Martin's, St Mary's and Samson

Native. A rare plant in Scilly that was previously only known from two sites (those also described by Lousley); the Plains on St. Martin's (SV924163) and a wall top at Borough, St Mary's (SV919117-SV921122) where it survives despite the wall being strimmed. In 2007 it was found on Samson (SV879131) during field surveys by botanists from CeC (Cornwall Environmental Consultants). Found on coastal grassland and on the top and sides of a stone hedge.

FABACEAE – Pea family
Melilotus officinalis (L.) Pall.
Ribbed Melilot
St Mary's, Tresco and St Martin's

Neophyte. Formerly occurred on Gugh where it grew in the fields below the houses until the late 1980s. The only recent records are from St Mary's, Tresco and St Martin's, as a constituent of 'conservation' and other seed-mixes. *Melilotus albus* also occurs occasionally in the same context.

FABACEAE – Pea family
Medicago lupulina L.
Black Medick
Inhabited islands

Native. Mainly scattered records from the inhabited islands. Probably the least common *Medicago* in Scilly. Found in cultivated and disturbed ground and around habitations.

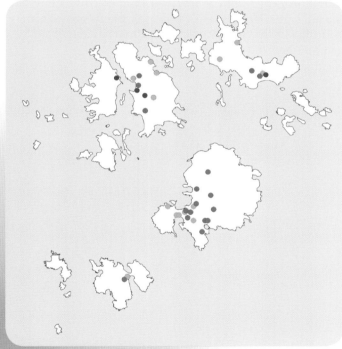

EU-DICOTS True Dicotyledons

FABACEAE – Pea family
Medicago sativa ssp. *sativa* L.
Lucerne
St Martin's and St Agnes

Neophyte. In Scilly probably an escape from cultivation or planted. Lucerne was still being grown as a crop when Lousley wrote his Flora. Since then it has virtually disappeared from the islands. In 1990 a plant was found between Covean and Bar, St Agnes and then another was found near the public toilets near Higher Town quay, St Martin's in 2009. It is likely the latter plants arose from 'wildflower' seeds scattered in the area to cover bare ground after the quay was rebuilt.

FABACEAE – Pea family
Medicago polymorpha L.
Toothed Medick
Inhabited islands

Native. Locally frequent on St Mary's, scattered elsewhere. Grows in sandy places, especially arable fields, bulbfields and sand dunes.

FABACEAE – Pea family
Medicago arabica (L.) Huds.
Spotted Medick
Inhabited islands

Native. A common plant on Scilly that can be locally abundant. Sometimes the black spots on the leaves are faint but more often they are very large – leaving only a thin margin of green around the edge. Besides arable fields it also grows on waste ground, hedgebanks and in gardens.

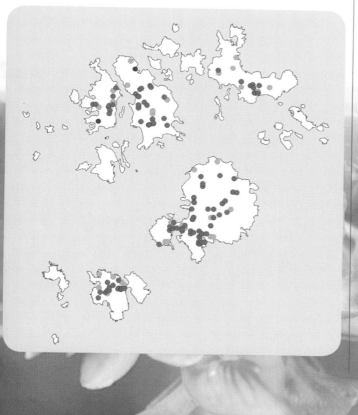

FABACEAE – Pea family
Trifolium ornithopodioides L.
Bird's-foot Clover
Inhabited islands, formerly Teän

Native. Mainly found at scattered sites around the coasts on the inhabited islands. Had been recorded from Teän in 1982. Often grows in similar places to *T. subterraneum* and *T. micranthum* but is easily overlooked when not in flower. In flower, the tiny white (occasionally pinkish) flowers stud the turf like little white teeth! Coastal grasslands and short turf.

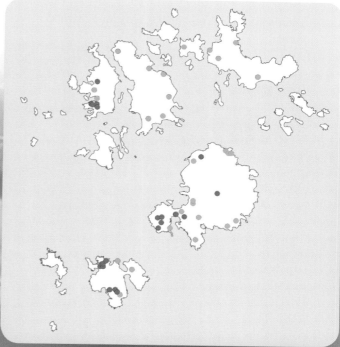

EU-DICOTS True Dicotyledons

FABACEAE – Pea family
Trifolium repens L.

White Clover

Inhabited islands, St Helen's, Teän, Samson, Nornour and Middle Arthur

Native. Abundant and widespread, found both on inhabited and some uninhabited islands. Larger flowered 'Dutch' clovers found in fields, grass leys and roadsides are relicts of cultivation. An attractive colour form, the variety *townsendii* Beeby, which has deep red or purple flowers and often purple to almost black leaves, is frequent in coastal grasslands. It was first discovered by the botanist Frederick Townsend when he visited Scilly in 1862. The variety was later named after him by William Hadden Beeby. A fine pressed specimen is held in the SLBI (South London Botanical Institute) herbarium that was collected by Beeby on Tresco in 1872. This variety is most common on Bryher and Tresco, but can be found on all the islands (Sell and Murrell, 2009). Variety *townsendii* is also found on the Channel Isles and may have been discovered there as early as 1821 (McClintock, 1975a). In cultivation the plant stays true to form although the colours may not be as intense. The native plant is common in coastal grassland, dunes, wall tops and on roadside verges.

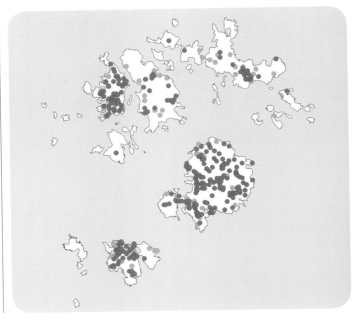

FABACEAE – Pea family
Trifolium occidentale Coombe
Western Clover
Inhabited islands, Teän, Samson and Middle Arthur

Native. This species was only recognised as a separate species from *T. repens* by Dr D.E. Coombe in 1961 (Coombe, 1961; Coombe and Morisset, 1967). According to Lousley (1971) the species was found (by Vercoe, in Coombe 1961) on Peninnis, St Mary's and Gugh in 1961 and by Lousley on Bryher in 1963, although the first published Tresco record I can find was in 1966 when Coombe recorded plants from around the Blockhouse on Tresco (Coombe and Morisset, 1967). Later it was shown to me on the Gugh by Jim Bevan, and has subsequently been found on all the inhabited islands and some uninhabited islands. It is not clear whether the plant is now more common than before or has just been better recorded, but it is now locally frequent mainly around the coast. *Trifolium occidentale* is quite a feature of the short coastal grassland where *T. repens* sometimes also grows (but flowers later). Although flowers have been found as early as January and February it is usually at its best in April and May, with sometimes a limited second flowering later in some years. Occasionally an odd flowering plant has been found as late as September or October, but this is unusual. Where *T. occidentale* and *T. repens* grow together the non-flowering plants can easily be determined. The leaves of *T. occidentale* are usually smaller, much thicker and glossy underneath and lack the translucent veins (use a lens and hold leaf up to light) of *T. repens*. Grows on short coastal grassland, dunes, wall tops and coastal heath.

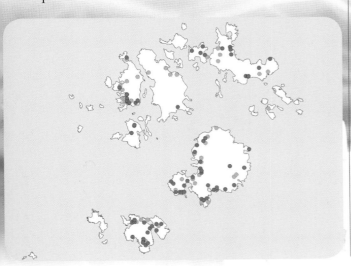

FABACEAE – Pea family
Trifolium hybridum L.
Alsike Clover
Bryher, St Agnes, St Mary's and St Martin's

Neophyte. Just a few scattered records, probably a relict from cultivation. Recent records are from Bryher in 2007 (SV88141528 and SV878155), St Mary's in 2005 (SV91871245), St Agnes in 2007 (SV886084) and from St Martin's in 1988 (SV917160) and 2012 (SV929156). Roadsides, grassy verges or waste ground.

EU-DICOTS True Dicotyledons

FABACEAE – Pea family
Trifolium glomeratum L.
Clustered Clover
St Mary's, St Martin's and Tresco

Native. An uncommon plant in Scilly; the only consistent records are as a weed in Tresco Abbey Gardens. On St Martin's it was found as a weed in the Vineyard in 2005, there are also earlier records for the Higher Town area. Found on St Mary's from SV90961137 in 2003 and near 'Nowhere', Old Town, St Mary's in 2004 and 2007. Last recorded on St Agnes in 1879. Cultivated ground, tracks and on wall-tops.

FABACEAE – Pea family
Trifolium suffocatum L.
Suffocated Clover
Inhabited islands, also Samson

Native. Although Suffocated Clover can be found in suitable places on all the inhabited islands it is easily overlooked; the tiny plants are frequently masked by sand that has blown over them. The characteristic form of the plant is a small rosette with tiny leaves with long petioles arranged in a circle. The flowers are clustered around the centre and are inconspicuous. Later the inflated pods are easier to see. A plant was found on Samson in 2008 near the ruined houses. Sandy ground, short sandy turf and along tracksides.

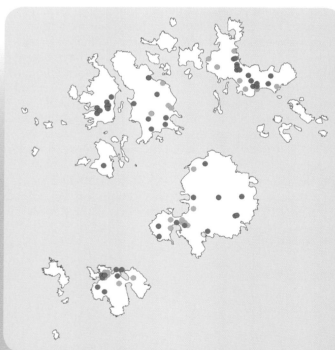

Trifolium fragiferum L.
Strawberry Clover
Extinct

Lousley includes records from Tresco (1870) and St Martin's (1904 and 1909). There are no recent records.

FABACEAE – Pea family
Trifolium campestre Schreb.
Hop Trefoil
Inhabited islands, also Teän

Native. Mainly scattered records from the inhabited islands. Also known from the east side of Teän, most recently in 2004. Found on the edges of fields, roadsides and sandy places.

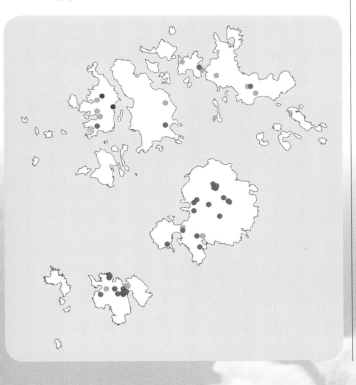

FABACEAE – Pea family
Trifolium dubium Sibth.
Lesser Trefoil
Inhabited islands, also Samson and Teän

Native. Much more common than *T. campestre*, abundant in some places. Refound on Teän in 2014. Frequent on cultivated and disturbed ground, sandy places and especially along road verges.

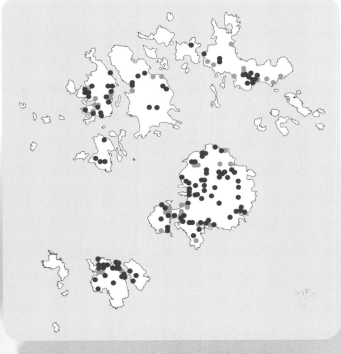

FABACEAE – Pea family
Trifolium micranthum Viv.
Slender Trefoil
Inhabited islands, also Nornour.

Native. Found at scattered locations around the inhabited islands; may occasionally be locally frequent. It was found on Nornour in 2002. Lousley had considered it rather rare, he had also recorded it from Teän. Can be easily overlooked when not in flower. Grows in short coastal turf and sandy areas.

FABACEAE – Pea family
Trifolium pratense L.
Red Clover
Inhabited islands

Native. Although it may be locally quite frequent on some of the inhabited islands it is only found in a discrete area of St Martin's and down the eastern side of Tresco. Lousley recorded it from Teän. The large and robust plants of the cultivated form var. *sativum* found in grasslands and verges are likely to be a relict of cultivation. Around the coast the plants are more typical of the native variety. Grows mostly in grasslands and along roadsides, also near the coast.

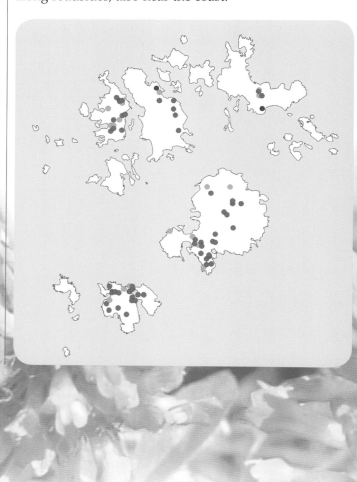

FABACEAE – Pea family
Trifolium medium L.
Zigzag Clover
Bryher, Tresco, St Agnes and St Mary's

Native. Nowhere common, there are just a few recent records from Bryher, Tresco, St Agnes and St Mary's. Lousley rejected the species after a misidentified specimen of *T. pratense* was sent to him from St Agnes, although *T. medium* had been recorded from there in 1893 and subsequently. In 1989 John Akeroyd recorded *T. medium* from Tresco, but there were no other records until 2003 when it was found that plants on Bryher previous dismissed as a cultivar of *T. pratense* were in fact this species. On St Agnes it has been found on the Meadow and a field above The Bar. It grows along tracks, grassland and field edges.

FABACEAE – Pea family
Trifolium incarnatum ssp. *incarnatum* L.
Crimson Clover
St Martin's and Tresco

Neophyte. Planted. A plant on sandy waste ground near Higher Town quay in 2008 seems to have originated from wildflower seeds spread in the area. This was the first record since Lousley and others reported that this species had sometimes been sown as a crop, for example in fields at Old Town Lane on St Mary's, and near Dial Rocks and New Inn on Tresco. In 2011 it was found that the clover was being grown on a smallholding just above Lawrence's Bay (SV9215) from where it may be spreading. Found in a small field on Tresco (SV899147) in 2014. Cultivated and disturbed ground.

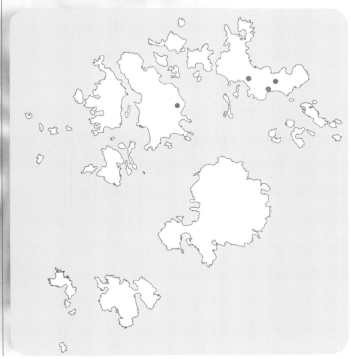

EU-DICOTS True Dicotyledons

FABACEAE – Pea family
Trifolium striatum L.
Knotted Clover
Inhabited islands

Native. A rare plant known from just a few places, with small concentrations of records from fields above The Bar on St Agnes and around Hugh Town and the Garrison, St Mary's. In 2011 there were records from Water Rocks Down, Tresco SV902107. Sandy ground and bulbfields.

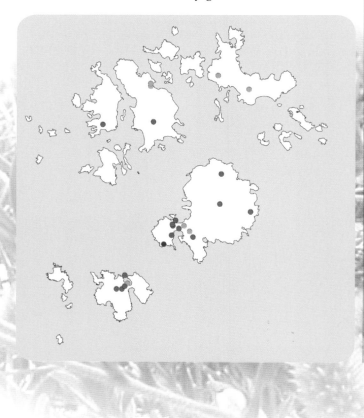

FABACEAE – Pea family
Trifolium scabrum L.
Rough Clover
Bryher, St Martin's, Tresco and St Mary's

Native. Also rare, there are scattered records from Bryher, St Martin's and St Mary's, formerly recorded by Lousley from St Agnes and Tresco. There was a record from Tresco (School Green, Old Grimsby) in 1975. In 2011 it was recorded at Rushy Bay, Bryher as well as from Garrison Arch and Porthcressa, St Mary's. Found in sandy places near the shore, by tracks, and on wall tops.

FABACEAE – Pea family
Trifolium arvense L.
Hare's-foot Clover
Inhabited islands, also Samson

Native. Recorded from the inhabited islands. On Samson it was rediscovered near the ruined houses (SV87741257 and SV87821246) in 2008. May be more common than formerly; especially on St Agnes. Here it had spread along the bank and path beside Big Pool, in an area that was heavily disturbed when rebuilding the sea defences. Sandy and disturbed soils, especially near the coast, often found on tops of walls and banks.

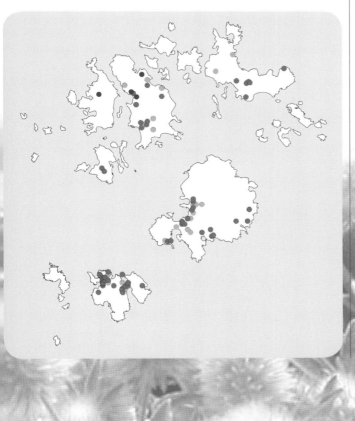

FABACEAE – Pea family
Trifolium subterraneum L.
Subterranean Clover
Inhabited islands, also Samson

Native. Although locally frequent on the inhabited islands, it seems surprising that the only record from uninhabited islands is from Samson (by the ruined houses in 2002). Often forms large patches that can be a feature of coastal grassland in early summer. The pods dig into the ground as they ripen. Grows in short turf on coastal areas, meadows and dune grassland.

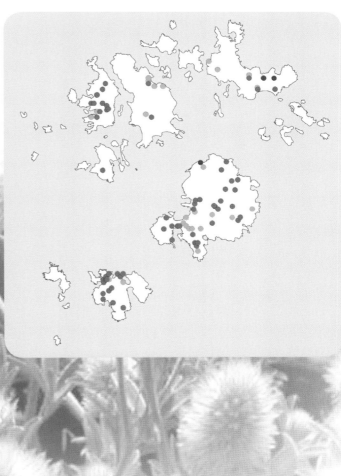

EU-DICOTS True Dicotyledons

FABACEAE – Pea family
Lupinus arboreus Sims
Tree Lupin
Inhabited islands

Neophyte. Originally a garden escape that first became established in dunes on Gugh, Tresco and St Martin's. Tree lupins are now found on all the inhabited islands and are still spreading. Mostly yellow-flowered but occasionally flushed purple. Found in dunes, sandy fields and waste ground.

FABACEAE – Pea family
Cytisus scoparius ssp. *scoparius* (L.)
Broom
St Mary's, Bryher, Tresco and St Martin's, also Samson

Native. Broom is common and widespread on St Mary's and Bryher. There are just a few individual records from Tresco, St Martin's and Samson. Records from St Martin's were not made until 2011, but it may have been overlooked earlier. The plants first found on North Hill, Samson in 2006 had not been recorded before, which suggests they may either have been missed earlier or had recently arisen from seed. They grow in, or close to the area where there had been heath fires in the past. It is also possible that now there are no longer rabbits on Samson it has allowed the plants to grow. As broom is frequently grown in gardens some plants found near habitations may be garden escapes. Found on cliff slopes, rough ground on heathland and a disused quarry.

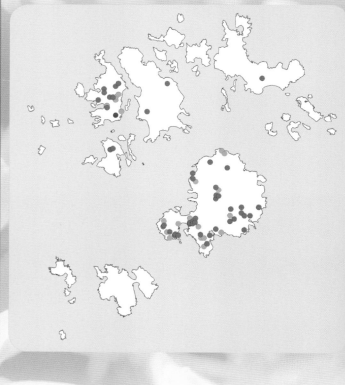

FABACEAE – Pea family
Spartium junceum L.
Spanish Broom
St Mary's

Garden escape near Longstone (SV917112).

Ulex europaeus L.
Gorse
Inhabited islands, also St Helen's, Northwethel, Teän, Samson, Great Ganilly and Little Arthur

Native. Gorse had been present on the islands since at least the Bronze Age when plant macrofossils were found at a cliff site at Bonfire Carn on Bryher. When Augustus Smith was laying out his garden on Tresco he planted gorse bushes (grown from seed brought from the mainland) around his estate to act as tree shelters. This intriguing fact suggests either that gorse was much less common at the time (c.1834), or that it was easier to get seed collected on the mainland. Another possibility is that the common gorse in Scilly at the time was *Ulex gallii* Western Gorse, a lower-growing species and unsuitable to act as a nurse to young trees? Historically gorse would have been utilised by the human population for fuel. Later gorse had been kept in check by grazing of stock and rabbits (the rabbit-grazing producing typically 'topiary' conical bushes). Gorse became dominant on the inhabited islands when the combination of less stock-grazing coincided with the crash in the rabbit population due to myxomatosis.

It now dominates some heathland areas and is very invasive. On most inhabited islands various methods have been used to control the spread of gorse; cutting, burning or free-range ponies and cattle. Gorse is found on heathland, coastal slopes and hedgebanks inland.

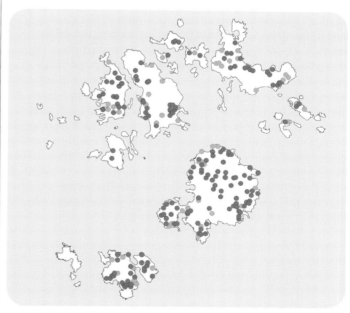

EU-DICOTS True Dicotyledons

FABACEAE – Pea family
Ulex gallii Planch.
Western Gorse
Inhabited islands, also Northwethel, St Helen's and Little Arthur

Native. Although Lousley described Western Gorse as very common, it is now only found in quantity on St Mary's, where it occurs around almost the whole coastline. There are scattered records from the other islands. *U. gallii* appears now to be absent from Wingletang Downs, St Agnes, where it is possible it has not survived burning, but it still occurs on Gugh. There are two records of plants that appeared to be intermediate between *Ulex europaeus* and *U. gallii* that may have been the hybrid. One of these was found on Wingletang Down, St Agnes in 2001, apparently now gone, the other on The Garrison, St Mary's in 2006. *Ulex gallii* plants on Scilly do not seem to grow as tall as their mainland compatriots, often occurring in low-growing communities with *Erica cinerea*. Coastal heathland, also occasionally on heathy sites inland.

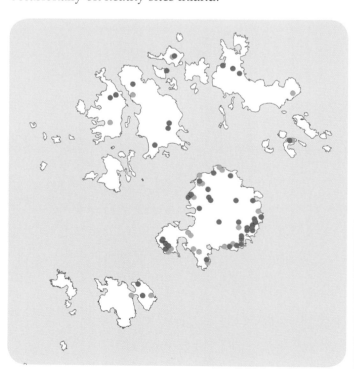

FABACEAE – Pea family
Albizia lophantha Benth.
Cape Wattle
Tresco

A rare garden escape on Tresco.

Acacia falciformis D.C.
Hickory Wattle
Tresco

Neophyte. A rare garden escape on Tresco.

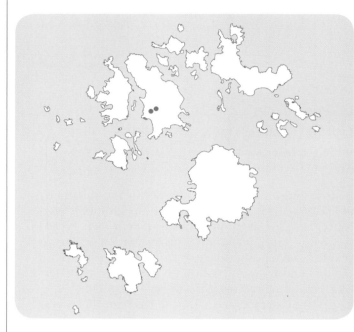

Acacia melanoxylon R. Br.
Australian Blackwood
Tresco and St Mary's

Neophyte. A rare garden escape on Tresco and St Mary's.

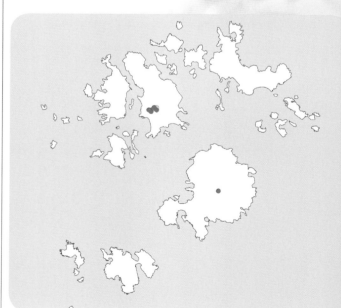

POLYGALACEAE – Milkwort family
Polygala vulgaris L.
Common Milkwort
Inhabited islands, also Northwethel

Native. The records for the species undoubtedly included some for the next species with which it is commonly confused. Grows on heathy and dune sites.

POLYGALACEAE – Milkwort family
Polygala serpyllifolia Hosé
Heath Milkwort
Inhabited islands, also Samson, White Island (off St Martin's), Great Ganilly and Northwethel

Native. The failure of many recorders to distinguish between the two species of *Polygala* in the field mean that the distribution maps do not accurately reflect the true situation. Heathy sites, often near the coast.

EU-DICOTS True Dicotyledons

ROSACEAE – Rose family

Spiraea salicifolia L.
Bridewort

Spiraea x pseudosalicifolia Silverside
Confused Bridewort

Spiraea douglasii Hook
Steeple-bush

Neophyte. Garden escape, near Newford Duck Ponds, 1992-1997, St Mary's. It is not clear which *Spiraea* grew there, as all three names have been used by recorders.

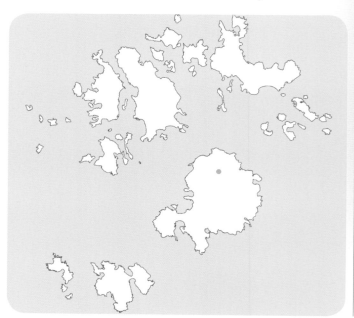

ROSACEAE – Rose family

Prunus cerasifera EhrH.
Cherry Plum

St Mary's

Neophyte. One in a hedge near the Airport in 1953. One in a roadside hedge near Sunnyside Farm was still there in 2013.

ROSACEAE – Rose family
Prunus spinosa L.
Blackthorn
St Mary's, St Martin's and Tresco

Native on the mainland but apparently planted or an escape from cultivation on Scilly. Just a handful of bushes in hedges and as individual trees, there is a group of about six near the Bakery on St Martin's (SV93051556).

ROSACEAE – Rose family
Prunus domestica L.
Wild Plum
St Mary's

Archaeophyte in the British Isles, but a neophyte on Scilly. One, possibly ssp. *insititia* recorded near Morning Point on the Garrison in 1998.

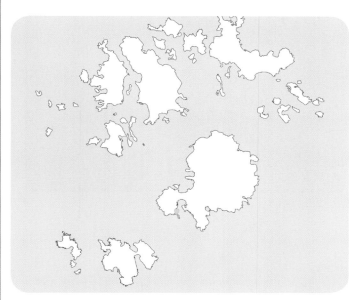

Prunus avium (L.) L.
Wild Cherry
St Mary's

Native on the mainland but planted in the Lower Moors extension.

Malus sylvestris (L.) Mill.
Crab Apple
St Martin's and St Mary's

Native on the mainland but neophyte on Scilly. A few records, possibly originated from garden throw-outs? Also planted in Lower Moors extension.

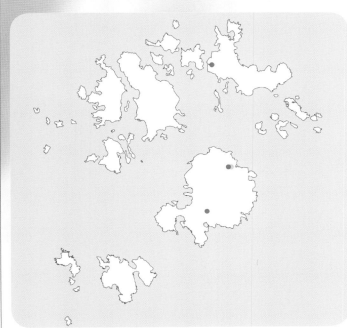

EU-DICOTS True Dicotyledons

ROSACEAE – Rose family
Malus pumila Mill.
Apple
Bryher, Tresco, St Mary's, St Martin's, also Samson

Archaeophyte. Planted, or originated from discarded apples. There is a small tree on Samson at SV87881315 that probably originated from an apple core. In hedges and near pathways, also Lower Moors extension.

ROSACEAE – Rose family
Sorbus aucuparia L.
Rowan
St Mary's

A native species on the mainland but only as a planted tree in Scilly. Also planted in the Lower Moors extension.

ROSACEAE – Rose family
Cotoneaster horizontalis Decne.
Wall Cotoneaster
Tresco and St Mary's

Neophyte. Occasional garden escape, probably birdsown.

ROSACEAE – Rose family
Cotoneaster simonsii Baker
Himalayan Cotoneaster
Tresco

Neophyte. Garden escape on Appletree Banks, probably birdsown.

Cotoneaster cambricus J. Fryer and B. Hylmö
Wild Cotoneaster
St Mary's

Neophyte. Garden escape on the Garrison in 1995.

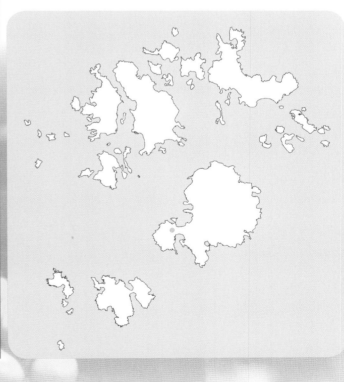

EU-DICOTS True Dicotyledons

ROSACEAE – Rose family
Crataegus monogyna Jacq.
Hawthorn
St Mary's, St Martin's and Tresco

Native in the British Isles but probably originally planted on Scilly. Found as a hedge or solitary tree mainly on St Mary's, less common on St Martin's and Tresco. Although there are a few hawthorn hedges some individual trees may have been birdsown. Most trees become gnarled, stunted and windswept. Hedges and field boundaries.

ROSACEAE – Rose family
Filipendula ulmaria (L.) Maxim.
Meadowsweet
St Mary's

Native. Only known from Holy Vale on St Mary's. Lousley had believed the species was extinct, not having been recorded since 1879, but it was found in a marshy area by the stream in Holy Vale (SV921114) in 1995 and again in the same area in 2002. It could still be there as the area is pretty inaccessible. Wetland.

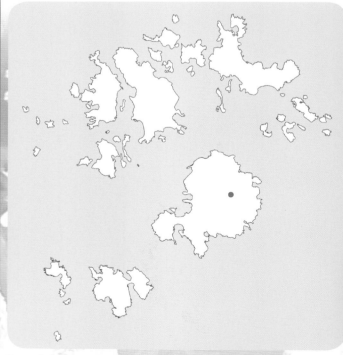

ROSACEAE – Rose family
Rubus caesius L.
Dewberry
St Mary's and Tresco

Native. Formerly found on St Mary's and also Tresco. Since Lousley's Flora there are just a few records from the 1990s. Two from Bar Point, St Mary's in 1996 (where it was also recorded in 1953 and 1963), one from Porth Mellon, also in 1996, and one from beside the pathway on the edge of the woods by the Abbey, Tresco (SV895145) in 1998. It does not ever have seemed to be common and was possibly restricted to a few dune and sandy places. Hedgebanks and dunes.

ROSACEAE – Rose family
Rubus fruticosus L. agg.
Blackberry
Inhabited islands, larger uninhabited islands

Native. Grows wherever the soils are deep enough to support it. Locally dominant especially on the Eastern Isles. The map shows the aggregate.

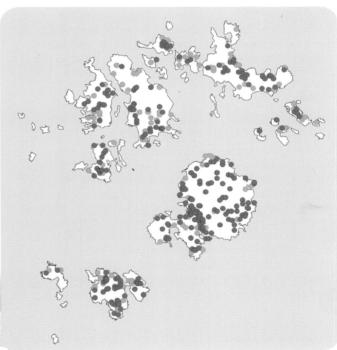

Rubus microspecies

According to Newton and Randall (2004) the majority of the brambles growing in Scilly are western species with distributions that often extends from NW France to SW England, the south of Wales and south and west Ireland. The following accounts are largely based on the papers and letters from D.E. Allen and the Atlas by Newton and Randall (2004).

EU-DICOTS True Dicotyledons

R. mollissimus Rogers
St Mary's and Tresco

Native. Allen found this much more plentiful on Scilly than the mainland. On St Mary's along the east coast from Pelistry Bay to Porth Hellick, among bracken. Also a bush on the Garrison. On Tresco he found four bushes on Middle Down and one on Abbey Hill. He further commented that the flowers were whitish in Scilly not the usual pink.

R. viridescens (Rogers) T.A.W. Davis
St Mary's

Native. A single plant on Lower Moors (Allen, 1997).

R. daveyi Rilstone
All inhabited islands

Native. Allen recorded this bramble from all the inhabited islands. Common on St Mary's; abundant on Wingletang and in a few hedges, St Agnes; a clump on Gugh in the gullery; frequent on Tresco; scarce on Bryher and frequent on St Martin's. What may be this species was recorded in open woodland in Holy Vale, St. Mary's growing right across the path at SV921113 by Tim Harrison in 2012.

R. dumnoniensis Bab.
Inhabited islands

Native. According to Allen this bramble is strongly maritime in its distribution and therefore particularly common on Bryher but occasional elsewhere. There are unconfirmed records by Harrison from Tresco, St. Mary's and St. Martin's in 2012.

R. iricus Rogers
St Mary's, Tresco, Bryher and St Martin's

Native. On St Mary's Allen recorded it as locally abundant, for example on The Garrison and Halangy Down. On Bryher he considered it local and on Tresco and St Martin's scarce, although on Tresco it was found among rhododendrons on Abbey Hill. This is a species with its main stronghold in the far west of Ireland although it also occurs on Alderney (Allen, 2001) but had long been suspected of being on Scilly. It is very interesting that the species should occur on two such disjunct, semi-maritime places. It is possible this may be the result of dispersal by birds (Allen, pers. comm.).

R. polyanthemus Lindeb.
St Mary's and Tresco

Native. Single bushes on Garrison Hill and Buzza Hill. On Tresco three bushes at Gimble Porth (Allen, 1997).

R. prolongatus Boulay and Letendre
St Mary's, St Agnes, Bryher and St Martin's

Native. Abundant on St Mary's, hedges on Barnaby Lane St Agnes, two places on Middle Down and a bush at the Abbey entrance, Tresco, recorded also from Bryher and from Tinkler's Hill, St Martin's (Allen, 1997).

R. riddelsdellii Rilstone
St Mary's and Tresco

Native. There are unconfirmed records by Tim Harrison from St Mary's (heathy areas), Tresco (woodland margin and heath). It is not mapped for Scilly in Newton and Randall (2004).

R. rubritinctus W.C.R. Watson
St Mary's and St Martin's

Native. A patch at the eastern end of Great Pool, Tresco and one or two bushes among bracken on The Plains, St Martin's (Allen, 1997).

R. sprengelii Weihe
St Mary's, Tresco and St Martin's

Native. A number of localities on St Mary's: Telegraph Road (SV915115), Bar Point, Innisidgen, west of Porth Wreck and Porth Hellick. On Tresco, Middle Down, Borough, east end of Great Pool, by Abbey Pool and Gimble Porth. On St Martin's Chapel Down and The Plains.

R. pydarensis Rilstone

Native. Shown as present in Scilly in the map in Newton and Randall (2004), but the source of the record is currently unknown.

R. ulmifolius Schott
Elm-leaved Bramble
All inhabited islands and the Eastern Isles

Native. The most often recorded bramble in Scilly. Probably the common species on the uninhabited islands especially on the Eastern Isles where it forms dense, impenetrable thickets.

R. leyanus Rogers
St Agnes

Native. A group of starved bushes in a bulbfield off Barnaby Lane (SV993082) (Allen, 1997).

R. newbouldianus Rilstone

Native. Also shown as present in Scilly in Newton and Randall (2004), but the source of the records is currently unknown.

R. peninsulae Rilstone
Tresco and St Martin's

Native. A clump in the nursery area of the Abbey Gardens and a patch of the edge of the *Salix* carr on Pool Road, Tresco. Abundant on St Martin's except on Tinkler's Hill (Allen, 1997).

R. angusticuspis Sudre
St Mary's and Tresco

Native. The first published record of this bramble on the Isles of Scilly was by Allen (2000). Previously only known from the southern end of the Welsh Marches, North Somerset and the Isle of Wight, it was found to be 'in local profusion' on both St Mary's and Tresco. Allen also found the Scilly plants to be more robust than their counterparts elsewhere.

R. rilstonei W.C. Barton and Riddelsd
St. Mary's, Tresco, St. Martin's

Native. Recorded from hedgerows and heaths on St. Mary's, Tresco and St. Martin's. The map in Newton and Randall (2004) only shows a pre-1988 record.

R. venetorum D.E. Allen
St Mary's and Tresco.
Native. This bramble, first described in 1998, is mainly known from France and was otherwise only relatively common in sheltered places on the north-eastern coastal area of St Mary's. There are now records from several places in Cornwall. It was recorded by Lousley in 1954 from two places on St Mary's, Low Pool (SV912108) and Carn Morval Downs (SV9012) (det. B.A. Miles in 1967 as *R. hastiformis*). Abundant in hedges and among bracken on slopes to the sea, Halangy Down, St Mary's, SV9012, 26 June 1995; wall-top, Back Lane, Tresco, (SV890156), 7 July 1995.

R. transmarinus D.E. Allen (tra)
Native. This species was formerly named *R. dumetorum* var. *ferox*. It is presently considered a synonym of *R. intensior* which has priority, although it has not been recorded from Scilly.

R. tuberculatus Bab.
St Mary's and Tresco
Native. A bush in the eroded pine plantation on Garrison Hill. Other, earlier records are considered errors by Allen. Tim Harrison recorded it as scattered in hedgerows on Tresco and St Mary's in 2012. The map in Newton and Randall (2004) only shows a pre-1988 record.

ROSACEAE – Rose family
Potentilla anserina L.

Silverweed
Inhabited islands, also Northwethel, Samson and Nornour

Native. Lousley recorded it from Teän where it may still be present. The form where both sides of the leaf are silver is common. Frequently grows in wet places around pools, the top of beaches, trampled places in gateways and waste places.

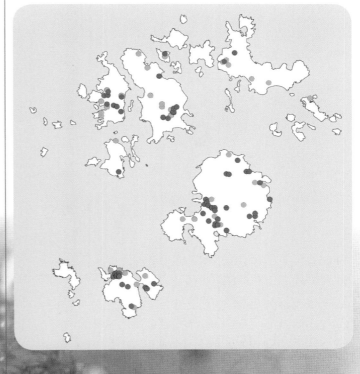

EU-DICOTS True Dicotyledons

ROSACEAE – Rose family
Potentilla erecta (L.) Raeusch.
Tormentil
Inhabited islands, also the larger uninhabited islands including St Helen's, Northwethel, Teän, Samson, Great Ganilly and Little Arthur

Native. Widespread on most islands but absent from Annet. Typically found on heathlands both inland and around the coast.

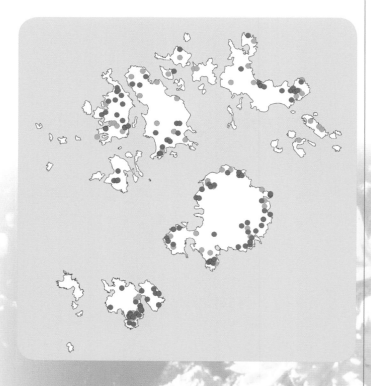

Potentilla anglica Laichard.
Trailing Tormentil
Unconfirmed

Native. Although Lousley records this species he comments his specimens were not typical. Occasionally recorded, but has not yet been satisfactorily confirmed.

ROSACEAE – Rose family
Potentilla reptans L.
Creeping Cinquefoil
St Agnes, St Mary's, Tresco and Bryher

Native. Only common in the south of St Mary's, otherwise there are a few recent records from St Agnes, Tresco and Bryher. Appears to be less frequent than in Lousley's time. Found on roadsides and waste ground.

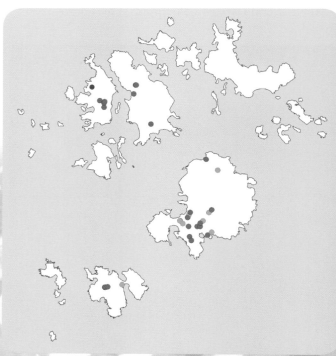

Poterium sanguisorba ssp. *balearicum* (Bourg. ex Nyman) Stace
Fodder Burnet
St Martin's

Neophyte. Found in a derelict field at Higher Town, St Martin's in 1996.

ROSACEAE – Rose family
Aphanes arvensis L.
Parsley-piert
Inhabited islands

Native. Scattered records on the inhabited islands. It is much less common than *Aphanes australis*, although they sometimes occur in the same fields. In cultivated fields and on disturbed ground.

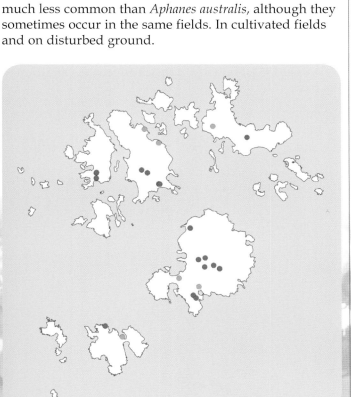

ROSACEAE – Rose family
Aphanes australis Rydb.
Slender Parsley-piert
Inhabited islands, Samson

Native. Found on all the inhabited islands, especially common on St Mary's. It is also known from Samson were it grows on the main path and around the ruined houses. Probably under-recorded, although it is more common than *A. arvensis*. Arable fields, disturbed ground and on sandy tracks in dunes.

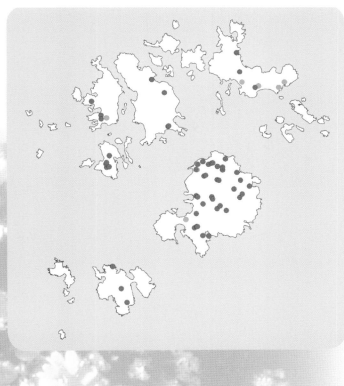

EU-DICOTS True Dicotyledons

ROSACEAE – Rose family
Rosa rugosa Thunb.
Japanese Rose
Inhabited islands

Neophyte. Occasional garden escape. A clump of bushes has been growing in the quarry at Porth Wreck, St Mary's for more than ten years.

ROSACEAE – Rose family
Rosa canina L.
Dog Rose
St Agnes (and Gugh), Tresco, St Mary's and St Martin's formerly St Helen's

Native. This rose has a patchy distribution on the islands and is apparently absent from Bryher. Has increased in numbers and range since Lousley wrote that there were less than twenty bushes in Scilly. Last recorded on St Helen's in 1983. Found on dunes, hedgebanks and along the coast.

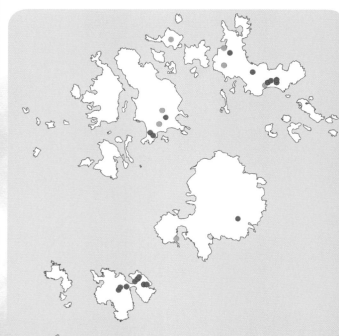

Rosa multiflora var. *cathayensis* (Rehder & E.H. Wilson) Bailey.
Multi-flowered Rose
St Mary's

Neophyte. The rose that has been known from a hedge on Salakee Farm (SV923107) for over 40 years was confirmed by Roger Maskew as this. A yellow-flowered rose found on Gugh (now lost?) has not been identified.

ROSACEAE – Rose family
Rosa rubiginosa L.
Sweet-briar
St Mary's, Tresco and Bryher

Native in the British Isles, but neophyte in Scilly as a garden escape. Near Garrison campsite, St Mary's in 1975; near Watch Hill, Bryher in 1993 and Gimble Porth, Tresco in 1975. Formerly found near the Penzance Road Gate and also opposite the Garden Gate on Appletree Banks, Tresco until 1967 (Lousley, 1971).

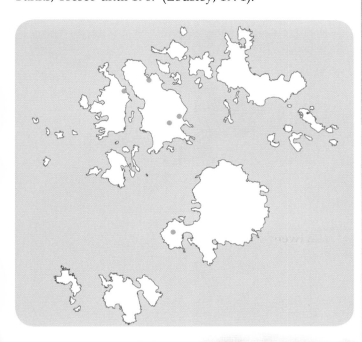

ROSACEAE – Rose family
Rosa micrantha Borrer ex Sm.
Small-flowered Sweet-briar
St Mary's and Tresco

Native in the British Isles, but neophyte in Scilly as a garden escape. Recorded by J.P. Bowman from Garrison Farm, St Mary's in 1974 and SW of Holy Vale in 1972, and from Appletree Banks and Abbey Hill, Tresco in 1972. In 2016 Christine Blackwell reported that the bush in Holy Vale (SV920115) was recovering after having been cut down.

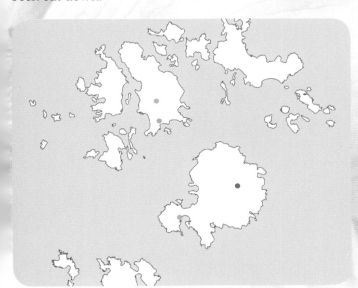

ELAEAGNACEAE – Sea-buckthorn family
Hippophae rhamnoides L.
Sea-buckthorn
St Mary's, Tresco

Native in the British Isles, but planted in Scilly. It was noted in 2004 that a few plants had been planted on Tresco on the edge of the dunes just outside the Gardens. It is to be hoped they are not allowed to spread further. A few saplings planted on St Mary's on the Garrison in 2010 were later removed to prevent them spreading.

Elaeagnus pungens Thunb.
Spiny Oleaster
Tresco

Neophyte. Found as a garden plant. There are no authenticated records of this shrub being naturalised. It grows outside Abbey Gardens or where it is planted as a hedge.

EU-DICOTS True Dicotyledons

ULMACEAE – Elm family
Ulmus L.
Elms
Inhabited islands

Native on the mainland, but originally planted in Scilly. The elm populations in Scilly are currently being investigated; they are a notoriously difficult group. The main species on St Mary's seems to be *Ulmus* x *hollandica* Dutch Elm. Elms are found on all the inhabited islands where they have been planted and have since spread, often by suckering, as can be seen in the extension to Lower Moors. Elms were first planted on St Mary's probably some time before 1600 as there is a reference to planted elms in Holy Vale in 1695 including one tree that was more than three feet in circumference, although other trees were not as large. Whether any descendants from those trees still exist is not known. There were reported to be 'fine' trees in Holy Vale in 1800, there are still fine trees there especially near Longstone but again their history is unknown. In 1813 Sir William Hooker visiting Scilly clearly had an unhappy time as he described 'wretched plants of elm and *Tamarix gallica*' (Lousley, 1971) in Holy Vale but this may not have been a fair description.

Ulmus minor ssp. *sarniensis* in lane by Content Lane, St Mary's

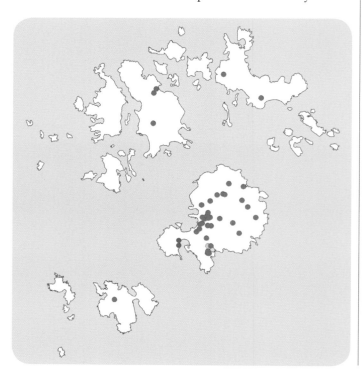

Ulmus minor ssp. *sarniensis*

Other elms in hedges are often old, but stunted by the elements. Some on St Agnes are reputed to be well over 100 years old but are still barely hedge height. Elms such as *U. procera* English Elm and *U. glabra* Wych Elm, have also been recorded but need confirmation. *Ulmus minor* has been recorded from St Mary's and St Martin's, with the subspecies *angustifolia* and *sarniensis* having been identified on St Mary's. Several *Ulmus minor* were planted in Holy Vale (SV92011152) as Cornish Elm *Ulmus minor* ssp. *angustifolia* and are now fine, large trees All the elms are believed to have originated from planted stock.

Fortunately Dutch elm disease has so far not reached the Isles of Scilly. But there are concerns about biosecurity when it has at times been proposed to bring timber with bark and even elm saplings into the islands. Occasional concerns about the health of some of the largest elms have been caused by trees apparently dying or being stressed either by extreme weather conditions, or as in the Holy Vale, where they are clearly growing on top of old walls and also have their roots in water. Some of the elms show signs of having been pollarded in the past, especially around field boundaries.

A visit by Dr Holger Thüs from the Natural History Museum in 2013 revealed that the elms had an important lichen flora associated with them including the species *Bacidia incompta*. Under the elm canopy is also an important foraging area for the local bats.

Holy Vale, St Mary's

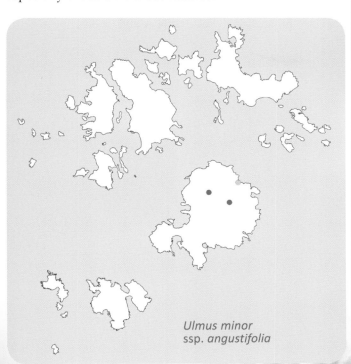

Ulmus minor ssp. *angustifolia*

EU-DICOTS True Dicotyledons

CANNABACEAE – Hop family
Humulus lupulus L.
Hop
Tresco, St Martin's and St Mary's

Native in the British Isles, but planted on Scilly. An occasional garden escape, found in lanes and along hedgebanks.

MORACEAE – Mulberry family
Ficus carica L.
Fig
Tresco, St Agnes, St Mary's and St Martin's

Neophyte. Occasional plants are found away from gardens, for example on Tresco and St Martin's where they are presumed to have self-seeded or have been bird sown.

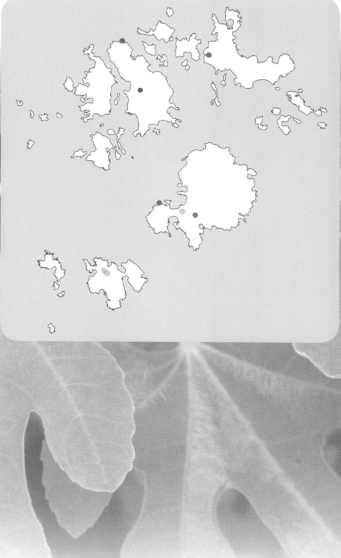

URTICACEAE – Nettle family
Urtica dioica L.
Common Nettle
Inhabited islands, many uninhabited islands

Native. Widespread on all the inhabited islands and also on Teän, St Helen's, Samson and Annet as well as on Puffin Island off Samson. There are no recent records from the Eastern Isles, where it was recorded by Lousley, or from the rarely visited Foreman's Island off Northwethel where it was recorded by Peter Clough in 1981. Usually associated with human activity; for example after there had been maintenance work around the ruined houses on Samson in 2007 there was a huge growth of nettles in response. There are records of nettle among plant remains from archaeological sites in Bronze Age Scilly. Also found where there is disturbance and enrichment of the ground from nesting seabirds. Fields, waste ground, rubbish dumps and other disturbed areas.

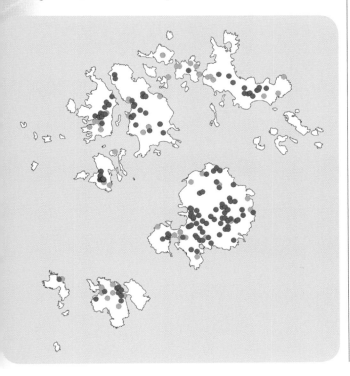

URTICACEAE – Nettle family
Urtica urens L.
Small Nettle
Inhabited islands

Archaeophyte. There are no recent records from any of the uninhabited islands, although Lousley had recorded this species growing around gull nests on Samson and Great Innisvouls. There are also pollen records from archaeological sites. Mainly found in cultivated habitats such as gardens, arable fields, allotments, and other disturbed ground.

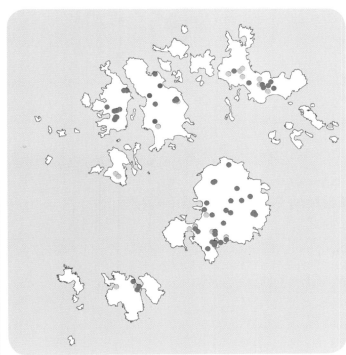

EU-DICOTS True Dicotyledons

URTICACEAE – Nettle family
Parietaria judaica L.
Pellitory-of-the-wall
St Mary's, Tresco, Bryher and St Martin's

Native. Although Lousley recorded this plant from all the inhabited islands, there are apparently no recent records from St Agnes. Occasionally found on banks or more natural sites, but most often on stone hedges and walls along roadsides and especially near habitations.

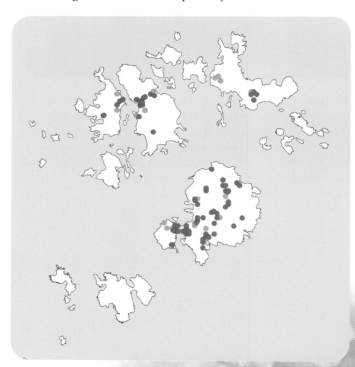

Parietaria officinalis L.
Eastern Pellitory-of-the-wall
Tresco

Neophyte. An unconfirmed plant found near Abbey Gardens by C. Westall in 2007 is presumed to have been an accidental introduction.

URTICACEAE – Nettle family
Soleirolia soleirolii (Req.) Dandy
Mind-your-own-business
Inhabited islands

Neophyte. Established on the inhabited islands. Found growing on damp walls and shady ground around gardens and habitations.

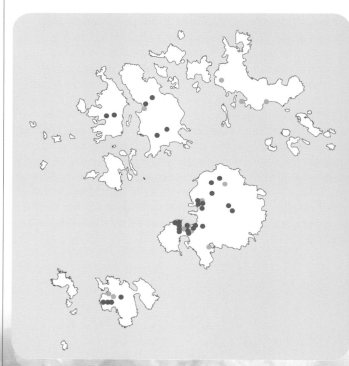

Tree planting on the Isles of Scilly

Most of the trees now found in the islands have been planted or are derived from planted trees. Tree planting was very popular in the past and many attempts were made to introduce trees to what at the time was a very exposed and open landscape. There is an account of sycamore, ash and elm being planted in Holy Vale about 1650 as well as cherry, pear and apples (Turner [1695] in *The Scillonian*, 1964). At that time there were apparently no trees and few shrubs, just 'Brambles, Furzes, Broom and Holly' when Robert Heath reported in 1750 (Heath, 1750).

It was found that trees such as *Pinus radiata* Monterey Pine and *Cupressus macrocarpa* Monterey Cypress were ideal subjects to grow in Scilly as they were resistant to the gales and salt spray that many British natives could not withstand. Elms, however grew well in more sheltered localities – for example along Holy Vale where they seem to have been planted in around 1600 although the present elms are more recent. Several plantations of British native trees have been established on St Mary's and elsewhere. Some of these, notably in the extension to Lower Moors and more sheltered places, appear to be thriving. Other plantations of deciduous species on exposed places have been less successful, although conifers thrived.

There are accounts of there being 'very thick stumps of oak, which evidently belonged to trees of extraordinary magnitude' in 1669 (Magalotti, 1821). Also, of the buried trunks of very large trees (having been found in the peaty soils sometimes five to six feet deep on Tresco and St Mary's) often from two to six feet in circumference and with remains of their branches and roots still present (Turner [1695] in *The Scillonian*, 1964).

Generally no attempt has been made to map planted trees, although sometimes they may be recorded where of interest – for example where they may have become naturalised, such as sycamore.

FAGACEAE – Beech family
Fagus sylvatica L.
Beech
Tresco and St Mary's

Native in the British Isles but a few have been planted on Scilly.

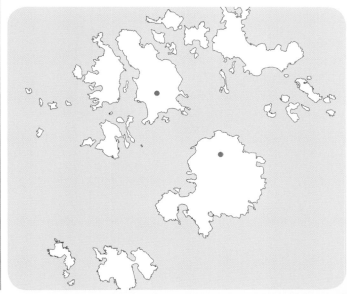

EU-DICOTS True Dicotyledons

FAGACEAE – Beech family
Castanea sativa Mill.
Sweet Chestnut
Tresco

Archaeophyte in the British Isles but planted in Scilly. Recorded on Abbey Road, Tresco by Tony Butcher in 1999, possibly no longer there?

FAGACEAE – Beech family
Quercus cerris L.
Turkey Oak
Tresco and St Agnes

Neophyte. Planted by the Abbey, Tresco and outside the Parsonage, St Agnes.

FAGACEAE – Beech family
Quercus ilex L.
Evergreen Oak
Tresco

Neophyte. Planted in several places in woodland, Abbey Hill and Abbey Drive.

FAGACEAE – Beech family
Quercus petraea (Matt.) Liebl.
Sessile Oak
St Mary's and Tresco

Although a native on the mainland, in Scilly all oaks have been planted. Abbey Hill, Tresco; several places on St Mary's, Mount Todden Down, as a hedge near Porthloo and at SV92661106.

EU-DICOTS True Dicotyledons

FAGACEAE – Beech family
Quercus robur L.
Pedunculate Oak
St Mary's and Tresco

Native on the mainland but usually planted in most places on Scilly. There are references to oak trees occurring in the islands historically and Lousley thought some may have survived on Tresco. A stunted tree that grew on Great Ganinick, known since at least 1938, has not been seen there since 1990.

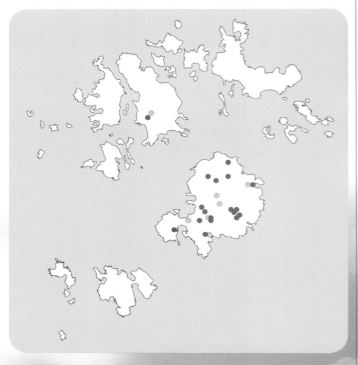

BETULACEAE – Birch family
Betula L. spp.
Birch
St Mary's

Native on the mainland but usually planted on Scilly. Some that appear to be *B. pendula* have been planted in Lower Moors extension, St Mary's and near Bar Point, St Mary's.

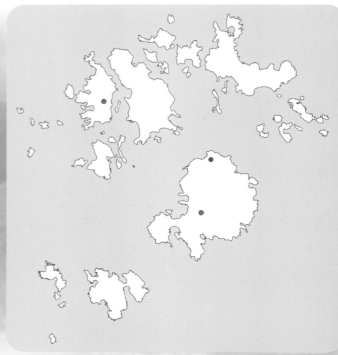

BETULACEAE – Birch family
Alnus glutinosa (L.) Gaertn.
Alder
St Mary's, St Martin's and Tresco

Native on the mainland but probably neophyte on Scilly. A few trees were recorded by Great Pool, Tresco in 1998 where they had been also recorded by Lousley. On St Mary's there are individual trees in Lower and Higher Moors. Although growing in typical wetland situations it is believed all these trees may have originally been planted, possibly by P.Z. McKenzie who was honorary Nature Conservancy warden at the time. There is a group of planted trees along a ditch on St Martin's. Some have recently been planted in the Lower Moors extension.

BETULACEAE – Birch family
Alnus incana (L.) Moench.
Grey Alder
Tresco and St Mary's

Neophyte. Planted on Abbey Hill and in the Lower Moors extension.

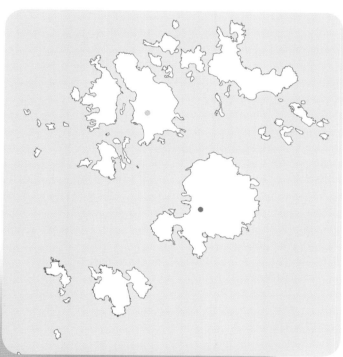

EU-DICOTS True Dicotyledons

BETULACEAE – Birch family
Corylus avellana L.
Hazel
St Mary's

Native in the British Isles but planted in Scilly. Planted in several places on St Mary's near Innisidgen, in the Lower Moors extension and as a hedge near Porthloo.

CELASTRACEAE – Spindle family
Euonymus japonicus Thunb.
Japanese Spindle
Inhabited islands

Neophyte. A frequent planted hedge species around bulbfields on the inhabited islands. Some plants may have occasionally self-seeded.

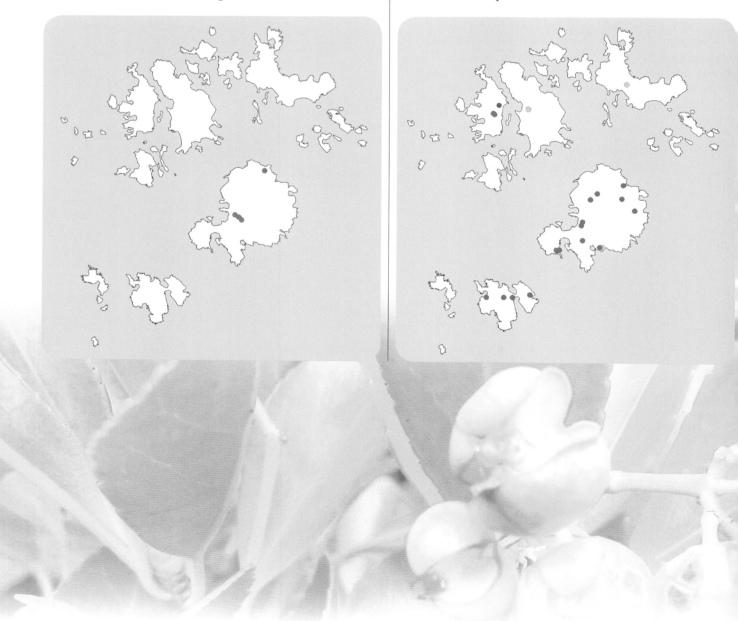

OXALIDACEAE – Wood-sorrel family

A number of species of *Oxalis* are found on Scilly, as either garden escapes or relicts of cultivation. Some of the pink-flowered species can be difficult to identify in the field as it is often necessary to examine the bulbils and rhizomes as well as details of leaf and inflorescence.

Oxalis rosea Jacq.
Annual Pink-sorrel
Tresco

Neophyte. Garden escape outside Abbey Gardens, Tresco. The latest record was from near the New Entrance in 2008.

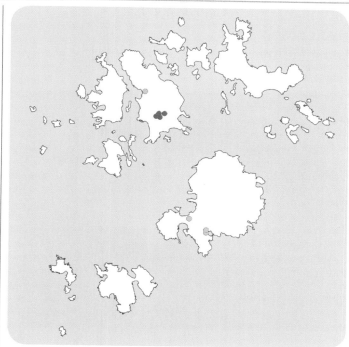

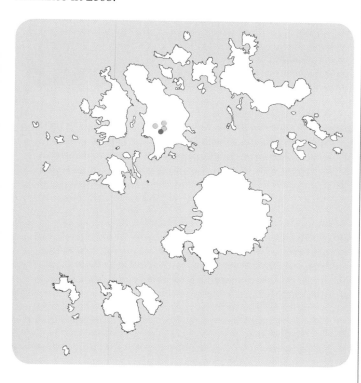

Oxalis corniculata L.
Procumbent Yellow-sorrel
Tresco and St Mary's

Oxalis corniculata var. *atropurpurea*

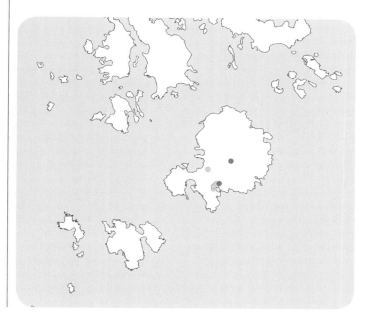

Neophyte. An established garden escape. The purple-leaved variety *atropurpurea* occurs in several places on St Mary's, including in Old Town churchyard.

EU-DICOTS True Dicotyledons

OXALIDACEAE – Wood-sorrel family
Oxalis exilis A. Cunn.
Least Yellow-sorrel
Tresco and St Mary's

Neophyte. A garden escape. Established in several places on Tresco and on St Mary's. This species is smaller in all its parts and was previously recorded as a variety of *O. corniculata*.

OXALIDACEAE – Wood-sorrel family
Oxalis megalorrhiza Jacq.
Fleshy Yellow-sorrel
Tresco, St Mary's, St Martin's and St Agnes

Neophyte. This is a native of Chile, where it apparently grows on coastal rocks. We know it was in cultivation in Tresco Abbey Gardens before 1879 because it is illustrated in painting number 7 by Mrs Le Marchant (King, 1985). Since then the plant has spread to walls on all the inhabited islands; besides being deliberately planted, it also seems to be spreading naturally by seed. It has been suggested the seeds may be spread by ants, but seedlings often appear in pot plants brought back from Tresco when the seed pods 'explode' shooting seeds in all directions. It is very frost sensitive and plants in exposed places are severely cut back some winters. Walls.

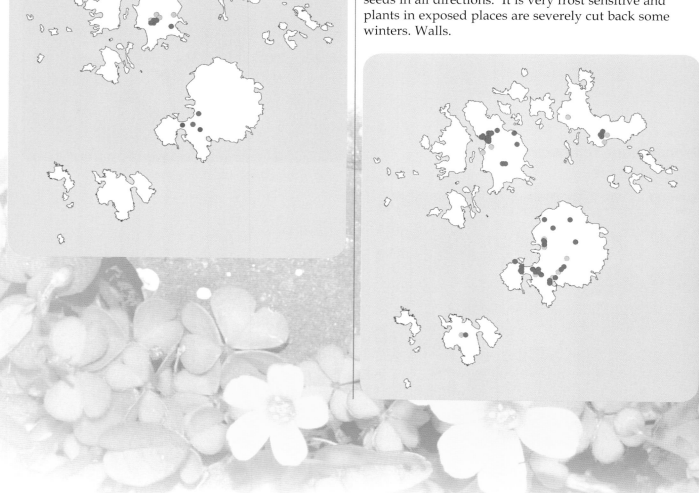

OXALIDACEAE – Wood-sorrel family
Oxalis articulata Savigny
Pink-sorrel
Inhabited islands

Neophyte. Locally frequent on all the inhabited islands. This escape from cultivation was first recorded by Lousley from Bar Point, St Mary's in 1939, but had been noted earlier on St Martin's in 1925. It is likely this is the *Oxalis* species that had originally been introduced from South America to be grown commercially as a garden plant in the 1890s. Unfortunately it was not successful and it seems to have become something of a pest in the bulbfields. Arable fields, waste places and gardens.

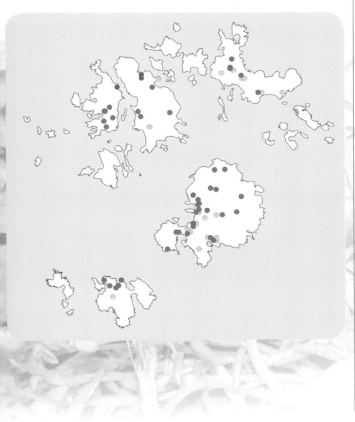

OXALIDACEAE – Wood-sorrel family
Oxalis debilis Kunth
Large-flowered Pink-sorrel
Inhabited islands

Neophyte. Occasional garden escape. This is var. *corymbosa* (DC.) Lourteig.

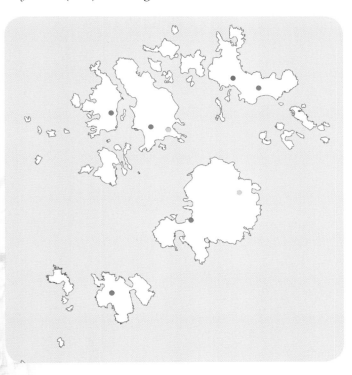

Oxalis acetosella L.
Wood-sorrel
Extinct/error?

There are records from the second half of the nineteenth century from St Mary's but without any woodland or other suitable habitat these seem doubtful. All recent records have proved to be errors for one of the introduced species.

EU-DICOTS True Dicotyledons

OXALIDACEAE – Wood-sorrel family
Oxalis latifolia Kunth
Garden Pink-sorrel
St Mary's, St Agnes and Tresco

Neophyte. Garden escape established in a few places. Found on waste land, roadsides and arable fields.

OXALIDACEAE – Wood-sorrel family
Oxalis tetraphylla Cav.
Four-leaved Pink-sorrel
Tresco and St Mary's

Neophyte. Rare garden escape only recorded from vicinity of Abbey Gardens and Old Grimsby on Tresco. On St Mary's it has been found in Carn Friars Road and recently (2011) at Thomas Porth. The leaves have only faint dark markings unlike those on some cultivars.

OXALIDACEAE – Wood-sorrel family
Oxalis pes-caprae L.
Bermuda Buttercup
Inhabited islands

Neophyte. As often stated 'neither a buttercup, nor from Bermuda'! Lousley first recorded this species on St Mary's in 1938. Probably originally introduced as a garden plant from South Africa it spread rapidly and is now a major, if attractive, pest in the bulbfields. The tiny bulbils are easily spread and are impossible to eradicate. In some fields it can be the main weed species and one of the earliest in flower, often co-dominant with *Allium triquetum*. Local children chew the flower stems as 'sorrel' for their lemony taste and leaves may be used in salads. Grows in arable fields, walls, roadsides and on disturbed ground.

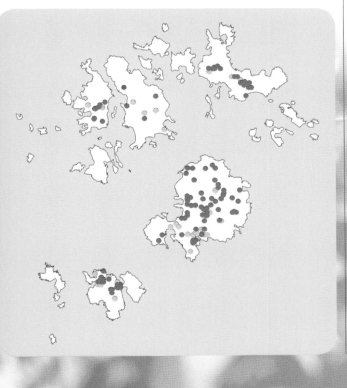

EUPHORBIACEAE – Spurge family
Mercurialis annua L.
Annual Mercury
Inhabited islands, also Samson

Archaeophyte. This is a common plant on St Mary's, but elsewhere on the inhabited islands there are just a few records. There are a cluster of records from Lower Town and another group at Higher Town on St Martin's. On St Agnes the only recent record was 1990 (SV884083); Bryher 2006 (SV881154); on Tresco it has been mostly recorded near the Abbey gardens and in 2007 several plants were found on the site of the new tennis courts then under construction. There is a record from South Hill, Samson near the ruined buildings in 1996. Found mainly in arable fields, dunes and disturbed ground.

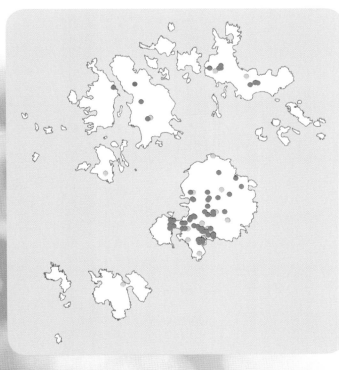

EU-DICOTS True Dicotyledons

EUPHORBIACEAE – Spurge family
Euphorbia mellifera Aiton
Honey Spurge
Tresco and St Agnes

Planted shrub. Self-seeded plants occur occasionally.

EUPHORBIACEAE – Spurge family
Euphorbia helioscopia L.
Sun Spurge
Inhabited islands

Archaeophyte. Scattered records, but nowhere particularly frequent. Found in arable fields and cultivated ground, including gardens.

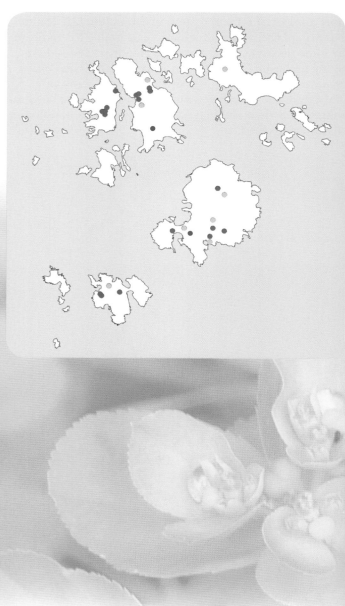

Euphorbia peplis L.
Purple Spurge
Extinct

Native. Known from St Agnes and St Mary's in late 1800s-early 1900s. Last record in 1936 (Lousley, 1971).

EUPHORBIACEAE – Spurge family
Euphorbia lathyris L.
Caper Spurge
St Mary's and St Martin's

Archaeophyte. Garden escape on St Mary's (SV918113) and St Martin's (SV924159).

EUPHORBIACEAE – Spurge family
Euphorbia peplus L.
Petty Spurge
Inhabited islands

Archaeophyte. There are concentrations of records on Tresco and St Mary's. Elsewhere there are very few recent records, just one from near the hotel on Bryher in 1995, from Higher Town, St Martin's (SV928156) in 1998 and a small population in bulbfields just above The Bar on St Agnes. Found mainly in bulbfields and other arable fields and gardens.

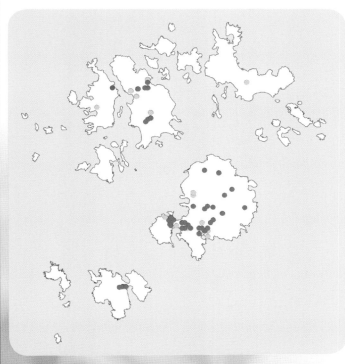

EU-DICOTS True Dicotyledons

EUPHORBIACEAE – Spurge family
Euphorbia portlandica L.

Portland Spurge

Inhabited islands, also Teän, Samson, White Island (off St Martin's), Great Ganilly and Nornour

Native. Sometimes found at the same sites as *E. paralias*. Widespread in suitable habitats around the islands, also more unusually found on a wall-top and sometimes in the gutters at Carn Thomas in Hugh Town, St Mary's. Mainly a coastal plant, growing in dunes, along sandy paths and on sandy or shingle beaches above HWM.

EUPHORBIACEAE – Spurge family
Euphorbia paralias L.

Sea Spurge

Inhabited islands, also Teän, St Helen's, Samson, Great Ganilly and Nornour

Native. Frequently found in the same sites as *E. portlandica*. It is generally more common than that species and inhabits some places where *E. portlandica* is not found. Sandy beaches and dunes.

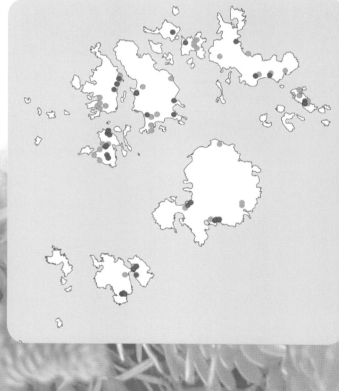

EUPHORBIACEAE – Spurge family
Euphorbia cyparissias L.
Cypress Spurge
St Mary's

Neophyte. Garden escape in several places on St Mary's, especially along Porthloo Lane.

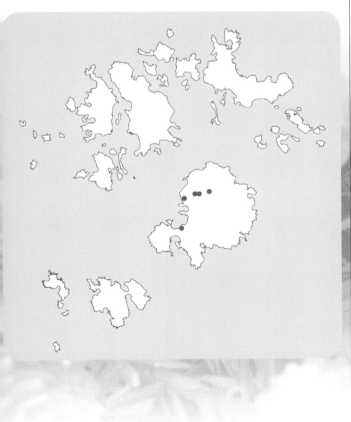

EUPHORBIACEAE – Spurge family
Euphorbia amygdaloides ssp. *amygdaloides* L.
Wood Spurge
Bryher, Tresco, St Martin's, also St Helen's, Great Ganinick and Great Ganilly

Native. Found in the north of the islands with concentrations on Bryher, Tresco and St Helen's. Coastal sites and heathland.

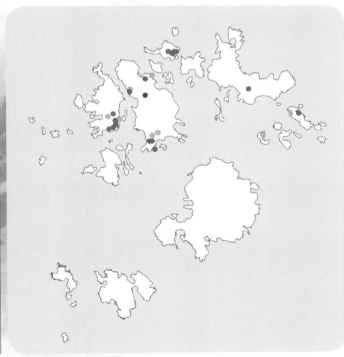

EU-DICOTS True Dicotyledons

ELATINACEAE – Waterwort family
Elatine hexandra (Lapierre) DC.
Six-stamened Waterwort
Tresco

Native. Only seen occasionally around Abbey Pool when the mud is exposed after a drought. A few plants were found on the drawdown edge of Abbey Pool in September 2016.

SALICACEAE – Willow family

If any tree is native in Scilly then it is possibly *Salix cinerea* ssp. *oleifolia*. However the comment by Borlase after his visit to the islands in 1752 (Borlase, 1756) that the wet ground at Porthellick which only had mire and flags would be an ideal place for a willow plantation does perhaps contradict this. But the name Porthellick translates in Cornish as porth (bay) of the willows, so there either had been or were some willows there? Other than *Salix cinerea* ssp. *oleifolia*, all the poplars and willows on Scilly have been planted. Some willows have been introduced for weaving lobster pots and lately for willow-weaving; many appear to be hybrids or cultivated varieties and the following list may be incomplete.

Populus alba L.
White Poplar
Tresco

Neophyte. Planted along Pool Road and near Great Pool, Tresco.

SALICACEAE – Willow family
Populus x *canescens* (Aiton) Sm.
Grey Poplar

Neophyte. Planted near Great Pool, Tresco.

Populus x *canadensis* Moench
Hybrid Black-poplar

Neophyte. Planted Tresco and St Mary's. Includes variety 'Serotina' on Tresco.

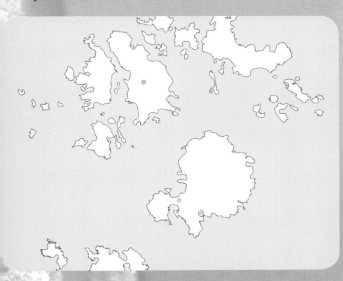

SALICACEAE – Willow family
Populus x *jackii* Sarg.
Balm of Gilead

Neophyte. Planted on Toll's Island and at Higher Moors, St Mary's.

EU-DICOTS True Dicotyledons

SALICACEAE – Willow family
Populus balsamifera L.
Eastern Balsam-poplar
Neophyte. One planted at Lunnon Farm on St Mary's.

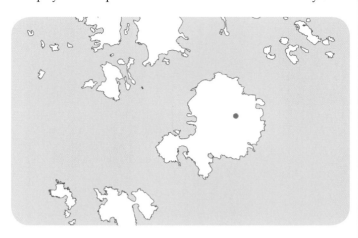

Salix fragilis L.
Crack-willow

Archaeophyte. Occasionally planted. Apparently introduced to St Mary's, St Martin's and Tresco.

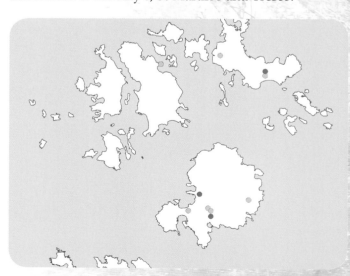

SALICACEAE – Willow family
Salix x rubens Schrank
Hybrid Crack-willow
Archaeophyte. Planted on St Mary's.

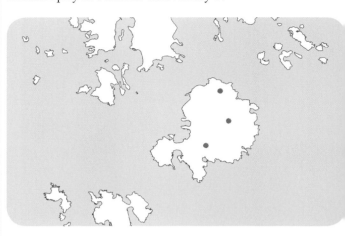

Salix alba L.
White Willow
Archaeophyte. Planted on St Mary's near Higher Moors and Porth Hellick.

Salix triandra L.
Almond Willow
Archaeophyte. One near the incinerator, St Mary's.

SALICACEAE – Willow family
Salix viminalis L.
Osier

Archaeophyte. Introduced to St Agnes, St Mary's and St Martin's for basketwork and willow-weaving.

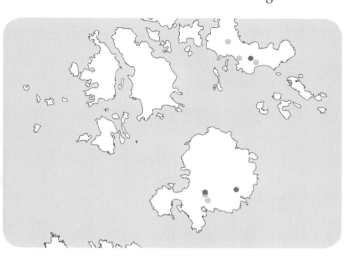

Salix x *smithiana* Willd.
Broad-leaved Osier

Native on the mainland, but planted on Tresco and St Mary's.

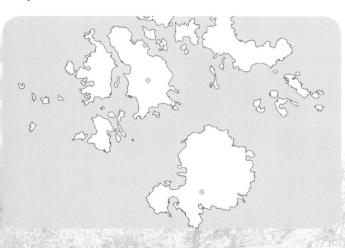

Salix caprea L.
Goat Willow

Native on the mainland, but planted on St Mary's near Higher Moors and Porth Hellick. There are some planted trees on Lower Moor that appear close to *S. caprea*, but as elsewhere may be the hybrid with *S. cinerea*, *S.* x *reichardtii* which J.E. Oliver considered to be more frequent than true *S. cinerea*. It is very variable in leaf shape and easily mistaken for *S. caprea*. Wetland.

SALICACEAE – Willow family
Salix cinerea ssp. *oleifolia* Macreight
Grey Willow

St Mary's, Tresco, Bryher, St Helen's and Great Ganilly

Native. This is a common willow in the Isles of Scilly. It is found in wetland sites on most islands although possibly Lousley overlooked it on Bryher. The stunted tree recorded on Great Arthur in 1968 has not been seen since. There are a few windswept trees on St Helen's and Great Ganilly. It is possibly these may have originated from twigs taken there by gulls as nesting material. See comments on possible hybrids with *S. caprea* above. Mainly wet areas, but also found in quite dry places.

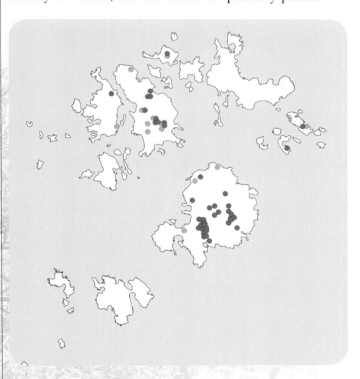

Salix x *multinervis* Döll
Native. Possibly this hybrid planted on Lower Moors.

EU-DICOTS True Dicotyledons

VIOLACEAE – Violet family
Viola odorata L.
Sweet Violet
St Mary's, St Martin's and St Agnes

Native in the British Isles, but probably neophyte in Scilly. Now well established on the above islands. It is unclear whether these plants originated from garden escapes and have reverted or whether they are the wild plant. Sweet violets were cultivated in Scilly but they are usually more robust plants. Usually not seen in flower (they flower in March so are over before most visitors arrive). At times this species has been mistaken for *Viola hirta* Hairy Violet. Occurs in scattered locations often near roadsides, sometimes forming quite large stands. Hedgebanks and verges.

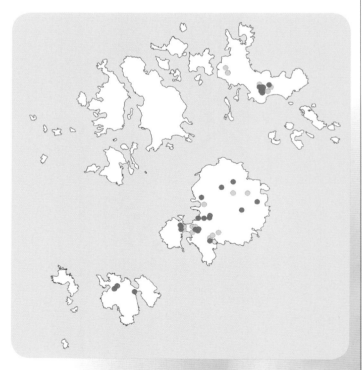

VIOLACEAE – Violet family
Viola riviniana Rchb.
Common Dog-violet
Inhabited islands, also Samson, Annet, St Helen's, Teän, Northwethel, White Island (off St Martin's), Nornour, Great Ganilly and Little Arthur

Native. This is the common violet in the islands, it is very variable with a tiny form attributable to var. *minor* sometimes found in exposed places. The colour of the spur is often white or pale but dark-spurred plants are not uncommon. The variability of the species seems to have led to the confusion over the distribution of this and the next two species on Scilly. Can be found in flower in virtually every month of the year but great drifts of the violets are a feature in late March to April on hedgebanks and also heathland where there has been grazing. Widespread on heathland, dunes and hedgebanks.

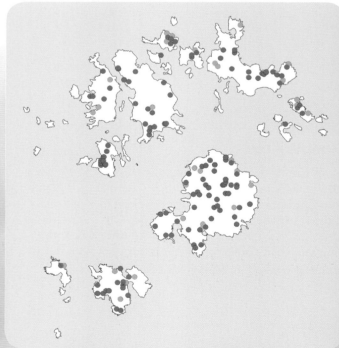

VIOLACEAE – Violet family
Viola reichenbachiana Jord. ex Boreau
Early Dog-violet
Tresco and St Agnes

Possibly native. Plants identified as this species in Scilly are being reviewed. Lousley recorded the species from most islands. The only confirmed plants found recently were on Tresco, as weeds in the Abbey Gardens. It now appears that most of the records of this species may have been errors due to confusion with *V. riviniana*.

Viola canina L.
Heath Dog-violet
Inhabited islands

Possibly native. Although Lousley records this species from St Mary's, St Agnes and Tresco, there have been no recent records. Without confirmation the status of this species on Scilly must be considered doubtful.

Viola x *wittrockiana* Gams ex Kappert
Garden Pansy
St Mary's

Neophyte. Occasional garden escape.

VIOLACEAE – Violet family
Viola tricolor L.
Wild Pansy
St Mary's, Bryher and Tresco

Native in the British Isles but probably neophyte in Scilly. The status of this species in the Isles of Scilly is unsatisfactory; it is possible all the records are of garden escapes, as there do not appear to be any reliable native records. The only recent records were from Tresco Abbey Gardens in 2000, Old Town, St Mary's in 1996 and from the Garrison in 2012. Disturbed ground.

VIOLACEAE – Violet family
Viola arvensis Murray
Field Pansy
Tresco, St Martin's, Bryher and St Mary's

Archaeophyte. Absent from St Agnes and not recorded on Bryher until 2011 (SV816145). It was not known from either island in Lousley's time. Plants with purple-flushed petals may have been mistaken for *V. tricolor*. Locally frequent in some cultivated fields, gardens and on disturbed ground.

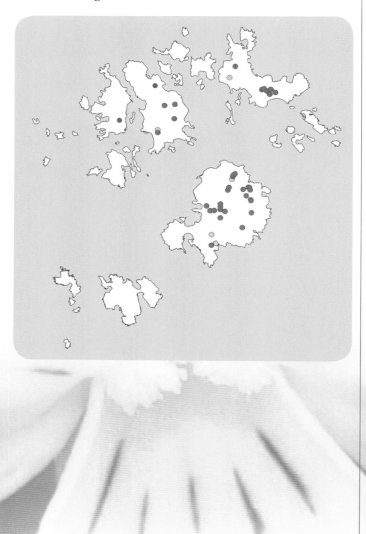

VIOLACEAE – Violet family
Viola kitaibeliana Schult.
Dwarf Pansy
Bryher, Tresco and Teän

Native. This is one of the botanical treasures of the Isles of Scilly. Only now known from the above three islands, with the main population being at Rushy Bay, Bryher, sometimes plants have been found on sandy ground almost as far as the Hotel area.

A site on sandy fields above Higher Town Bay, St Martin's at SV937154 has not been seen since 1990. It was first discovered in Scilly in 1873 by the botanist J. Ralfs. The original Ralfs site at New Grimsby was lost under buildings at the seaplane base during the 1914-18 war (Raven, 1950). Lousley found it in a sandy field at Pelistry between 1953 and 1957, a site that has apparently since also been lost. There are recent records for populations on Teän and Tresco. On Teän, Clare and John O'Reilly counted sixty-one and 227 plants respectively at two sites (SV909164 and SV911163) in 2004. On the Tresco site at Appletree Banks (SV89921393) the O'Reillys counted twenty plants in 2004. Plants were present on Teän in 2010 but after 2014 the sites were covered in deep sand. Only one plant was found at the Tresco site (although they may have been over).

The Rushy Bay site is subject to inundation from the sea in severe storm surges. During winter 1989-1990 storms deposited debris, rocks and sand over the whole site. This was cleared away by islanders and in spring 1990 the pansy appeared in astounding numbers. This population explosion continued for the next few years with around 20,000 plants in 1992 according to estimates by Pat Sargeant (Sargeant, 1993).

Rushy Bay dune heath Dwarf Pansy site

The O'Reillys only found 200 plants there in 2004 although the species was doing well on Teän in the same year. It was suggested the decline in numbers may have been associated with the reduction in numbers of rabbits (see below). After the storms during the winter of 2007-2008 around ten plants appeared in spring 2008 instead of the thousands that appeared in 1992. Over the next few years there was some recovery with several hundred flowering plants in April 2011. By late May that year after several weeks of drought most plants had gone to seed and flowering plants were difficult to find. In May 2012 numbers appeared to be increasing although no count was made, but in May 2013 there were again hundreds of plants although most had gone to seed making them difficult to count again. Another severe storm in February 2014 caused sand and rocks to be deposited over the area and changes made to the Rushy Bay site, but by April the pansy was appearing in hundreds again, as again in 2015. A few hundred plants appeared in 2016, but the vulnerability of the area leaves their long-term survival uncertain.

An interesting observation by David Mawer relates to how the seeds are spread across the site. In 2003 they had been collecting seed for the Millennium Seed Bank at Wakehurst, and had put the capsules out to dry in trays on the floor. Next day they found seeds everywhere – having been flung out by the popping capsules over a metre in some cases.

Attempts have been made to reinforce the sea bank at Rushy Bay, but this low dune and insubstantial fencing were unable to prevent the recent storm incidents. The disturbance caused by the storms may to some extent compensate for the loss of rabbits. In the past digging by rabbits exposed bare sand which seemed to encourage germination, but rabbits are now gone from Bryher. Grazing by ponies or cattle also keeps the vegetation short and may help to produce suitable areas of bare sand. Additionally the turf gets very parched in summer and also wind-pruned which reduces competition from taller plants. Sandy ground, especially dune grassland.

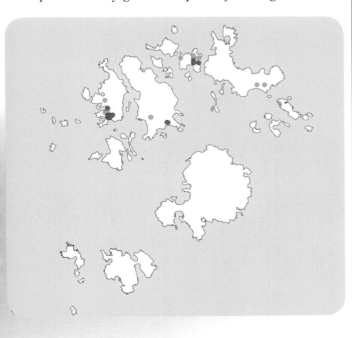

LINACEAE – Flax family
Linum bienne Mill.
Pale Flax
St Mary's

Native. Apparently lost from almost all its former sites, except from within a small area from Holy Vale to Maypole (SV9211) where about 75-100 plants were found in 2013 growing on a grassy headland, a verge and sometimes individual plants are found in the lane. Lousley recorded the species from a number of places on St Mary's: Pelistry, Old Town, Telegraph and Bar Point. There was a specimen from Bryher in the herbarium of the Misses Dorrien-Smith, 1922-1940. Roadsides and a grassy bank.

Linum usitatissimum L.
Flax
St Mary's and Tresco

Neophyte. Casual or from bird-seed. Occasional.

EU-DICOTS True Dicotyledons

LINACEAE – Flax family
Linum catharticum L.
Fairy Flax
St Martin's

Native. The only recent records are from the Great Bay area (SV923165), St Martin's in 1995 and in 2002. This is close to where Lousley recorded it in 1936. Lousley also recorded it near Tremelethen, St Mary's, possibly in 1939. The St Martin's site is near where *Pilosella officinarum* also grows, in an area where there may be deposits of shell sand. Dune grassland.

LINACEAE – Flax family
Radiola linoides Roth
Allseed
Inhabited islands, also Toll's Island, St Mary's, White Island (off St Martin's), St Helen's and Great Ganilly

Native. Widely spread throughout the islands with a concentration of records from St Agnes and the Gugh. Can be locally frequent in damp and sandy places, heathlands and grassy areas.

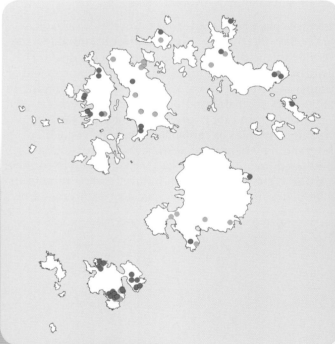

HYPERICACEAE – St John's-wort family
Hypericum androsaemum L.
Tutsan
St Mary's and Tresco

HYPERICACEAE – St John's-wort family
Hypericum x *inodorum* Mill.
Tall Tutsan
St Mary's and Tresco

Native in the British Isles, but a neophyte in Scilly as a garden escape recorded recently from three places on St Mary's, Holy Vale (SV920115), track to Bar Point (SV923126) and Old Town churchyard (SV911101). Also from near the Gardens on Tresco.

Garden escape recorded from several places on St Mary's, most recently on The Garrison and Bar Point and also from near the Abbey Gardens on Tresco.

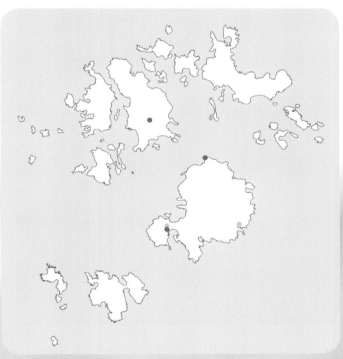

EU-DICOTS True Dicotyledons

HYPERICACEAE – St John's-wort family
Hypericum hircinum L.
Stinking Tutsan
St Mary's

Neophyte. Garden escape recorded from several places on St Mary's, Holy Vale (SV920116) in 1994, near Bant's Carn (SV909122) in 1996 and Holy Vale (SV921114) in 1997.

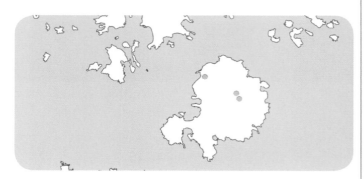

Hypericum perforatum L.
Perforate St John's-wort
St Mary's

Native in the British Isles but not on Scilly. A plant found in the restored area at the east of Porthcressa beach in 2000 and one at Porthloo in 2011 are believed accidental introductions with building or garden materials.

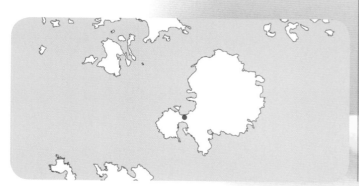

HYPERICACEAE – St John's-wort family
Hypericum humifusum L.
Trailing St John's-wort
Inhabited islands

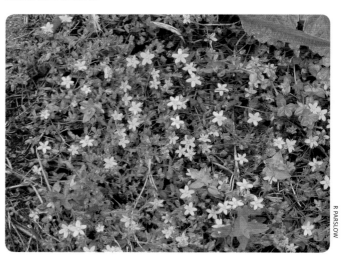

Native. Widespread and common on St Mary's, less so on the other inhabited islands. Found on shallow soils on heathland, in arable fields and man-made habitats such as gutters and between kerbstones along roadsides.

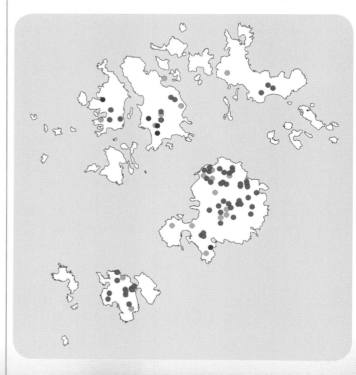

HYPERICACEAE – St John's-wort family
Hypericum pulchrum L.
Slender St John's-wort
St Mary's, St Martin's, also Great Ganilly and Little Arthur

HYPERICACEAE – St John's-wort family
Hypericum elodes L.
Marsh St John's-wort
St Mary's

Native. An uncommon plant on Scilly that is mainly found on sites in the eastern half of the islands. It is now known from St Martin's, an addition to the distribution shown on the map in Lousley's Flora. Not recorded from Teän possibly since before 1952. Grows on coastal heathland and cliff grassland often among taller vegetation.

Native. Lousley describes this species as ' Higher Moors – decreasing, limited to a small bog near Tremelethen, and exceedingly scarce by 1967'. Since then there had been just one record from Lower Moors (SV912106) by Mrs Harvey in 1994. The species was then believed to have become extinct until a small number of plants were found on boggy ground near Shooter's Pool, St Mary's (SV91301078) in July 2011. May be still there. Marshy ground.

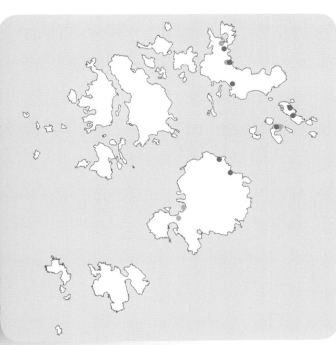

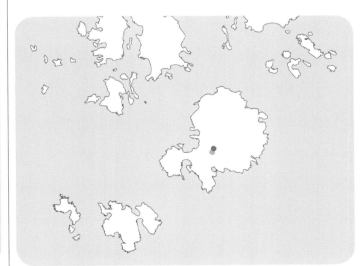

EU-DICOTS True Dicotyledons

GERANIACEAE – Crane's-bill family
Geranium x *oxonianum* Yeo
Druce's Crane's-bill
St Mary's

Neophyte. Garden escape, Old Town, St Mary's 2012.

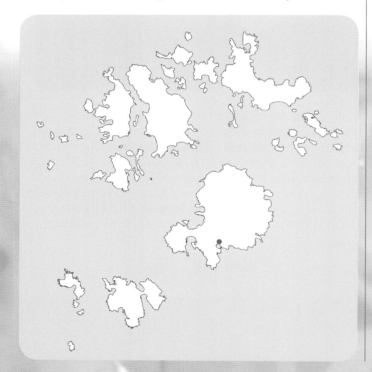

GERANIACEAE – Crane's-bill family
Geranium versicolor L.
Pencilled Crane's-bill
St Mary's and St Martin's

Neophyte. Garden escape, Lower Moors, St Mary's (SV913106), 2005.

GERANIACEAE – Crane's-bill family
Geranium rotundifolium L.
Round-leaved Crane's-bill
St Martin's

Native in the British Isles but probably neophyte in Scilly. One casual record from west of Middle Town, St Martin's by Rose Murphy in 1995.

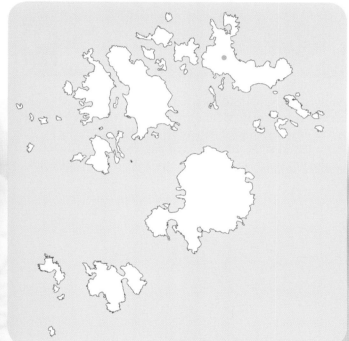

GERANIACEAE – Crane's-bill family
Geranium dissectum L.
Cut-leaved Crane's-bill
Inhabited islands

Archaeophyte. Can be locally abundant. White-flowered plants occur occasionally. Arable fields, roadsides, grassy areas and disturbed ground.

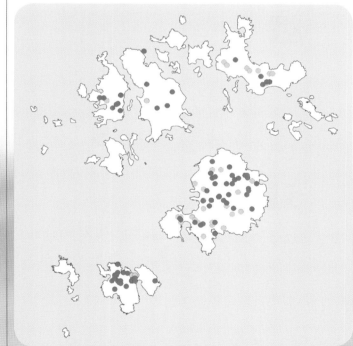

EU-DICOTS True Dicotyledons

GERANIACEAE – Crane's-bill family
Geranium molle L.
Dove's-foot Crane's-bill
Inhabited islands, also Teän, St Helen's and Great Ganilly

Native. This species is locally abundant in places. There is also a 1981 record from Foreman's Island (near Northwethel). Diminutive plants can be found in places such as Rushy Bay, Bryher growing in almost pure sand. White-flowered plants are also quite common. Grows in a range of habitats including arable fields, roadsides, dune grassland and coastal habitats.

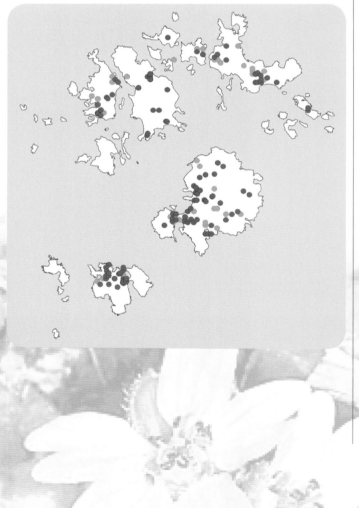

GERANIACEAE – Crane's-bill family
Geranium pusillum L.
Small-flowered Crane's-bill
St Mary's

Native. Lousley describes this as rare, he only cites one record from Porth Mellon in 1938. In May 2013 a plant was found by Mark Spencer in a bulbfield at Rocky Hill Farm (SV914112). It may have been overlooked due to confusion with small specimens of *G. molle*. Bulbfield.

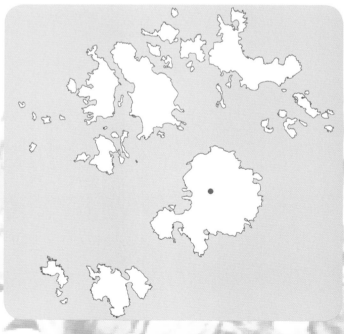

GERANIACEAE – Crane's-bill family
Geranium lucidum L.
Shining Crane's-bill
St Mary's

Native in the British Isles but a Neophyte in Scilly, with a deliberate introduction on The Garrison, Hugh Town. The plants have not spread far from where they were originally planted. Clare Harvey found two plants in a flower pot in 1973 and planted them in her garden outside Benhams on the Garrison. It initially spread along the road but has not been seen since 1995, so may have died out.

GERANIACEAE – Crane's-bill family
Geranium robertianum L.
Herb-Robert
St Mary's and St Martin's

Native. This is a rare plant on Scilly. Lousley did not accept a record by Curnow and Ralfs in 1879 but considered it a misidentification for *G. purpureum*. The first accepted recent record appears to have been from Old Town churchyard, St Mary's in 1973 where it still grows. Although not in Lousley the first records appear to have been from Garrison Lane and Old Town Churchyard (where it still is) in 1973. It appears to have since spread along the road from Old Town to Watermill. In 2011 a plant was found on St Martin's near Lower Town (SV91691615). White-flowered plants have occurred in Old Town churchyard. Waste land, walls and gardens.

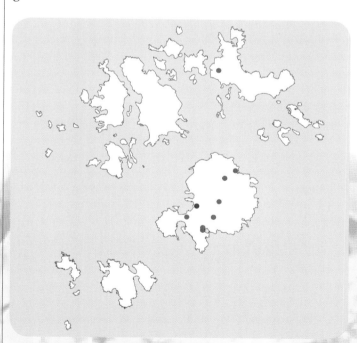

Geranium purpureum Vill.
Little-Robin
Unconfirmed

Native. Lousley refers to a specimen collected in Scilly in 1878 that is in the Herbarium of the National Museum of Wales (Baker, 1955). There have been no other records.

EU-DICOTS True Dicotyledons

GERANIACEAE – Crane's-bill family
Geranium yeoi Aedo and Muñoz Garm.
Greater Herb-Robert
Tresco, Bryher and St Mary's

Neophyte. Garden escape found in a few places. Roadsides, waste ground and a churchyard.

GERANIACEAE – Crane's-bill family
Geranium maderense Yeo
Giant Herb-Robert
Inhabited islands

Neophyte. Originally a native of Madeira that was first introduced to the Abbey Gardens in 1929. It was noted growing near gardens on Bryher in the 1980s and on other islands subsequently. Now widespread on all the inhabited islands to the extent it has become an accepted part of the Scillonian landscape. Roadsides, waste ground and coastal areas.

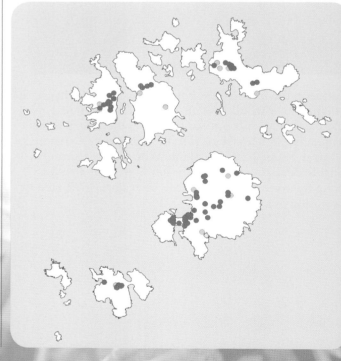

GERANIACEAE – Crane's-bill family
Erodium maritimum (L.) L'Hér.
Sea Stork's-bill

Inhabited islands, also Teän, Samson, White Island (St Martin's), Nornour and Great Ganilly

Native. Scattered sites around the islands. Often found in man-made habitats such as buildings, pathways and in pavement cracks. There are good populations on some of the ancient monuments such as The Garrison and the castles on Tresco. White-flowered plants are not uncommon. Shallow soils over granite, sandy ground, dunes, walls and other structures.

GERANIACEAE – Crane's-bill family
Erodium moschatum (L.) L'Hér.
Musk Stork's-bill

Inhabited islands

Archaeophyte. A large, handsome plant with impressive, spiky fruits. Can often be a feature, or a major weed (according to your perspective), of bulbfields on the inhabited islands. Frequent in arable fields and on disturbed ground.

EU-DICOTS True Dicotyledons

GERANIACEAE – Crane's-bill family
Erodium cicutarium (L.) L'Hér.

Common Stork's-bill

Inhabited islands, also Teän, Samson, Nornour and Great Ganilly

Native. Frequent in suitable habitats. The ssp. *dunense* is recorded occasionally but may just be a dwarf form that occurs in places such as Rushy Bay, Bryher. White-flowered plants are not uncommon. Found in sandy places near the coast, dunes, bulbfields and field edges; occasionally grows in cracks in pavements and other man-made habitats.

GERANIACEAE – Crane's-bill family
Erodium lebelii Jord.

Sticky Stork's-bill

Gugh, St Martin's, St Mary's and Great Ganilly

Native. There are very few records of this species on Scilly, and none recently. Lousley recorded it from Great Ganilly. The most recent records are from Church Street, Hugh Town where a plant persisted in a pavement crack from 1993 to 1996. In 1994 there was a record from English Island Point, St Martin's, and in 1995 it was recorded on Gugh near The Bar. Other than the plant in the pavement crack, all the records are from bare sandy areas near the coast.

GERANIACEAE – Crane's-bill family
Pelargonium tomentosum Jacq.
Peppermint-scented Geranium
Tresco and St Martin's

Neophyte. Garden escape naturalised in woodland on Tresco since at least 1971 (Clement, 2006) and near Great Pool. In a hedgerow near Carron Farm (SV928155) St Martin's in 2011.

GERANIACEAE – Crane's-bill family
Pelargonium x *hybridum* Aiton
Geranium
Inhabited islands

Neophyte. Occasional records from the inhabited islands are almost certainly of garden escapes. Not believed to be naturalised.

EU-DICOTS True Dicotyledons

LYTHRACEAE – Purple-loosestrife family
Lythrum salicaria L.
Purple-loosestrife
Tresco and St Mary's

Native. Mainly found in larger wetlands and beside pools; Higher and Lower Moors and Holy Vale, St Mary's also Great and Abbey Pools, Tresco. Also found on St Mary's in several of the small water-filled pits on Salakee Down, pools in a field near Maypole and in wet fields on Lunnon Farm adjoining Higher Moors. Grows on margins of pools and in wetlands.

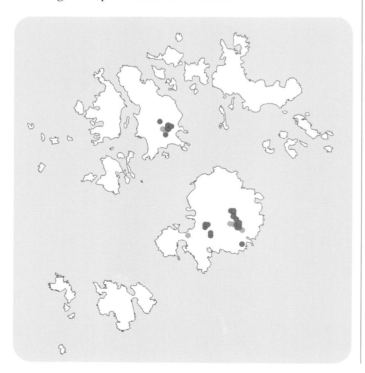

LYTHRACEAE – Purple-loosestrife family
Lythrum hyssopifolia L.
Grass-poly
St Martin's (and St Mary's?)

Archaeophyte. A casual record from an arable field below Arthur's Farm Café on St Martin's in 1984. A second report from an arable field on St Mary's (SV92021189) in 2008 could not be confirmed.

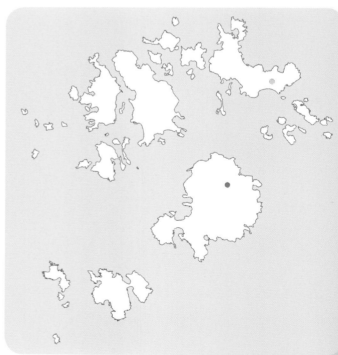

LYTHRACEAE – Purple-loosestrife family
Lythrum portula (L.) D. A. Webb
Water-purslane
Tresco and St Agnes

Native. A rare plant in Scilly, only known from around the Abbey Pool (SV896141- SV897142) and Abbey Hill (SV893144) on Tresco and Wingletang, St Agnes (SV883074). Formerly recorded from St Mary's, most recently in 1956. Ssp. *longidentata* was recorded from Middle Down, Tresco (SV8915) in 1975 by David Allen. Wet areas beside pools and in freshwater seepages.

ONAGRACEAE – Willowherbs
The smaller willowherbs can often be very abundant in arable fields. Many appear to be hybrids and identifications are not easy to confirm except when made by experts such as Geoffrey Kitchener.

Epilobium hirsutum L.
Great Willowherb (or Codlins and Cream)
St Mary's, Tresco and Bryher

Native. Scattered records from St Mary's; from Abbey Pool and both sides of Great Pool, Tresco and Veronica Farm area, Bryher. Has clearly increased since Lousley's time. From wetland sites, ditches and waste ground.

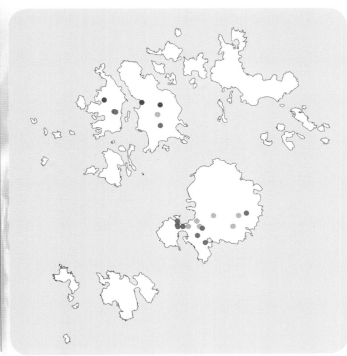

EU-DICOTS True Dicotyledons

ONAGRACEAE – Willowherb family
Epilobium parviflorum Schreb.
Hoary Willowherb
St Mary's and Tresco

Native. A scatter of records from St Mary's and records from Tresco, Borough Farm, Abbey Gardens and a dump at SV895150. The hybrid *E.* x *palatinum* (*E. parviflorum* x *E. tetragonum*) was recorded by G. Kitchener at Halangy, St Mary's (SV91131263) in 2009. Arable fields and waste land.

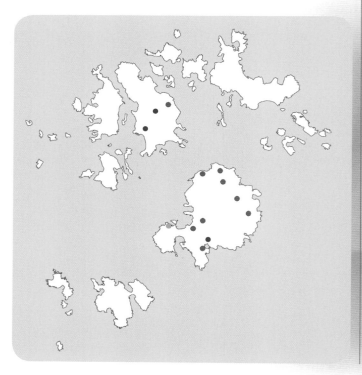

ONAGRACEAE – Willowherb family
Epilobium montanum L.
Broad-leaved Willowherb
St Mary's, St Agnes, Tresco

Native. Recent records from a few places on St Mary's, St Agnes and Tresco Abbey Gardens. Found in cultivated fields and disturbed ground.

ONAGRACEAE – Willowherb family
Epilobium lanceolatum Sebast. and Mauri
Spear-leaved Willowherb
St Mary's and St Agnes

Native. Only known from St Mary's (SV908102, SV912102 and SV926112) and St Agnes (SV881082). On disturbed ground.

ONAGRACEAE – Willowherb family
Epilobium tetragonum L.
Square-stalked Willowherb
Inhabited islands

Native. Found on all the inhabited islands, ssp. *lamyi* has been recorded occasionally. On arable fields and disturbed ground.

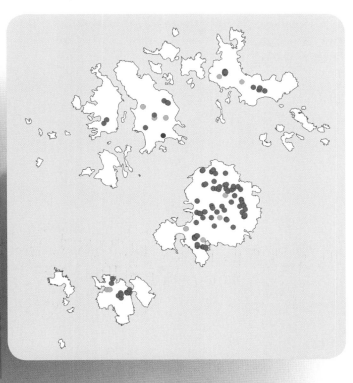

EU-DICOTS True Dicotyledons

ONAGRACEAE – Willowherb family
Epilobium obscurum Schreb.
Short-fruited Willowherb
St Agnes, St Mary's, St Martin's and Tresco

Native. Common on all the inhabited islands with the apparent exception of Bryher. The hybrids *E.* x *semiobscurum* (*E. obscurum* x *E. tetragonum*) and *E. dacicum* (*E. obscurum* x *E. parviflorum*) have been recorded by G. Kitchener in 2009, the first at Halangy (SV91131263) and the second in Pungies Lane (SV913121). Cultivated fields and waste ground.

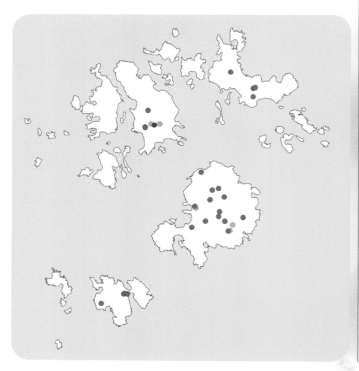

ONAGRACEAE – Willowherb family
Epilobium ciliatum Raf.
American Willowherb
Tresco, St Mary's and St Agnes

Alien. Common, probably under-recorded due to confusion with other arable willowherbs. G. Kitchener recorded the hybrid *E. ciliatum* x *E. obscurum* from several sites on all the above islands in 2009. Waste and cultivated ground.

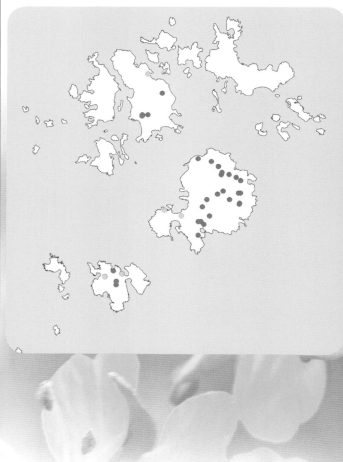

ONAGRACEAE – Willowherb family
Epilobium palustre L.
Marsh Willowherb
St Mary's

Native. Very rare, Lousley recorded it from both Higher and Lower Moors, but there have only been two recent records, three plants in 2007 on Lower Moors, St Mary's (SV912106) and a plant in a ditch (SV91281029) in 2015. Wetland.

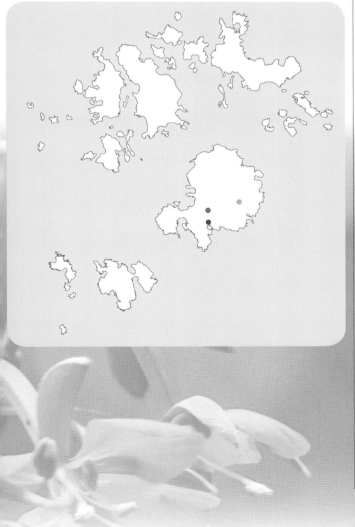

ONAGRACEAE – Willowherb family
Chamerion angustifolium (L.) Holub
Rosebay Willowherb
St Mary's

Native in the British Isles, but probably neophyte in Scilly. One casual record from The Garrison, St Mary's in 1989.

EU-DICOTS True Dicotyledons

ONAGRACEAE – Evening-primroses

Due to the difficulty in identifying species and hybrids of Evening Primroses only recent, accepted records are included here. The following records have been confirmed by R.J. Murphy. It is likely that plants formerly identified to species are now believed to be hybrids. Garden escapes are frequent. (Rostański, 1982; Sell and Murrell, 2009; Stace, 2010; Murphy, 2016).

Oenothera glazioviana P. Micheli
Large-flowered Evening-primrose
St Mary's, Tresco, St Agnes and St Martin's

Neophyte. Recorded from all the inhabited islands except Bryher. On waste land and roadsides.

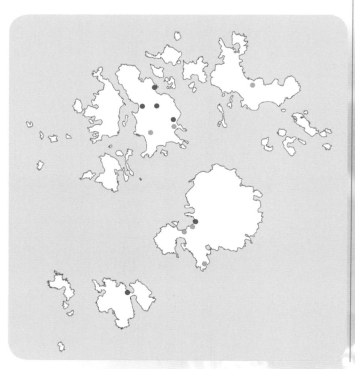

ONAGRACEAE – Willowherb family
Oenothera x *fallax* Renner
Intermediate Evening-primrose
St Mary's

Neophyte. Recorded from Old Town Churchyard, St Mary's in 2003.

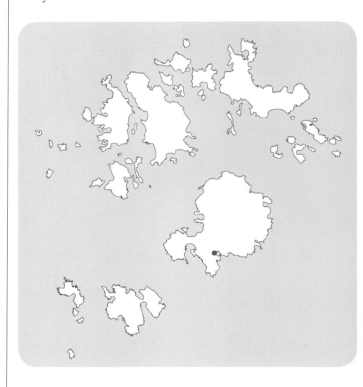

Oenothera cambrica Rostański
Small-flowered Evening-primrose
St Mary's

Neophyte. Known from one site near Trenoweth, St Mary's (SV918123) in 2006. Waste land.

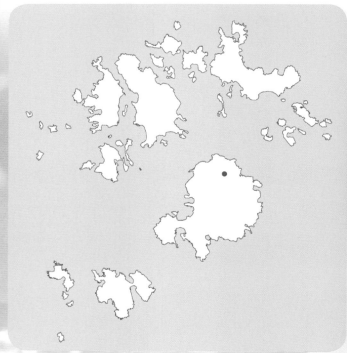

ONAGRACEAE – Willowherb family
Oenothera biennis L.
Common Evening-primrose
Tresco and St Mary's

Neophyte. Found on the rubbish dump near Abbey Garden, Tresco (SV892141). A few records from St Mary's, not all confirmed. On wasteland and roadsides.

ONAGRACEAE – Willowherb family
Fuchsia magellanica Lam.
Fuchsia
St Agnes, Bryher, Tresco, St Mary's

Neophyte. Occasional garden escape. Planted in gardens and as an informal hedge on the inhabited islands from where it may occasionally escape. The larger-flowered cultivar 'Corallina' has been recorded from Tresco, although there is no evidence it has become established. In 2015 a large, flowering bush of the cultivar was found on a field margin near Bant's Carn (SV91021220).

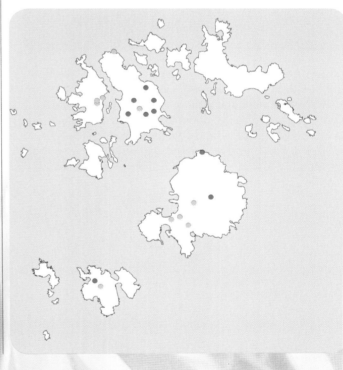

EU-DICOTS True Dicotyledons

ONAGRACEAE – Willowherb family
Circaea lutetiana L.
Enchanter's-nightshade
St Mary's, Tresco

Native. Recorded from The Garrison, St Mary's where it appears to be spreading as a garden weed. First recorded in 1952 from a cottage wall on Star Castle Hill, and from the garden at Benhams in 1970. It has since spread locally to the gardens and roadsides near the Sallyport. In 2011 it was found as a weed in Tresco Abbey Garden, presumably also a recent introduction. Cultivated places.

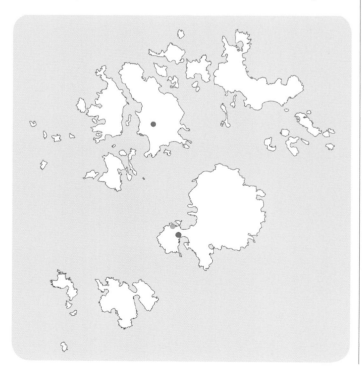

MYRTACEAE – Myrtle family
Leptospermum scoparium J.R. and G. Forst
Broom Tea-tree
Tresco

Neophyte. Established escape from cultivation Abbey Hill area, Tresco. *L. lanigerum* Woolly Tea-tree has also been reported from the same area.

MYRTACEAE – Myrtle family
Eucalyptus globulus Labill.
Southern Blue-gum
Tresco

Neophyte. Planted Abbey Hill and woodland area, Tresco.

MYRTACEAE – Myrtle family
Eucalyptus viminalis Labill.
Ribbon Gum
Tresco

Neophyte. Planted Abbey Hill and woodland area, Tresco.

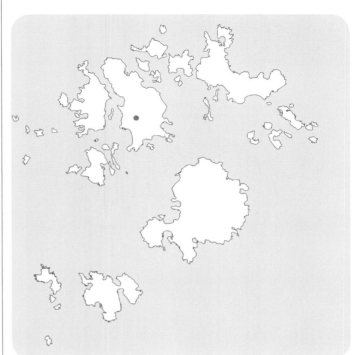

EU-DICOTS True Dicotyledons

MYRTACEAE – Myrtle family
Eucalyptus pulchella Desf.
White Peppermint Gum
Tresco, St Martins

Neophyte. Planted Abbey Hill and woodland area, Tresco, also St Martin's.

Eucalyptus spp.
Other species of *Eucalyptus* are planted on Tresco and may become established. *E. viminalis* and *E. urnigera* have been recorded, but it is not clear whether they are established away from plantings.

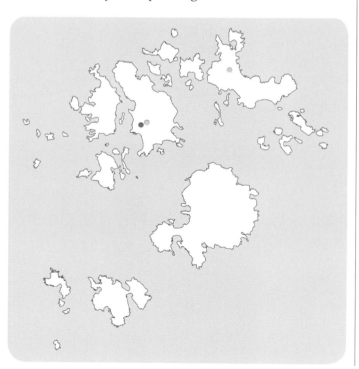

MYRTACEAE – Myrtle family
Ugni molinae Turcz.
Chilean Guava
Tresco and St Martin's

Neophyte. Garden escape self-sown on Abbey Hill, 2007 and 2008. Spreading on slopes above Great Bay, St Martin's, but may be cleared by IoSWT. Probably originally birdsown.

MYRTACEAE – Myrtle family
Luma apiculata (DC.) Burret
Chilean Myrtle
Tresco, St Martin's and St Mary's

Neophyte. Established escape from cultivation in a few places. May still be present on Toll's Hill, St Mary's.

ANACARDIACEAE – Sumach family
Rhus typhina L.
Stag's-horn Sumach
St Mary's

Neophyte. Garden escape near Longstone and near Newman House, Hugh Town.

EU-DICOTS True Dicotyledons

SAPINDACEAE – Maple family
Aesculus hippocastanum L.
Horse-chestnut
Tresco

Neophyte. Planted tree near Borough Farm, Tresco. There may be others.

SAPINDACEAE – Maple family
Acer pseudoplatanus L.
Sycamore
Inhabited islands

Neophyte. There are accounts of sycamore being planted about 1650 (Turner [1695] in *The Scillonian*, 1964) in Holy Vale and it was also one of the trees planted in the first shelterbelts during the inception of the Abbey Gardens. A tree planted in the Parsonage garden on St Agnes is still there and there are trees at Higher Down, St Martin's and near Carron Farm and the road to the quay. There are sycamores in a number of places on St Mary's, for example near The Garrison campsite and along the lane from Borough to Watermill Lane. Sycamore appears to be one of the few broadleaved trees that tolerate the heavy salt spray that is blown over the islands during storms. Some trees have arisen from self-seeding from planted trees.

SAPINDACEAE – Maple family
Acer campestre L.
Field Maple
St Mary's

Native in the British Isles, but a neophyte in Scilly. Planted tree, known from a hedge west of Porthloo Lane. May have been included in recent plantings on St Mary's.

RUTACEAE – Rue family
Correa backhouseana Hook.
Tasmanian-fuchsia
Tresco and St Mary's

Neophyte. Frequently used as hedging or as an ornamental shrub. Naturalised in woodland on Tresco and on the Garrison, St Mary's.

EU-DICOTS True Dicotyledons

MALVACEAE – Mallow family
Malva moschata L.
Musk-mallow
St Mary's and St Martin's

Native. Lousley also noted the erratic appearance of this species. There have been occasional records; St Martin's (SV917160) in 1988; St Mary's (SV9110) and near Bant's Carn (SV909125) in 1993, but they do not appear to be associated with gardens.

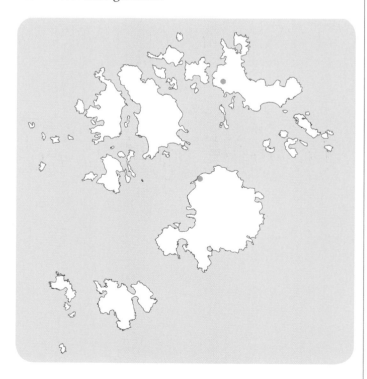

MALVACEAE – Mallow family
Malva sylvestris L.
Common Mallow
Inhabited islands

Archaeophyte. Widespread and common on the inhabited islands. Lousley recorded the species on Teän where it was probably a relict of former human occupation. There can be confusion with *M. pseudolavatera* where it occurs, as the plants are superficially alike. The best feature to check is the epicalyx lobes which are narrow and free in *M. sylvestris* and broad, blunt and joined below for about a third of their length in *M. pseudolavatera*. Grows around buildings, along roadsides, waste ground and field edges.

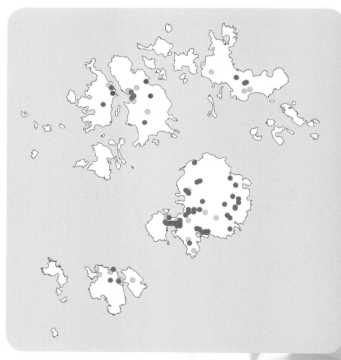

Malva pusilla Sm.
Small Mallow
Extinct

Last recorded as a casual near New Quay, St Martin's in 1936 and 1939.

MALVACEAE – Mallow family
Malva neglecta Wallr.
Dwarf Mallow
Inhabited islands

Archaeophyte. Widespread and common on all the inhabited islands, but has not been found on any of the uninhabited islands since 1938 when Lousley cites a record from Great Ganilly by J. Dallas. Found in similar places to *M. sylvestris*, disturbed and waste ground.

MALVACEAE – Mallow family
Malva arborea (L.) Webb & Berthel.
Tree-mallow
Inhabited islands, most uninhabited islands including small islets

Native. Essentially coastal, but is also found away from the coast. Typically found on some of the smaller rocky islets where it may be the only tall vegetation. On Rosevear in Western Rocks, and on some of the Norrard Rocks, it is important for providing shelter and nesting places for both breeding and migrating birds. A short-lived perennial or biennial, there often appear to be alternating generations of flowering plants with new growth appearing at the base of the previous year's dead white stems. Grows on beaches above HWM, coastal rocks and occasionally along roadsides inland.

EU-DICOTS True Dicotyledons

MALVACEAE – Mallow family
Malva pseudolavatera Webb & Berthel.
Smaller Tree-mallow
Inhabited islands

Neophyte. Lousley and others have considered this to be native to Scilly. It was first discovered by William Curnow in 1876. Lousley (1971) only recorded it from St Mary's, St Agnes and Tresco. It is still common on St Mary's, Tresco and St Agnes, but there are now some records from St Martin's and Bryher. It is uncommon on St Martin's, records include at SV930150 in 1979, SV923166 in 1989, SV921162 in 1999 and SV919162 in 2005. It has only recently been found on Bryher; on the coastal path in the south of Bryher (SV87791409) in 2008, and at SV881152 in 2010 and near Anneka's Quay (SV881150) in 2007 (still there and spreading 2014). These recent records on Bryher suggest it may have accidentally spread there from Tresco as there is frequent daily boat traffic back and forth between the two islands. At one time the population of the plant on St Agnes appeared to be declining, but recovery apparently coincided with less herbicide use. As it can sometimes be easy to confuse this species with the similar *Malva sylvestris* it is best to check the epicalyx to confirm identification (see note under *M. sylvestris*). Occurs in bulbfields and other cultivated ground, waste ground and roadsides.

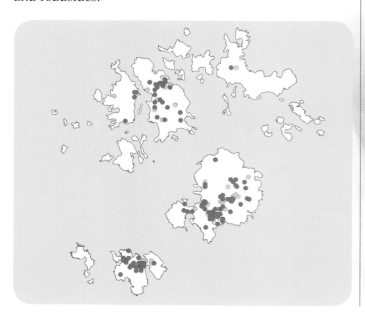

MALVACEAE – Mallow family
Hibiscus trionum L.
Bladder Ketmia
Tresco

Neophyte. There is a casual record from Tresco in 1984.

TROPAEOLACEAE – Nasturtium family
Tropaeolum majus L.
Nasturtium
Inhabited islands

Neophyte. Frequent garden escape that is found on scattered sites on the inhabited islands. It persists on a number of places where garden waste has been dumped including on the coast near dwellings. The Moorwell dump on St Mary's often had a fine show of flowering nasturtiums!

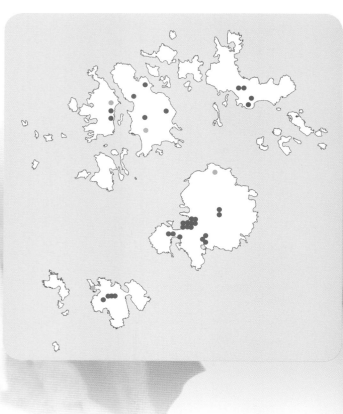

RESEDACEAE – Mignonette family
Reseda alba L.
White Mignonette
St Mary's

Neophyte. Was at one time common around Hugh Town, but was becoming rare and decreasing by 1952 (Lousley, 1971). The area where it grew was lost when it was built over and the largest colony destroyed by dumping of road metal. The last record was from Porth Mellon (SV908107) in 1993.

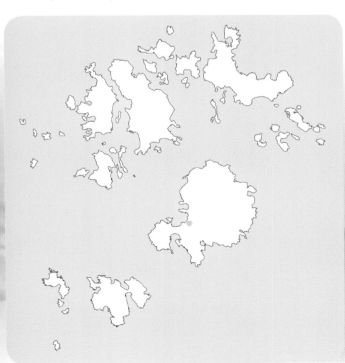

RESEDACEAE – Mignonette family
Reseda lutea L.
Wild Mignonette
Gugh

Neophyte. Had been known from the field below The Gugh houses and bulbfield for many years. The last record was of one flowering plant in the sand pit near the house in 2000.

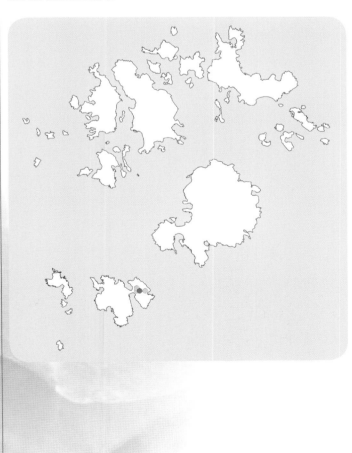

EU-DICOTS True Dicotyledons

BRASSICACEAE – Cabbage family
Erysimum cheiranthoides L.
Treacle-mustard
St Mary's and Bryher

Archaeophyte. Only recorded from two places on St Mary's; Green Lane, Pelistry (SV926118) in 1995 and the track to Salakee Farm (SV919108) in 2004. In 2004 it was also found in two places on Hillside Farm, Bryher (SV87701470) and (SV87761471). The records were all from cultivated and waste land.

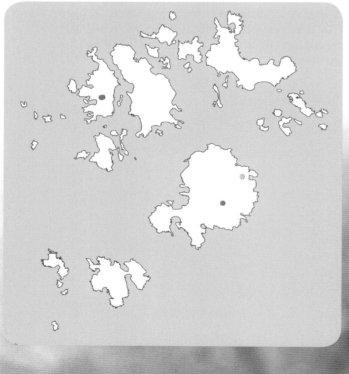

BRASSICACEAE – Cabbage family
Erysimum cheiri (L.) Crantz
Wallflower
Tresco, St Martin's and St Mary's

Archaeophyte. An occasional garden escape. May be established on The Garrison.

BRASSICACEAE – Cabbage family
Arabidopsis thaliana (L.) Heynh.
Thale Cress
Inhabited islands

Native. Lousley considered this species to be very rare; he only recorded it from St Mary's, Tresco and St Martin's. It is difficult to say whether it was really a recent arrival in 1940 and rare when Lousley was compiling his Flora, or whether it had just been under-recorded. It is very easily overlooked among bulbfield communities so may be more common than the records suggest. There is also some evidence that its appearance can be sporadic. Grows in bulbfields, other cultivated land and waste places.

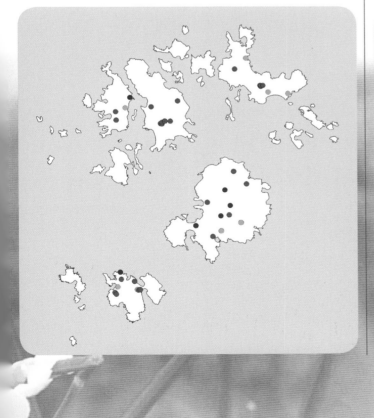

BRASSICACEAE – Cabbage family
Capsella bursa-pastoris (L.) Medik.
Shepherd's-purse
Inhabited islands

Archaeophyte. Although a common weed it may be under-recorded on Scilly. Mainly in arable fields, roadsides, gardens, allotments and disturbed ground.

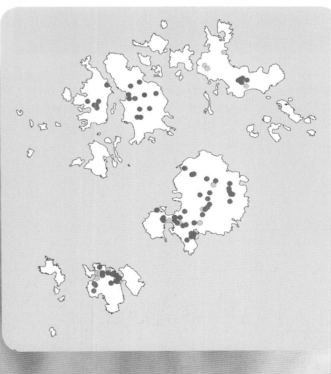

EU-DICOTS True Dicotyledons

BRASSICACEAE – Cabbage family
Barbarea vulgaris W.T. Aiton
Winter-cress
Tresco, St Agnes and St Mary's

Native in the British Isles, but neophyte in Scilly. Possibly a garden escape. Known from a wall near the church on Tresco (SV893154) until 2000, from both St Agnes (SV87900806) and St Mary's (SV90091032) in 2010.

BRASSICACEAE – Cabbage family
Barbarea verna (Mill.) Asch
American Winter-cress
Tresco and St Mary's

Neophyte. No recent records. Formerly recorded from Tresco, Dolphin Town (SV893154) in 1986, Old Grimsby, (SV892155) in 1987, and from near Juliet's Garden, St Mary's (SV909116) in 1998.

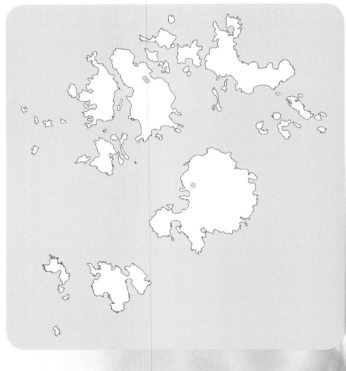

BRASSICACEAE – Cabbage family
Rorippa palustris (L.) Besser
Marsh Yellow-cress
St Mary's

Native in the British Isles, but neophyte in Scilly. Only one record, a number of plants growing in a sandy street in Hugh Town in 2011.

BRASSICACEAE – Cabbage family
Rorippa sylvestris (L.) Besser
Creeping Yellow-cress
St Mary's

Native in the British Isles, but neophyte in Scilly. Known for a number of years from fields at Trenoweth (SV91871245). Still there in 2011.

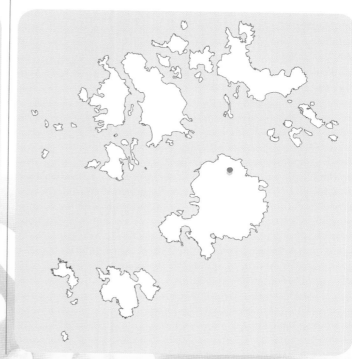

EU-DICOTS True Dicotyledons

BRASSICACEAE – Cabbage family
Nasturtium officinale W.T. Aiton
Water-cress
St Mary's and Tresco

Native. Restricted to a few wetland sites where it is abundant. Plants on St Mary's, at Lower and Higher Moors and in the stream through Holy Vale are exceptionally large and robust. It also grows along the Watermill stream. On Tresco is found on the edge of Great Pool. Possibly reduced in distribution since Lousley's time as many of the roadside gutters are no longer suitable for wetland plants since the roads were asphalted. Ditches and marshes.

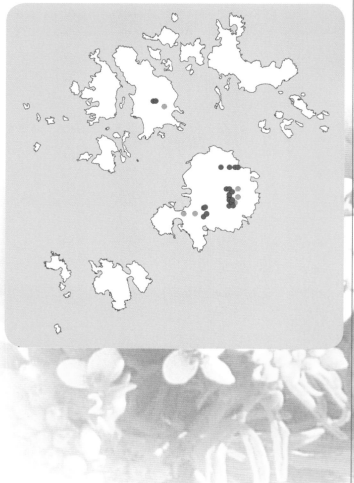

BRASSICACEAE – Cabbage family
Armoracia rusticana P. Gaertn, B. Mey & Scherb.
Horse-radish
St Mary's and Tresco

Archaeophyte. Recorded from two places, School Green, Tresco (SV893156) where it had been present since before 1957 and near Porthloo, St Mary's (SV909113).

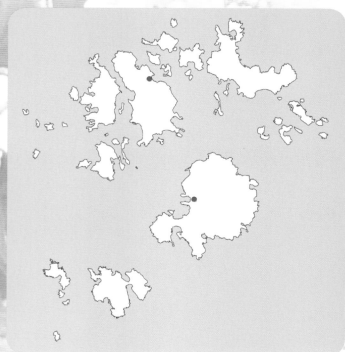

BRASSICACEAE – Cabbage family
Cardamine pratensis L.
Cuckooflower
St Mary's and Tresco

Native. Restricted to just a few sites, but even there tends to be sparse. Occurs on Higher Moors, Lower Moors and wet fields near Porthloo, St Mary's and on Tresco is found near Abbey Pool, Great Pool and at Borough Farm. A field at Old Town, St Mary's where it used to be found was used to store materials from the building of the new school so it may have been lost from there. Wetlands and damp fields.

BRASSICACEAE – Cabbage family
Cardamine flexuosa With.
Wavy Bitter-cress
St Mary's, Tresco and St Agnes

Native. Scattered mostly in damp, shady places such as Holy Vale, Rose Hill, Old Town and the Garrison on St Mary's; from Lower Town, St Agnes and around the gardens, Great Pool and woodland on Tresco. Cultivated and disturbed ground.

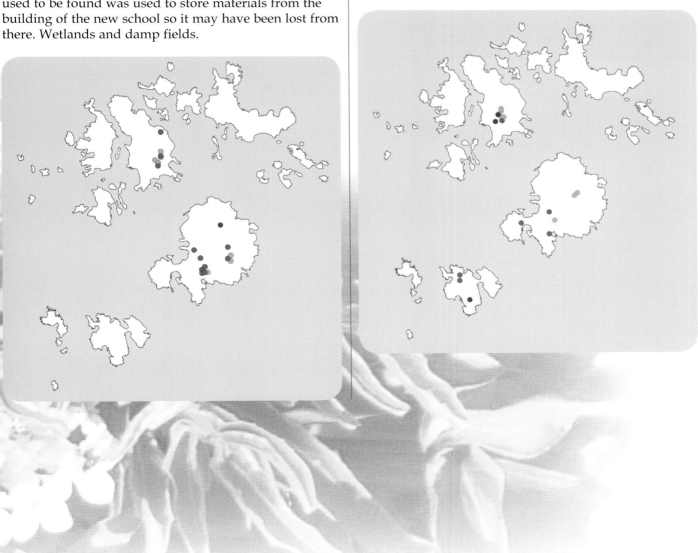

EU-DICOTS True Dicotyledons

BRASSICACEAE – Cabbage family
Cardamine hirsuta L.
Hairy Bitter-cress
Inhabited islands also Gugh

Native. More common than *C. flexuosa* although found in similar places, may be under-recorded in the arable fields. Cultivated fields, gardens and disturbed ground.

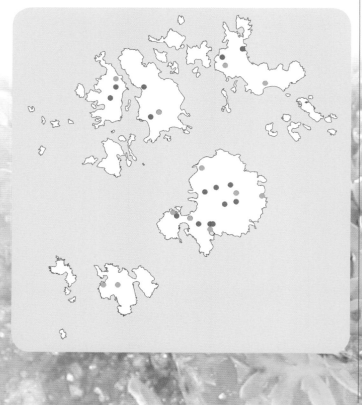

BRASSICACEAE – Cabbage family
Lepidium heterophyllum Benth.
Smith's Pepperwort
St Mary's

Native. Although Lousley cites a few records from St Mary's these were in 1939 and the only Tresco record was in 1879. Now only found near Morning Point on the Garrison (SV899099 - SV901099), St Mary's, where a small number of plants persist on disturbed ground and on top of The Garrison wall.

BRASSICACEAE – Cabbage family
Lepidium coronopus (L.) Al-Shehbaz
Swine-cress
Inhabited islands

Archaeophyte. Widespread on all the islands, but fewer records from St Martin's. Less common or less often recorded than *L. didymum*. Cultivated land, waste and trampled places.

BRASSICACEAE – Cabbage family
Lepidium didymum L.
Lesser Swine-cress
Inhabited islands

Neophyte. A very common plant on all the inhabited islands. Frequently found growing together with *L. coronopus*. Cultivated fields, gateways, gardens and other areas of disturbed ground.

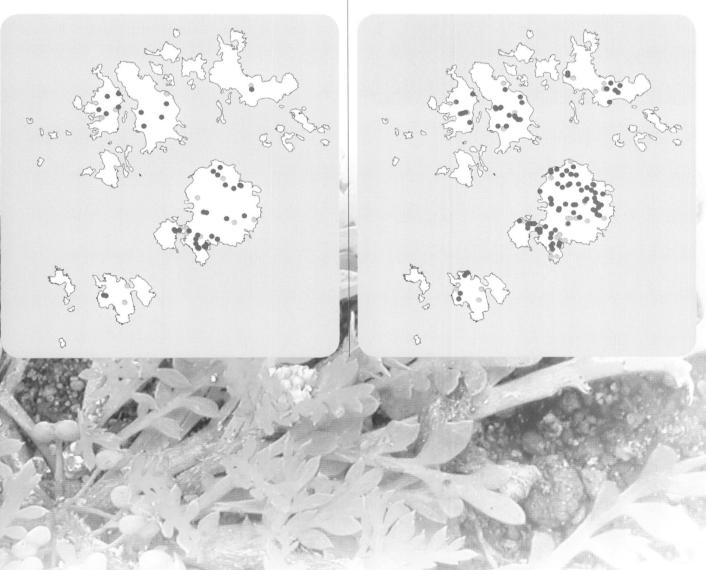

EU-DICOTS True Dicotyledons

BRASSICACEAE – Cabbage family
Lunaria annua L.
Honesty
Bryher, St Martin's, St Agnes and St Mary's

Neophyte. Occasionally recorded as a garden escape. Apparently absent from Tresco.

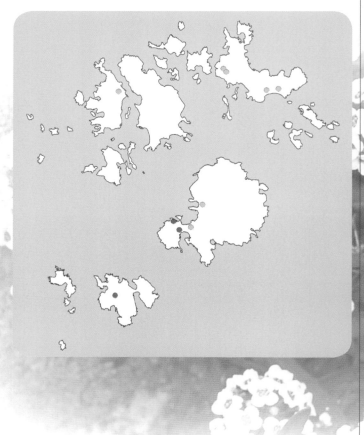

BRASSICACEAE – Cabbage family
Lobularia maritima (L.) Desv.
Sweet Alison
St Mary's, St Martin's and Bryher

Neophyte. A common garden plant long established around Hugh Town where it was first found by Townsend (Lousley, 1971) in 1864. Lousley considered it reached its greatest abundance in the 1920s before Hugh Town became more built up. Occurs occasionally as a garden escape on the other inhabited islands (Lousley had recorded it from Tresco), but apparently does not persist. Sandy ground, roadsides and pavements near the sea.

Diplotaxis tenuifolia (L.) DC
Perennial Wall-rocket
Extinct

Formerly on Tresco, the last record from near the Abbey in 1938.

BRASSICACEAE – Cabbage family
Diplotaxis muralis (L.) DC
Annual Wall-rocket (or Stinkweed)
St Martin's, St Mary's and Tresco

Neophyte. Grows in a few discrete areas including New Grimsby, Tresco; near the hotel on St Martin's and several places on St Mary's including Porthloo and in Hugh Town. Last recorded at Old Grimsby, Tresco in 1987. Associated with habitations, gardens and on sandy ground.

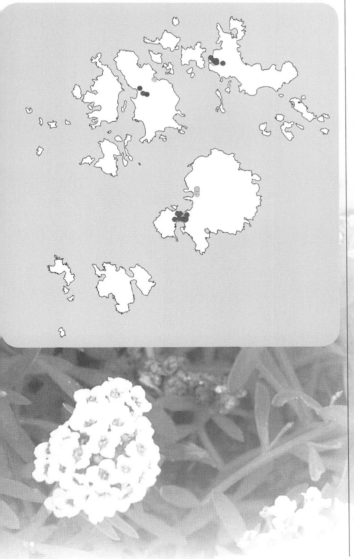

BRASSICACEAE – Cabbage family
Brassica napus ssp. *oleifera* (DC) Metzg.
Oil-seed Rape
St Mary's

Neophyte. Only recorded from Porthloo (SV909115) in 1997 and Old Town (SV913102) in 2012; also St Agnes in 1997, presumed an accidental introduction possibly with birdseed.

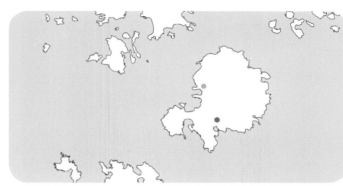

Brassica rapa L.
Turnip
St Mary's and St Martin's

Archaeophyte. Occasional escape from cultivation; from St Martin's in 1992 and 1995 (SV919162) and on St Mary's, Star Castle in 1975 (SV913117) in 2000 and Hugh Town in 2003.

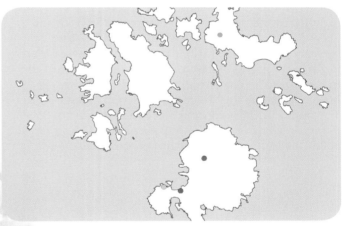

Brassica rapa ssp. *oleifera* (DC) Metzg.
Turnip-rape
St Mary's

Neophyte. One record from Hugh Town, St Mary's (SV905105) in 2000.

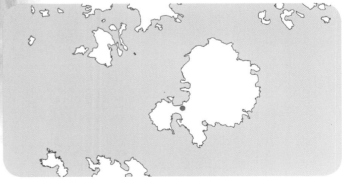

EU-DICOTS True Dicotyledons

BRASSICACEAE – Cabbage family
Brassica nigra (L.) W.D.J. Koch
Black Mustard
Bryher, Tresco and St Mary's

Native or alien in the British Isles, probably introduced to Scilly. Single records from Bryher (SV882155) and Hugh Town, St Mary's (SV905105) in 1997. More recently from Tresco (SV893156) in 2007; from St Mary's, Porth Mellon (SV908108) in 2003. Believed by Lousley to have been an old introduction to Hugh Town, St Mary's where it was described as 'locally plentiful' by Townsend in 1864. A common roadside and coastal plant in West Cornwall so it is surprising it is so rare on Scilly. Near the shore on disturbed ground.

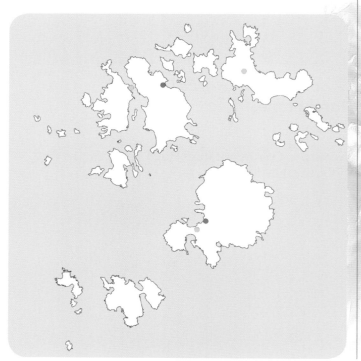

BRASSICACEAE – Cabbage family
Sinapis arvensis L.
Charlock
Inhabited islands

Archaeophyte. Appears to have been more frequent in the past according to Lousley's account. Now recorded from just a few places on all the inhabited islands. At times may be locally common in places, but does not persist. Cultivated and disturbed land.

BRASSICACEAE – Cabbage family
Sinapis alba L.
White Mustard
Tresco

Archaeophyte. Recorded from fields near Great Pool, Tresco (SV894146) in 1995 and Pentle, Tresco in 2015. A former crop previously recorded on Tresco in 1952 and 1967, and St Agnes in 1953. Recorded from a field on St Martin's in 2007. Arable fields.

BRASSICACEAE – Cabbage family
Cakile maritima Scop.
Sea Rocket
Inhabited islands, also Teän, Samson, Great Ganilly and Little Arthur

Native. Populations fluctuate markedly in numbers and locality from year to year. The plants are variable in leaf shape and flower colour. Our plants are ssp. *integrifolia* (Hornem.). Usually occurs on sandy or shingle shores, but can also be found inland especially in dunes and sandy areas.

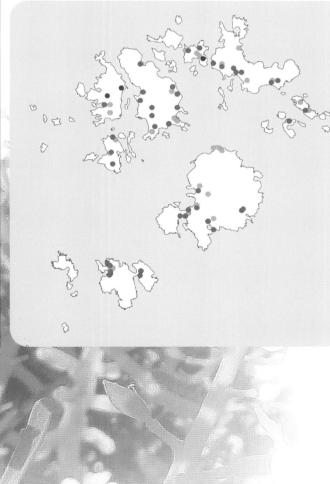

EU-DICOTS True Dicotyledons

BRASSICACEAE – Cabbage family
Crambe maritima L.
Sea-kale
Inhabited islands, also Great Ganilly

Native. Occurs on a few sites where it has been known for many years although numbers can fluctuate from year to year. Although vulnerable to storms that wash away the top growth the deep roots often survive and also the buoyant seeds are washed back onto the shore. On one occasion an islander was seen collecting the plants in sacks – presumably to feed animals. Grows at the top of sandy or shingle beaches above HWM.

BRASSICACEAE – Cabbage family
Raphanus raphanistrum ssp. *raphanistrum* L.
Wild Radish
St Mary's, St Martin's, Tresco and St Agnes

Archaeophyte. Locally abundant on St Mary's, otherwise scattered localities on the other inhabited islands. Was thought absent from Tresco until found there in 2010. Appears to have spread to more areas since Lousley's time when he considered it 'rather rare'. Yellow, white and violet-veined varieties occur, often growing together. Can be confused with Sea Radish or Charlock when not in fruit where they grow nearby. Found in cultivated fields and on waste ground.

BRASSICACEAE – Cabbage family
Raphanus raphanistrum ssp. *maritimum* (Sm.) Thell.
Sea Radish
All inhabited islands, also Teän

Native. Records from Bryher and Tresco are recent: plants were found on the coastal path on Bryher (SV87911481) in 2009 and at Pentle bay, Tresco in 2008. Principally a coastal plant but does occur at a few more inland sites such as near the Garrison camp site on St Mary's. Can be easily confused with ssp. *raphanistrum*, from which it can only reliably distinguished when the fruits are ripe. Sandy shores and waste ground near the sea.

Raphanus raphanistrum ssp. *maritimum* fruit

Raphanus sativus L
Garden Radish
St Mary's and Tresco

Neophyte. Casual. An escape from cultivation Holy Vale (SV921114) in 2002 and near Pentle House 2015.

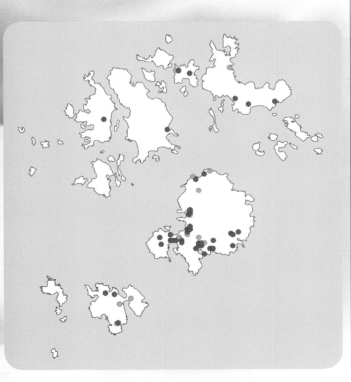

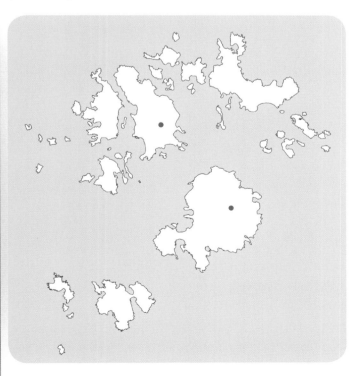

EU-DICOTS True Dicotyledons

BRASSICACEAE – Cabbage family
Sisymbrium officinale (L.) Scop.
Hedge Mustard
Inhabited islands

Archaeophyte. Frequently recorded on all the inhabited islands. Common along roadsides, waste places and borders of cultivated fields.

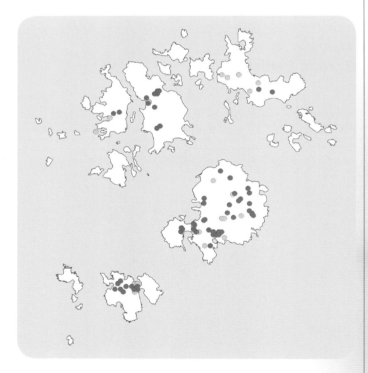

Sisymbrium orientale L.
Eastern Rocket
Extinct

Not seen since 1940. Was recorded from several sites on St Mary's, St Agnes and Tresco between 1932 and 1940, but apparently did not persist. It is a distinctive plant due to the extremely long fruits so unlikely to have been overlooked.

BRASSICACEAE – Cabbage family
Thlaspi arvense L.
Field Penny-cress
Inhabited islands

Archaeophyte. Although recorded from all the inhabited islands, the records are few and scattered and it is nowhere very common save for a small concentration around Old Town, St Mary's. Although Lousley described the plant as 'common' it appears to have declined since his day. As it is such a distinctive plant in fruit, the paucity of records appears to be genuine. Mainly found in arable fields and on wasteland.

BRASSICACEAE – Cabbage family
Matthiola incana (L.) W.T. Aiton
Hoary Stock
St Mary's, St Agnes, Tresco and Bryher

Neophyte. Garden escape on Scilly. Naturalised in a few places; apparently absent from St Martin's. The flowers are wonderfully perfumed, especially at night. Grows on walls and verges mostly near habitations and the coast.

BRASSICACEAE – Cabbage family
Matthiola longipetala (Vent.) DC.
Night-scented Stock
St Mary's and St Agnes

Neophyte. Occasional garden escape.

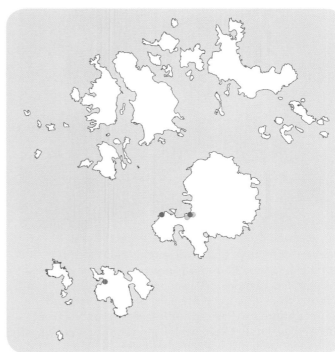

EU-DICOTS True Dicotyledons

BRASSICACEAE – Cabbage family
Cochlearia officinalis L.
Common Scurvygrass
Inhabited islands, also most of the uninhabited islands and islets

Native. A common and widespread plant found mainly around the coast. Found on some of the most exposed uninhabited islands and islets including in the Norrards, Western Rocks, Teän, St Helen's, Northwethel and Round Island. In 1989 ssp. *scotica* was recorded from Toll's Island by J. Akeroyd, J. David and S. Jury. Usually grows among maritime rocks and shingle, but also occurs inland in a few places such as bases of walls and in Hugh Town.

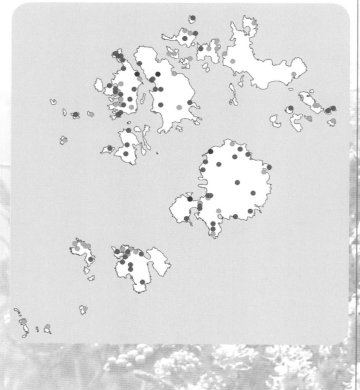

BRASSICACEAE – Cabbage family
Cochlearia danica L.
Danish Scurvygrass
Inhabited islands, also Teän, Annet, Northwethel, Samson and some Eastern Isles

Native. Generally occurs on the more exposed coastal areas although may be more widespread than the records indicate, but the plant flowers early in the year before the uninhabited islands are generally accessible. Lousley refers to a possible hybrid between *C. officinalis* and *C. danica* that was collected by a Miss M. Jaques in 1957 from Turfy Hill, St Martin's. In 2014 the hybrid was found by Fred Rumsey and Mark Spencer on St Mary's (SV9239412215). Besides the usual habitat among coastal rocks, cliffs and on beaches, the plant is found inland on stone walls, and also in pavement cracks and wall bases in Hugh Town.

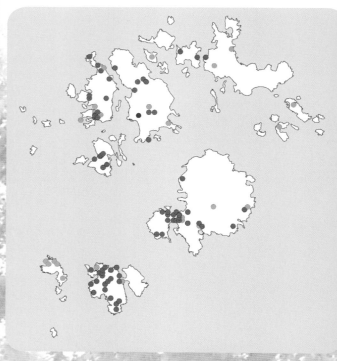

BRASSICACEAE – Cabbage family
Iberis sempervirens L.
Perennial Candytuft
St Mary's

Neophyte. Garden escape. St Mary's SV915115 in 2009.

Iberis umbellata L.
Garden Candytuft
St Mary's

Neophyte. Garden escape, Old Town, St Mary's (SV915102) in 2004.

TAMARICACEAE – Tamarisk family
Tamarix gallica L.
Tamarisk
Inhabited islands, Samson

Neophyte. Sir William Hooker visited Scilly in 1813, but apparently without having had a very good time, he mentioned the presence of 'wretched plants of elm and *Tamarix gallica*' (Lousley, 1971) in Holy Vale. Tamarisk was one of the first hedging plants introduced to Scilly early in the century. It later fell out of favour when glossy-leaved evergreen shrubs were found to make far superior windbreaks. There are still some fine old trees, especially on St Agnes and occasionally some apparently naturalised plants. A tree on Samson by the ruined houses is thought to have been planted there by the former inhabitants so could be around 170 years old. Hedging.

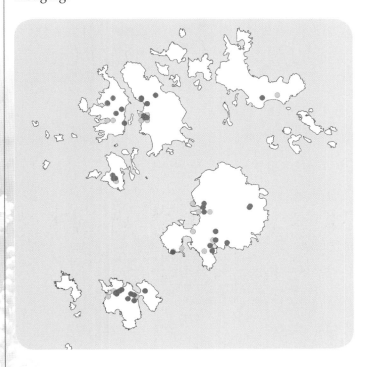

Tamarix africana Poir.
African Tamarisk
Tresco

Neophyte. One identified in a hedgerow near the sea at Old Grimsby, Tresco in 1992/3 may still be there.

EU-DICOTS True Dicotyledons

PLUMBAGINACEAE – Thrift family
Armeria maritima (Mill.) Willd.
Thrift
inhabited islands, also all uninhabvited islands other than the Western Rocks and most of the Norrards.

Native. Abundant on most islands, other than the most rocky and exposed. Thrift is almost exclusively coastal, although on Scilly it is still within the influence of the sea where it is found at a few inland sites, such as on heathland and the top of stone fences on St Mary's. On Annet enormous, metre square thrift tussocks dominate large areas in the north of the island. In May the whole island glows pink seen from the distance. The storms in 2014 severely damaged the thrift tussocks but it is expected they will eventually recover. Found on all kinds of coastal habitats, rocks, wall-tops, shingle, cliff-top grassland and coastal heath.

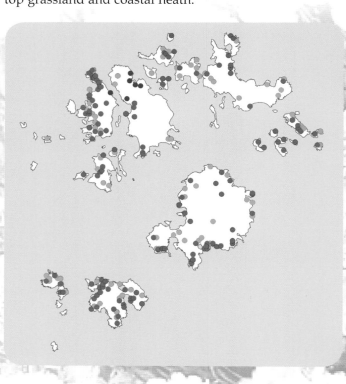

POLYGONACEAE – Knotweed family
Persicaria amphibia (L.) Delarbre
Amphibious Bistort
Tresco

Native. Occurred by the Abbey Pool (SV897142) between 1968 and 1999. Although there are no recent records it may still persist in less accessible areas of the Pool.

POLYGONACEAE – Knotweed family
Persicaria maculosa Gray
Redshank
Inhabited islands, also Northwethel

Native. Although recorded from all the inhabited islands is only relatively widespread on St Mary's and Tresco. It was also recorded on Northwethel in 1992. Occurs on cultivated ground, roadsides, and beside pools and ditches.

POLYGONACEAE – Knotweed family
Persicaria lapathifolia (L.) Delarbe
Pale Persicaria
Tresco and St Mary's

Native. A rare plant in Scilly. There have only been a handful of records of this species. Tresco (SV890149) in 1995 and St Mary's; Longstone (SV91811145) in 1984, Pelistry (SV926118) in 1995, a wet field near Lower Moors (SV91251032) in 2003.

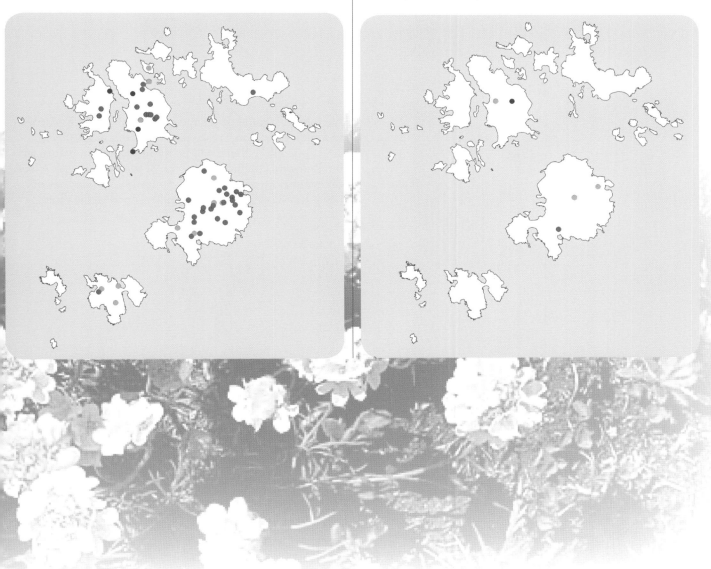

EU-DICOTS True Dicotyledons

POLYGONACEAE – Knotweed family
Persicaria hydropiper (L.) Delarbe
Water-pepper
St Mary's, Tresco, St Martin's and Samson

Native. An uncommon plant that is associated with wet places such as beside the Holy Vale stream. Found along muddy tracks, beside pools and a damp seepage near the shore on Samson (SV880122) in 1996.

POLYGONACEAE – Knotweed family
Fagopyrum esculentum Moench
Buckwheat
Tresco and St Martin's

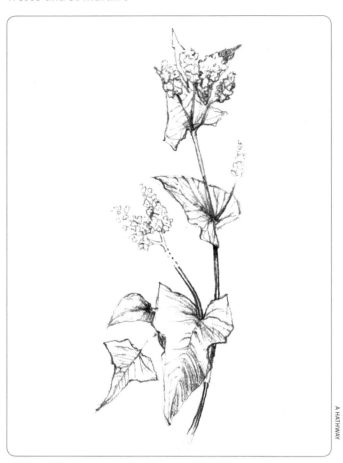

Neophyte. Casual or escape from cultivation. Recorded in a maize field on Tresco (SV89591502) in 2010 and 2011. It is likely there will be more records in future as buckwheat is included in some of the 'conservation' seed mixtures now used in Scilly. So far buckwheat has only been noted on St Martin's and Tresco in game crops.

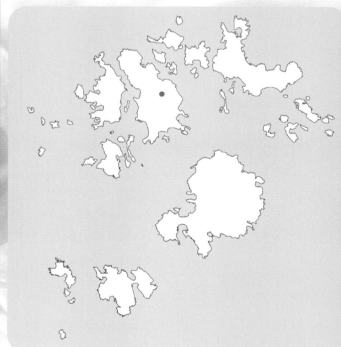

POLYGONACEAE – Knotweed family
Polygonum maritimum L.
Sea Knotgrass
Bryher and St Martin's

Native. Rare and sporadic. There have been records from just two places; Bryher, from Great Porth (SV875145) in 2003, 2008, 2010, 2012 and St Martin's, Higher Town Bay, (SV936152) in 2007. The Bryher plants apparently did not persist from year to year but reappeared at intervals, presumably from buried seed. All the records were of plants growing on sandy beaches just above HWM.

POLYGONACEAE – Knotweed family
Polygonum oxyspermum C.A. Mey. & Bunge
Ray's Knotgrass
Bryher and Tresco

Native. Another rare plant whose appearances are erratic. There are just two recent records; Tresco (SV90121388) in 2004 and Bryher (SV87631450) in 2009. Our plant is ssp. *raii* (Bab.). From sandy beaches above HWM.

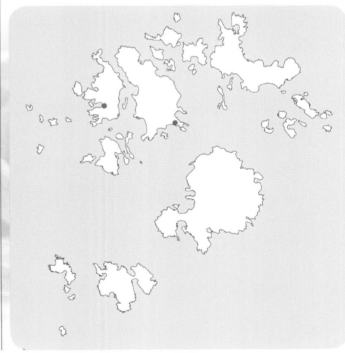

EU-DICOTS True Dicotyledons

POLYGONACEAE – Knotweed family
Polygonum arenastrum Boreau
Equal-leaved Knotgrass
Inhabited islands

Archaeophyte. Apparently less common than *P. aviculare* but probably overlooked. Scattered records from all the inhabited islands. Waste places, roadsides, field edges and sandy wet places.

POLYGONACEAE – Knotweed family
Polygonum aviculare L.
Knotgrass
Inhabited islands and Northwethel

Native. Found on all the inhabited islands and also on Northwethel. Apparently more common than *P. arenastrum* and found in similar places. Plants on shores often have fleshy leaves.

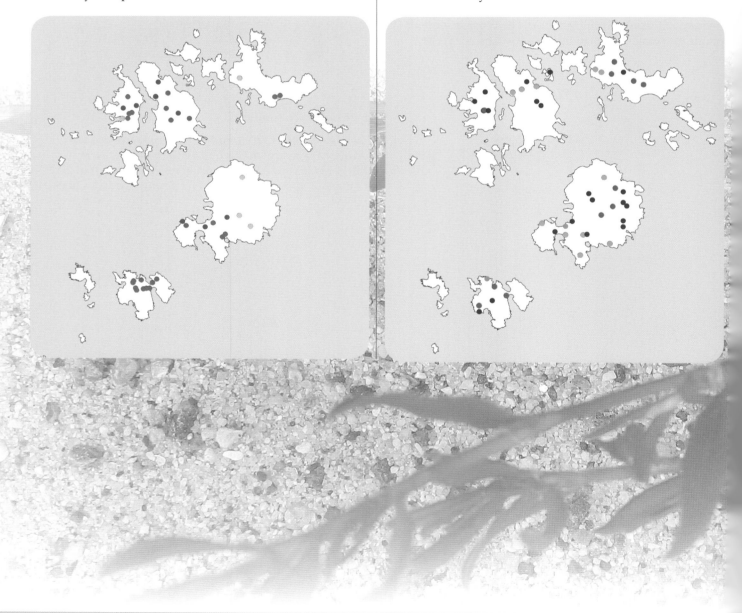

POLYGONACEAE – Knotweed family
Fallopia japonica (Houtt.) Ronse Decr.
Japanese Knotweed
Tresco, St Martin's and St Mary's

Neophyte. The plant has been subject to an aggressive extermination programme by the Isles of Scilly Council and is no longer common. It appeared to have already gone from St Agnes before the herbicide programme started. Last recorded on St Martin's in 1995. A few patches still persist in places on St Mary's and Tresco. Roadsides and waste ground.

POLYGONACEAE – Knotweed family
Fallopia baldschuanica (Regel) Holub
Russian-vine
Bryher and St Mary's

Neophyte. Garden escape that has been recorded in a few places; north of Old Town (SV915105) in 1996 and on The Garrison, St Mary's where it persists, also on Bryher from The Town (SV882153) in 1997, Timmy's Hill and Watch Hill in 2007. Hedges and waste ground.

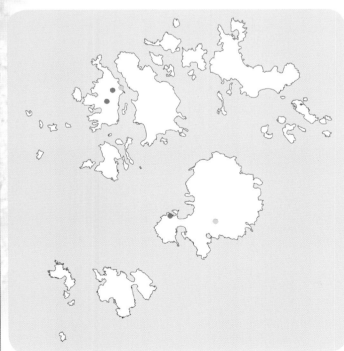

EU-DICOTS True Dicotyledons

POLYGONACEAE – Knotweed family
Fallopia convolvulus (L.) Á Löve

Black-bindweed
Inhabited islands, formerly also Northwethel

Archaeophyte. There are only scattered records from the inhabited islands. This is one of the arable weed species found at Bronze Age archaeological sites. Lousley considered it 'rather common', but this is clearly no longer so. It may be under-recorded as it is much less showy than other arable plants. The most recent records are from Hillside Farm, Bryher in 2005, near the former Post Office (SV881154) in 2007 and Great Porth in 2012; from Tresco as a weed in the vegetable garden at Tresco Abbey in 2010 and the churchyard in 2012; St Marys from Bar Farm in 2005, Porthloo in 2005, Porth Mellon in 2005 and Lunnon Farm (SV924111) in 2013; St Martin's (SV932157 and SV93181562) in 2011. The only record from St Agnes was in 1997 (a field that is now under grass); and there was a 1992 record from Northwethel. Cultivated fields and disturbed ground.

POLYGONACEAE – Knotweed family
Muehlenbeckia complexa (A. Cunn.) Meisn.

Wireplant
Tresco and St Mary's

Neophyte. Garden escape, that originates from New Zealand. Locally very abundant, covering large areas of hedges and hillsides on discrete areas of Appletree Banks on Tresco and many places on St Mary's. Frequently completely obliterates all other vegetation reducing it to a sort of mad topiary. Apparently has completely gone from St Agnes where Lousley recorded it near the Parsonage in 1936. Some garden centres in Cornwall still offer it as an 'attractive climbing plant'!

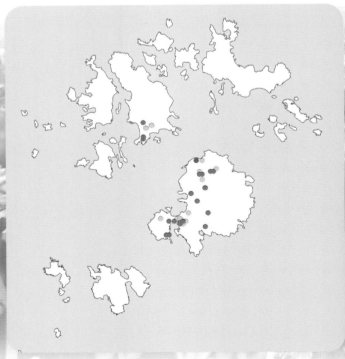

POLYGONACEAE – Knotweed family
Rumex acetosella L.
Sheep's Sorrel
Inhabited islands, Samson, Annet, St Helen's, Teän and the Eastern Isles

Native. A widespread and common plant on most of the larger islands. In the Eastern Isles it has been recorded on Menawethan, Great Ganilly, Great Innisvouls, Nornour, Great Ganinick and Middle Arthur. May still be present on Annet where it was last recorded in 1983. In places may be so abundant as to form large red-coloured patches, for example on sandy ground on some of the uninhabited islands. The plant is found in a range of habitats; heathland, grasslands, cultivated ground, wasteland and on the tops of stone hedges.

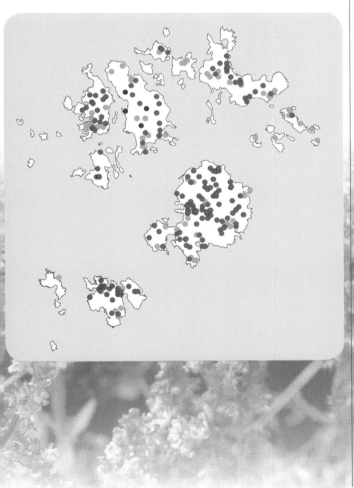

POLYGONACEAE – Knotweed family
Rumex acetosa L.
Common Sorrel
Inhabited islands, also most uninhabited islands except the Western Rocks and Norrads

Native. Widespread and very common on all the inhabited, as well as most of the uninhabited islands with the exception of the very bare rocky ones. On the seabird islands the plants are usually large and fleshy. Ssp. *hibernicus* has been found at Rushy Bay, Bryher and may occur at other dune sites. Found in a range of habitats from dunes, cliffs and other coastal habitats to cultivated land, wasteland, roadsides, grassland and heaths.

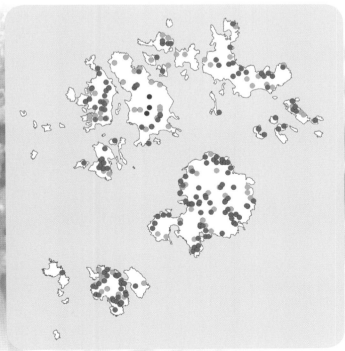

EU-DICOTS True Dicotyledons

POLYGONACEAE – Knotweed family
Rumex frutescens Thouars
Argentine Dock
Gugh

Neophyte. Naturalised only at one site on The Gugh (between SV888084 - SV891083). It was apparently first recorded in 1995, but had clearly been present for some time. Grows among dense bracken and brambles at the back of dunes.

POLYGONACEAE – Knotweed family
Rumex crispus L.
Curled Dock
Inhabited and most of the uninhabited islands including many of the small islets

Native. The most common and widespread dock in Scilly. There are few places where this plant cannot be found. It is only absent from the smaller, rocky islets, although it was found on Rosevear in the Western Rocks in 1990 and may still be there. Inland plants are ssp. *crispus*, but on the coast ssp. *littoreus* is the common plant. They can easily be distinguished when in fruit by the tubercles on the achenes; usually the three are different sizes or there may be only one in ssp. *crispus* and in ssp. *littoreus* the three are more or less equal in size. Until the differences were recognised ssp. *littoreus* was frequently misrecorded as *R. rupestris*. Found in both cultivated and coastal habitats, fields, dunes and roadsides.

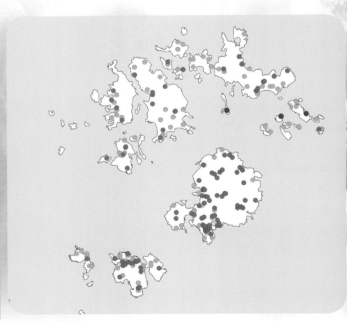

Rumex hydrolapathum Huds.
Water Dock
Extinct or error?

A record from Higher Moors, St Mary's is believed to be an error.

POLYGONACEAE – Knotweed family
Rumex conglomeratus Murray
Clustered Dock
Tresco, St Mary's, St Agnes, Samson

Native. Only found in any amount on Tresco and a few localities on St Mary's. The record from St Agnes in 1997 was from a field that is now under grass. On Samson the plant was found by the well on South Hill in 1984 and 1994 and from Southward Well in 2002. Typically found around pools and in wet grasslands, although occasionally occurs in cultivated habitats.

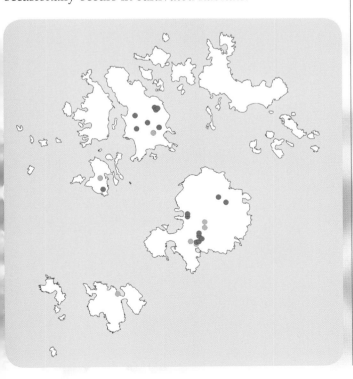

POLYGONACEAE – Knotweed family
Rumex sanguineus L.
Wood Dock
Tresco

Native. Only known from the wooded area near the Abbey (SV893144 – SV895145) where other woodland plants have been recorded.

EU-DICOTS True Dicotyledons

POLYGONACEAE – Knotweed family
Rumex rupestris Le Gall

Shore Dock

Samson, Teän and Annet

Native. Although Lousley recorded the species as 'common' and from all the inhabited and many of the uninhabited islands it is very probable his records date from before the Second World War. By 1982 there were populations on only four islands, although there was a brief resurgence on St Agnes from buried seed in 1984 and 1993 (neither site being typical or ideal). Since then the population of this species in the Isles of Scilly has declined drastically. Small, relict colonies existed on the above three islands until 2014 when it is believed they were swept away in the severe storm surges; the last plant on Tresco had already gone by 2013. Whether the species will reappear in future remains to be seen.

Since 1982 this species had been monitored almost annually and the decline has therefore been recorded. *Rumex rupestris* has been the subject of an English Nature's Species Recovery Programme; in Cornwall plants were reintroduced to sites, but this was not attempted on Scilly. There are plenty of apparently suitable sites for the plant here and the decline appears to be mainly due to increased storminess, coastal squeeze and occasionally drought (as in 1996 when freshwater seepages dried up killing some young plants).

In many places a common associate of *R. rupestris* was *R. crispus* ssp. *littoreus*, although there is no evidence of competition between the two species and very few purported hybrids have been found, so introgression does not seem to be a factor. In the past this coastal form of *R. crispus* had frequently been confused with and misrecorded as *R. rupestris*. *Carpobrotus* is another plant which has been blamed for loss of the dock where they occur together, but on Scilly that does not seem to have been significant. Neither is there any evidence that rats eat the seeds.

In more accessible places there has been some evidence of trampling or damage by people; on one occasion we found dry seed heads had been used on a barbeque and another time a tent was erected over a site on Samson. But these were just single incidents and not believed to be a factor in the plant's decline. Despite producing plenty of good seed this is washed away into the sea with apparently not much being washed back onto suitable places for germination. The seed can survive submergence for weeks in seawater, but again this does not seem to result in new plants being recruited. On the Teän site seedlings were often produced, but it is low-lying and vulnerable as the sea frequently scours the area at the highest tides.

Rumex rupestris grows just above HWM on rock platforms, at the base of cliffs and in freshwater seepages. It has been found that freshwater is essential for germination.

POLYGONACEAE – Knotweed family
Rumex pulcher L.
Fiddle Dock
Inhabited islands, formerly Teän and Little Ganilly

Native. Widespread on the inhabited islands although may still occur on some of the uninhabited islands; it was recorded on Teän by Lousley (without details) and on Little Ganilly in 1983. It may be more widespread than the map suggests; the fiddle-shaped leaves are obvious and easily recognised in spring, but may be overlooked later in the year. Frequent on grassland, roadsides, waste places and cultivated fields.

POLYGONACEAE – Knotweed family
Rumex obtusifolius L.
Broad-leaved Dock
Inhabited islands

Native. A common plant known from all the inhabited islands. Along ditch-sides on the Moors, St Mary's the plants grow very large. Found mostly along roadsides, waste places, wet places and edges of fields.

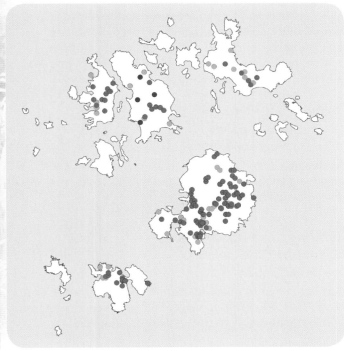

EU-DICOTS True Dicotyledons

POLYGONACEAE – Knotweed family

Rumex crispus x *R. rupestris*
Samson and St Agnes

Native. There are few records of this hybrid despite the species frequently growing together. It was recorded on The Gugh in 1992 and 1993, and near Beady Pool, St Agnes in 1996. A plant found by the well on Samson in 1994 that appeared to be the hybrid did not persist.

Rumex x *abortivus* Ruhmer
(*R. conglomeratus* x *R. obtusifolius*)
St Mary's

Native. Two records by Geoffrey Kitchener in 2009: Old Town (SV91311034) and Newford Ponds (SV9121).

Rumex x *dufftii* Hausskn.
(*R. sanguineus* x *R. obtusifolius*)
Tresco

Native. Recorded from between Abbey Wood and Great Pool (SV89351447) by Geoffrey Kitchener 2009.

Rumex x *muretii* Hausskn.
(*R. conglomeratus* x *R. pulcher*)
St Mary's

Native. Only one record, in 1993 from St Mary's (SV910105) by Alan Underhill.

Rumex x *ogulinensis* Borbás
(*R. pulcher* x *R. obtusifolius*)
St Martin's

Native. Recorded by Lousley from St Martin's in 1936.

Rumex x *pratensis* Mert. & W.D.J. Koch
(*R. crispus* x *R. obtusifolius*)
Tresco, St Martin's and St Mary's

Native. Recorded from a few sites on the above islands. From Lower Moors and Peninnis Head, St Mary's by Colin French in 1995 and from sites on all three islands by Geoffrey Kitchener in 2009. This hybrid has apparently been generally overlooked in the past.

Rumex x *pseudopulcher* Hausskn.
(*R. crispus* x *R. pulcher*)
Tresco, St Mary's and St Martin's

Native. Recorded from Tresco by Lousley in 1939. Geoffrey Kitchener recorded the hybrid from sites on all three islands in 2009.

Rumex x *trimenii* E.G. Camus
(*R. rupestris* x *R. pulcher*)
Tresco and Samson

Native. The only relatively recent record is from Carn Near, Tresco in 1984. Lousley collected specimens from both Samson and Tresco (Lousley & Kent, 1981).

CARYOPHYLLACEAE – Pink family
Arenaria serpyllifolia L.
Thyme-leaved Sandwort
Inhabited islands

Native. Known from a few discrete areas on the inhabited islands, although although the only very recent records are from Rushy Bay, Bryher, the latest in May 2015. Our plant is ssp. *serpyllifolia* Although there are no new records from former localities on either Samson or Teän it is very easily overlooked. Sandy areas, dune grassland and open grassland.

CARYOPHYLLACEAE – Pink family
Honckenya peploides (L.) Ehrh.
Sea Sandwort
Inhabited islands, Samson, Teän, Northwethel, and the Arthurs

Native. A bright green, succulent little plant that is locally frequent at coastal sites on most islands. Widespread at the top of sandy and shingle beaches; it is typically found growing from just above HWM to the top of the beach and sometimes spreading a little way inland into dunes and sandy banks. Beaches.

CARYOPHYLLACEAE – Pink family
Stellaria media (L.) Vill.
Common Chickweed
Inhabited islands, Annet, Menawethan, Great and Little Innisvouls

Native. A widespread weed of cultivation. On the uninhabited islands it is mainly associated with seabirds, often growing where their guano is deposited. Had been recorded from Samson and Teän by Lousley in 1936, but not since. Cultivated and disturbed land.

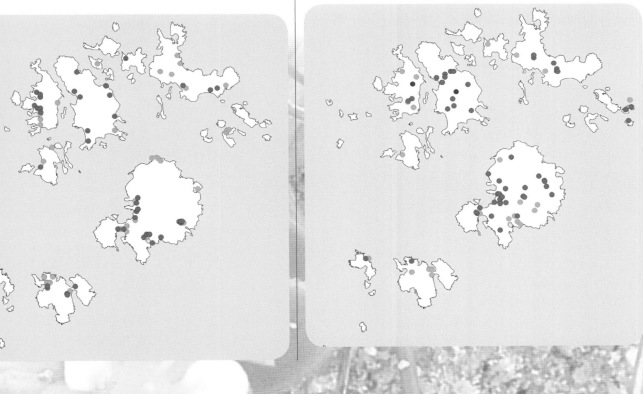

EU-DICOTS True Dicotyledons

CARYOPHYLLACEAE – Pink family
Stellaria pallida (Dumort.) Crép.
Lesser Chickweed
St Agnes, Bryher, Tresco and St Martin's, Annet and Teän

Native. Few records; usually the plant flowers very early and can be over before most botanists are around, although there are some records for June and one from St Martin's in October. Found on Annet (SV862088) in 2002 and St Agnes (SV882080) in 2005. It was last recorded from Samson in 1954 and Northwethel 1957. Most records are from coastal sites although there are some from inland sites such as arable fields. Coastal sites, cultivated and disturbed ground.

CARYOPHYLLACEAE – Pink family
Stellaria neglecta Weihe
Greater Chickweed
St Mary's

Native. Although common and widespread on the mainland, *Stellaria neglecta* is very rare in Scilly. The only recent records were from Little Porth, Porthcressa in 2005 and Shooter's Pool in 2011. Damp and waste ground.

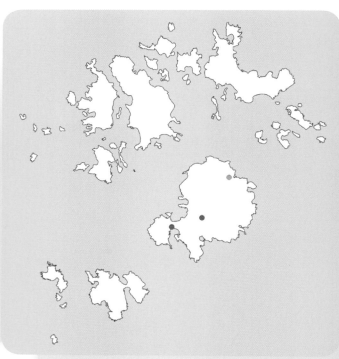

Stellaria holostea L.
Greater Stitchwort
St Mary's

Though native in the British Isles, it is introduced to Scilly. There is only one record of this species in Scilly, a plant growing in the Lower Moors extension, St Mary's in 2005 where it was almost certainly introduced with the planted trees.

CARYOPHYLLACEAE – Pink family
Stellaria alsine Grimm
Bog Stitchwort
St Mary's

Native. Formerly considered locally common by Lousley, but it is now quite rare. The most recent records are from wetlands on St Mary's: Lower Moors 1996, Higher Moors in 2002 and Holy Vale 2000. Wet ground and ditches.

CARYOPHYLLACEAE – Pink family
Cerastium tomentosum L.
Snow-in-summer
St Agnes and St Mary's

Neophyte. Occasional garden escape recorded from Higher Town, St Agnes and near Telegraph, St Mary's.

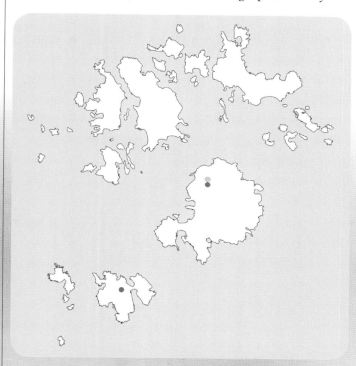

EU-DICOTS True Dicotyledons

CARYOPHYLLACEAE – Pink family
Cerastium fontanum Baumg.
Common Mouse-ear
Inhabited islands

Native. Widespread and common on the inhabited islands except St Martin's where it has not been seen recently. Recorded on Middle Arthur and Great Ganilly in the Eastern Isles in 1994 where it may still be present. The common subspecies in Scilly is ssp. *vulgare* although ssp. *holosteoides* has been identified from Hillside Farm on Bryher. Found in a range of habitats including cultivated land, waste places, roadsides, dunes, cliffs and around pools.

CARYOPHYLLACEAE – Pink family
Cerastium glomeratum Thuill.
Sticky Mouse-ear
Inhabited islands

Native. Found on all the inhabited islands where it is very common and widespread. Often appears to be larger and more robust than mainland plants. Lousley recorded it from Teän as a possible relict from former cultivation, but it has not been seen there recently. Cultivated fields and roadsides.

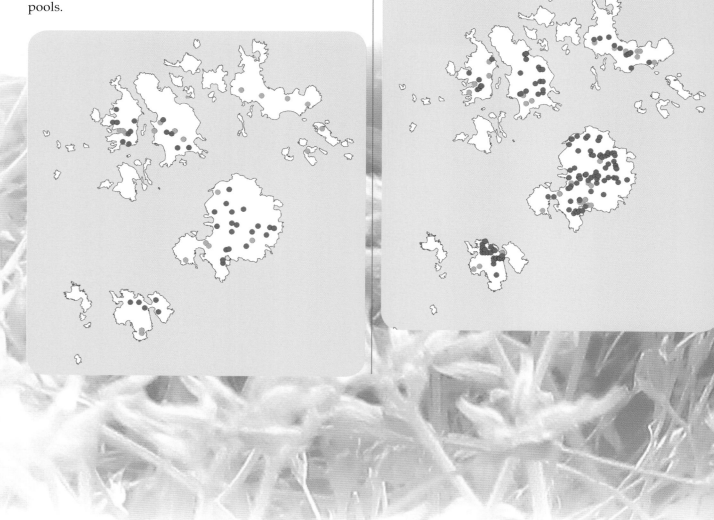

CARYOPHYLLACEAE – Pink family
Cerastium diffusum Pers.
Sea Mouse-ear
Inhabited islands, also St Helen's, Teän, Annet, Great Ganilly and Nornour

Native. Scattered records from all the inhabited and some larger uninhabited islands. It usually flowers early and may therefore be overlooked. A variable species mainly found growing in thin soils over granite around the coast and on wall tops.

CARYOPHYLLACEAE – Pink family
Cerastium semidecandrum L.
Little Mouse-ear
Bryher, St Mary's, St Martin's and St Agnes

Native. There are only a few records of this species. This may also be partly due to the plant flowering very early and being missed by most visiting botanists. It grows on sandy, short grassland on dunes, grassland and heaths.

EU-DICOTS True Dicotyledons

CARYOPHYLLACEAE – Pink family
Sagina procumbens L.
Procumbent Pearlwort
Inhabited islands, also Samson and St Helen's

Native. Widespread and common from scattered places over all the inhabited islands and also recorded from St Helen's and Samson. Not recorded from Great Ganilly since 1936. Grows mainly on heathland, margins of bulbfields and other cultivated places, on tracks and in short, wet grassland.

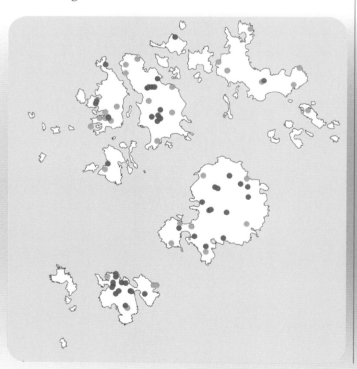

CARYOPHYLLACEAE – Pink family
Sagina apetala Ard.
Annual Pearlwort
Inhabited islands, also Samson

Native. Recorded from all the inhabited islands although there are only two records from St Martin's. Lousley describes it as rare so it appears to have increased substantially since 1971. Although two subspecies *apetala* and *erecta* have been recorded in Scilly, they may only reflect slight variations of the same species. Grows on bare ground, paths and walls on heathland, also on arable land and in gardens.

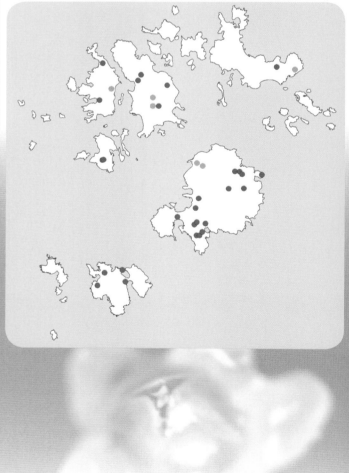

CARYOPHYLLACEAE – Pink family
Sagina maritima Don.
Sea Pearlwort
Tresco, St Agnes, St Mary's and Bryher

Native. Suprisingly there are fewer records of this species than of some other pearlworts. Lousley described it as very common and lists it from all the main uninhabited island groups. Although there are records from Annet (1983) and Foreman's Island (1981) there are none from there recently. It is not known whether this is a genuine change in distribution, or that it is being under-recorded. Usually found on bare ground on cliffs, walls, paths in dunes and heaths, usually coastal, but occasionally inland.

CARYOPHYLLACEAE – Pink family
Polycarpon tetraphyllum (L.) L.
Four-leaved Allseed
Inhabited islands, Teän and Northwethel

Native. Abundant on all the inhabited islands and has also been recorded from Teän (1990) and Northwethel (1992). In 2009 it was recorded on Great Ganilly. The var. *diphyllum* has been recorded occasionally. Although *Polycarpon* is now mostly associated with man-made habitats on Scilly it does occur in dunes and sandy ground near the coast. Also found in arable and bulb fields, along wall bases and frequently in cracks in pavements.

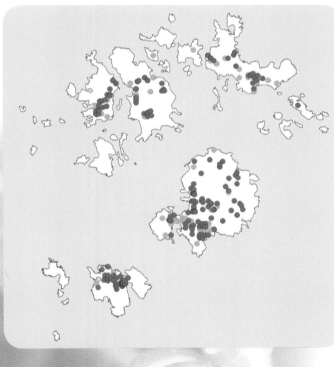

EU-DICOTS True Dicotyledons

CARYOPHYLLACEAE – Pink family
Spergula arvensis L.
Corn Spurrey
Inhabited islands

Native. A common plant on the inhabited islands. It has been found both in the Bronze Age and Iron Age archaeological remains. Occasionally can appear in extraordinary profusion and be the dominant weed, especially in newly opened land (see photo on page 70). It is believed the seeds can remain viable in the seedbank for a very long time. Usually in arable fields but can also occur in any disturbed or cultivated ground.

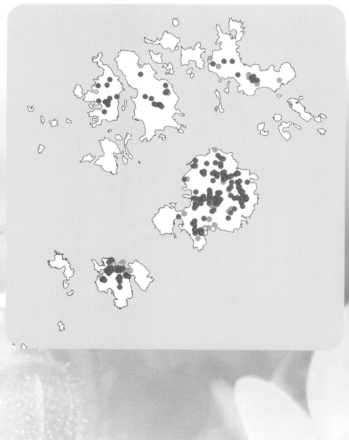

CARYOPHYLLACEAE – Pink family
Spergularia rupicola Lebel ex Le Jol.
Rock Sea-spurrey
Inhabited islands, the Norrard Rocks, Eastern Isles, Western Rocks and the islands around St Helen's

Native. Recorded on almost every island in Scilly than can support vegetation, this is perhaps the most widespread species throughout. It is one of the few plants that can survive on some of the most exposed rocky islets such as those in the Western Rocks, Norrards and Eastern Isles. It was described by Lousley as an 'outstanding botanical feature of the Isles of Scilly'. It grows on coastal rocks, inland walls and heathland carns, very occasionally on the ground.

CARYOPHYLLACEAE – Pink family
Spergularia marina (L.) Besser.
Lesser Sea-spurrey
Bryher and St Mary's

Native. On Bryher it is only known from along the leat and the fringes around the brackish Pool. A plant was found in Lower Moors, St Mary's in 2009, was still there in 2010 and by 2011 there were five plants. The last record previously from this area had been in 1953. The Lower Moors plants perhaps may be a relict of when the Moors were more brackish. Wet ground and ditches in brackish areas.

CARYOPHYLLACEAE – Pink family
Spergularia rubra (L.) J. & C. Presl.
Sand Spurrey
Inhabited islands

Native. More widespread than formerly, possibly still expanding its range and is far more frequent than a couple of decades ago. Lousley recorded the species from St Mary's, St Martin's and Tresco, but suggested it was increasing as 'it spread into habitats made available by human activity'. It is now common on St Agnes and has been found on Bryher, including at Rushy Bay in 2014. The plant apparently thrives on compacted ground such as the headlands in bulbfields, as well as on paths and other bare ground.

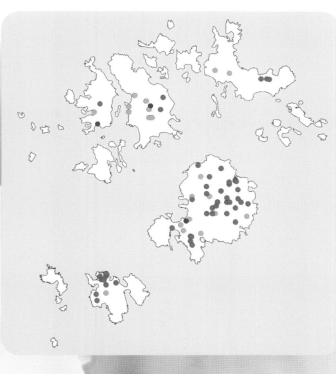

EU-DICOTS True Dicotyledons

CARYOPHYLLACEAE – Pink family
Spergularia bocconei (Scheele) Graebn.
Greek Sea-spurrey
St Mary's

Neophyte. Originally only known from a site at Porth Hellick in 1953 from which it disappeared and which Lousley and others later were unable to relocate. In 2004 a plant was found on a track at Tolman Point by Ian Bennallick, but was not found there the following year or since. Then ten plants were found near the desalination plant at Mount Todden in 2005 and three plants in 2006, growing among a mass of *S. rubra* in a saltwater spill. No plants of either *Spergularia* species could be found there in 2014. As the plant is an annual and usually found on sandy, bare and compacted ground such as pathways, it is likely the seeds of this plant may be carried on feet or by vehicles to suitable areas. Bare ground.

CARYOPHYLLACEAE – Pink family
Silene vulgaris (Moench) Garcke
Bladder Campion
Tresco

Native. Only known from one site, a wall on a verge at New Grimsby SV89341557 where it was first found by Rosi and Jim Bowyer in 2009, it was still there in 2014. The plant appears to be ssp. *vulgaris*. This species was recorded on several occasions in the 1800s, but the records were rejected by Lousley as errors. It is not known whether the plant above was an accidental arrival or possibly introduced as a garden plant. But it does not seem to have been planted. Wall.

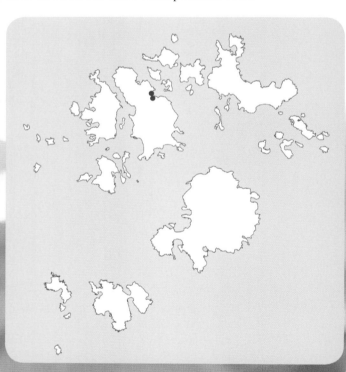

Agrostemma githago L.
Corncockle
Extinct

A record from the allotments, St Mary's in 1988 almost certainly had been accidentally introduced with garden material or seeds.

CARYOPHYLLACEAE – Pink family
Silene uniflora Roth
Sea Campion

Inhabited islands, Annet, St Helen's, Teän, Guther's Island, most of the Eastern Isles (although apparently not Great Ganilly) and Samson (including Puffin Island)

Native. Occurs as discrete patches on all the inhabited and many uninhabited islands. Most sites are near the coast, but occasionally found inland. Rocks, walls and banks.

CARYOPHYLLACEAE – Pink family
Silene latifolia ssp. *alba* (Mill.) Greuter & Burdet
White Campion

St Agnes, St Mary's, St Martin's and Tresco

Archaeophyte. Recorded from a just a very few localities, of which only a few are recent or have been verified due to confusion with the common white-flowered form of *Silene dioica*. Found in hedgerows or edges of bulbfields.

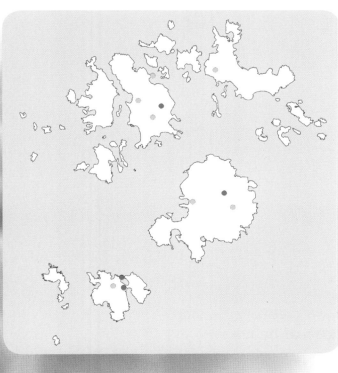

EU-DICOTS True Dicotyledons

CARYOPHYLLACEAE – Pink family
Silene dioica (L.) Clairv.
Red Campion
Inhabited islands, Annet, St Helen's, Northwethel, Teän, Nornour, Great Ganilly and Little Innisvouls

Native. A widespread and common plant found on all the inhabited and larger uninhabited islands. Often found with pink or white flowers as well as deep red. Often found on disturbed ground, for example, on heathland after gorse has been burned. Grows on hedgebanks, cliff slopes, road verges and along field edges.

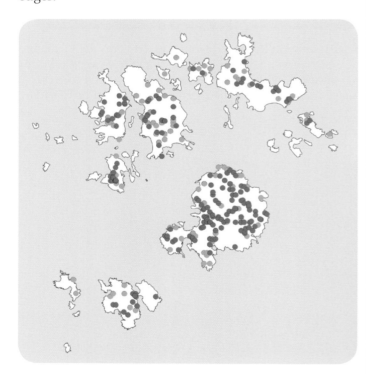

Silene coeli-rosa (L.) Godr.
Rose-of-heaven
Tresco

Neophyte. A garden escape recorded on St Mary's in 1973.

CARYOPHYLLACEAE – Pink family
Silene noctiflora L
Night-flowering Catchfly
Tresco

Archaeophyte. The flowering plant photographed on Tresco (SV89091590) by Charles Campbell in 2012 was believed to be an accidental introduction.

Silene gallica L.
Small-flowered Catchfly
Inhabited islands

Archaeophyte. Widespread and locally common on all the inhabited islands. Where it occurs in the bulb fields it can at times be abundant. On The Garrison exceptionally large plants have been found on the top of the wall towards Morning Point since 2010. Several colour varieties have been recognised: var. *gallica* is pale pink, var. *anglica* off-white and var. *quinquevulnera* has a crimson spot on each white petal. This latter variety is very attractive, rather like a small Sweet-William *Dianthus barbatus,* but is now believed extinct in the wild. It has been conserved in a few private gardens on St Mary's from where it occasionally escapes – for example on Tolman Point from a nearby vegetable garden. Found on roadsides and disturbed land, cultivated habitats including arable fields, allotments and gardens.

CARYOPHYLLACEAE – Pink family
Silene flos-cuculi (L.) Clairv.
Ragged Robin
St Mary's

Native. Lousley recorded the species as locally plentiful at both Higher and Lower Moors. There are records since 2011 from both the Moors but usually only a few discrete patches. In 2012 there were more plants in Higher Moors and in 2013 Martin Goodey photographed a good display there. Wetlands.

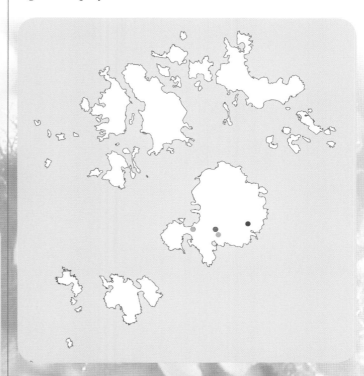

Saponaria officinalis L.
Soapwort
Extinct

Formerly occurred on Tresco, probably before 1940, as there are no recent records. Lousley considered this was another of the plants associated with the monks of the former Priory on Tresco.

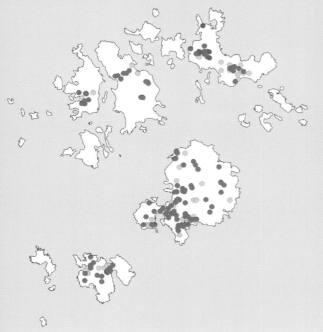

EU-DICOTS True Dicotyledons

CARYOPHYLLACEAE – Pink family
Dianthus plumarius L
Pink
St Mary's

Neophyte. Hugh Town, St Mary's 1997 and 2012. Occasional escape from cultivation. Pinks are grown commercially as a summer crop.

CARYOPHYLLACEAE – Pink family
Dianthus barbatus L.
Sweet-William
St Mary's

Neophyte. Holy Vale in 1996 and 1997 and Porth Mellon, St Mary's in 1997.

AMARANTHACEAE – Goosefoot family
Chenopodium rubrum L.
Red Goosefoot
Inhabited islands, also Teän, Samson, Great Arthur, Northwethel and Great Ganilly

Native. This species has apparently spread since Lousley published his map in 1971. There is a record of the species on Teän in 1972. This is another plant whose remains have been found at Bronze Age archaeological sites. *Chenopodium rubrum* is now widespread and locally frequent in a range of habitats; from arable and disturbed land, as well as typically around the margins of pools, wetland and coastal sites.

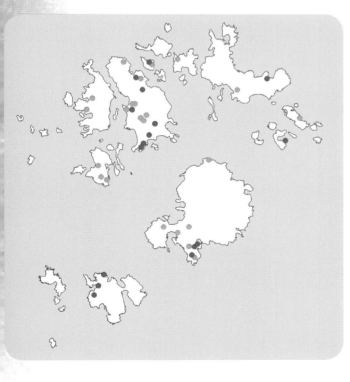

AMARANTHACEAE – Goosefoot family
Chenopodium polyspermum L.
Many-seeded Goosefoot
Tresco, St Mary's, Great Ganilly

Archaeophyte. An uncommon plant, with just a few recent records mainly from Tresco in 2004 (SV893142, SV89371541 and SV89211409). The last record from St Mary's was from near Bar Point (SV912127) in 1998. The record from Great Ganilly was a plant photographed by Ren Hathway on the strandline there in 2005. Arable and disturbed land.

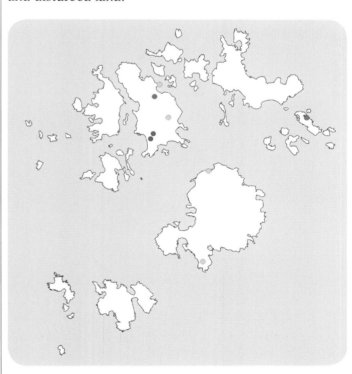

Chenopodium bonus-henricus L.
Good King Henry
Extinct

Records from a farm on Tresco in 1960 and from an unknown place on St Mary's in 1967.

EU-DICOTS True Dicotyledons

AMARANTHACEAE – Goosefoot family
Chenopodium hybridum L.
Maple-leaved Goosefoot
Tresco, St Mary's and Bryher

Archaeophyte. There are only three records of this species, each from a different island; Tresco (SV900144) in 1992, St Mary's (SV911100) in 2005 and Bryher (SV880152) in 2008. All from cultivated or disturbed ground.

AMARANTHACEAE – Goosefoot family
Chenopodium murale L.
Nettle-leaved Goosefoot
Inhabited islands

Archaeophyte. There are scattered records from all the inhabited islands most frequently on Tresco; from Bryher it was recorded from Veronica Farm in 2005 and near Fraggle Rock in 2014. Found in arable fields, other cultivated and disturbed ground.

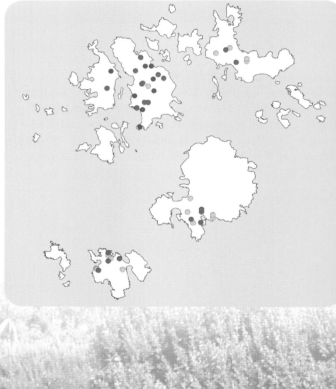

AMARANTHACEAE – Goosefoot family
Chenopodium ficifolium Sm.
Fig-leaved Goosefoot
St Agnes

Archaeophyte. Only known from St Agnes. Recorded from bulbfields near Covean in 1995 and near the Island Hall in 2007. G. Kitchener also recorded it from SV8808 in 2009. Cultivated fields.

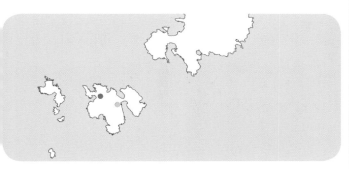

Chenopodium quinoa Willd.
Quinoa
St Martin's and Tresco

Neophyte. Currently only known as a planted crop. As a conservation mixture/wild bird food on St Martin's and Tresco.

AMARANTHACEAE – Goosefoot family
Chenopodium album L.
Fat-hen
Inhabited islands, Samson, Puffin Island off Samson, Round Island, Teän and Middle Arthur

Native. Although there are only scattered records for this species around the islands it can be locally frequent, for example as a weed in arable fields. There are records from Bronze Age archaeological sites when it was probably used as food. It also occurs on other cultivated land, disturbed ground and occasionally in places where gulls have been nesting or roosting.

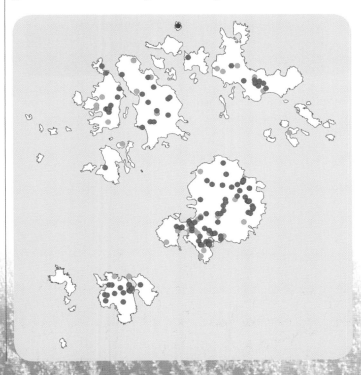

EU-DICOTS True Dicotyledons

AMARANTHACEAE – Goosefoot family
Atriplex hortensis L.
Garden Orache
St Mary's

Neophyte. Casual. In garden at Little Porth after storms deposited sand and debris in a garden by the beach. 2014.

AMARANTHACEAE – Goosefoot family
Atriplex prostrata Boucher ex DC.
Spear-leaved Orache
Inhabited islands, also most uninhabited islands

Native. One of the most widespread coastal plants in the islands; it is found on even the the smallest islands that can support vegetation, for example Rosevear, Round Island, Mincarlo and most of the Eastern Isles. Hybrids between this and other *Atriplex* species can cause confusion. Usually found on coastal sites but also occasionally inland on pathways, the back of dunes and on rubbish dumps.

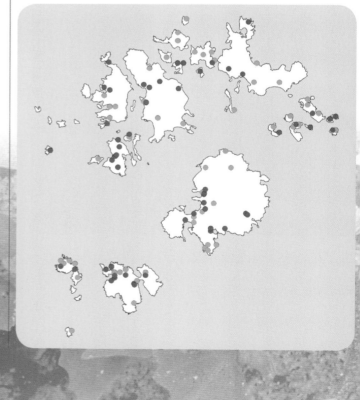

AMARANTHACEAE – Goosefoot family
Atriplex prostrata x *Atriplex glabriuscula*
Hybrid Orache
Bryher

Native. Recorded by Owen Mountford near Great Porth (SV874148) in 2011. Although *Atriplex* frequently hybridise, this is the first record of this hybrid for Scilly.

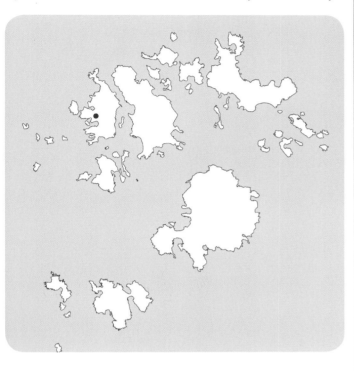

Atriplex x *gustafssoniana* Tascher.
(*A. prostrata* x *A. longipes*)
Kattegat Orache
Teän

Native. This distinctive hybrid was recorded by John Sproull in 2008 from Didley Point, Teän (SV903171). Although unconfirmed there is no reason to doubt the identification which was supported by a sketch of the leaf. Only one of the parents occurs on Scilly.

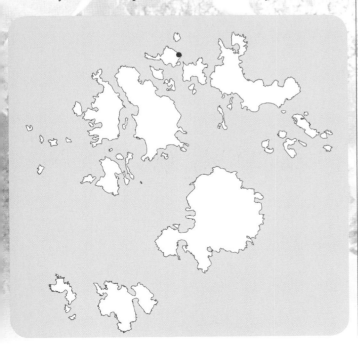

AMARANTHACEAE – Goosefoot family
Atriplex glabriuscula Edmondston
Babington's Orache
Inhabited islands, also Samson, Teän and Eastern Isles

Native. This orache is probably much more common than the records suggest, many botanists seem to ignore the *Atriplex* species on the strandline. There are scattered records from both inhabited and many uninhabited islands. These include a 1981 record from Foreman's Island (near Northwethel) and 1984 record from Gugh Bar, with more recent records from Samson, Teän, and from Menawethan, Great and Little Innisvouls and Ragged Island (all in 1987) and Middle Arthur (1994). Beaches around the coast.

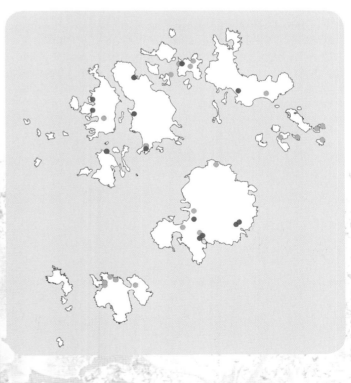

EU-DICOTS True Dicotyledons

AMARANTHACEAE – Goosefoot family
Atriplex patula L.
Common Orache
Inhabited islands

Native. There is a scatter of records from St Mary's. Elsewhere it is nowhere common. On Bryher it has been recorded in the Hotel area in 1995; also from Priglis, St Agnes in 1995; on Tresco from Abbey Road in 1997 and the path to Cromwell's Castle in 2007; and in a beet field at Higher Town, St Martin's in 1989. Cultivated fields, coastal and waste places.

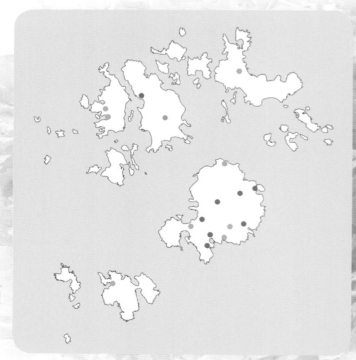

AMARANTHACEAE – Goosefoot family
Atriplex laciniata L.
Frosted Orache
Inhabited islands, Samson, St Helen's, Teän, White Island (St Martin's), Middle and Little Arthur

Native. This species is a feature of strandlines around the island beaches. Plants are also occasionally found inland. Other than current records, there are also records from Annet (1983), Little Ganilly ((1983) and Menawethan (1984). Beaches around the coast.

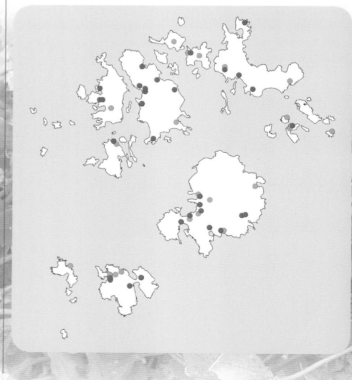

AMARANTHACEAE – Goosefoot family
Atriplex halimus L.
Shrubby Orache
St Mary's and St Agnes

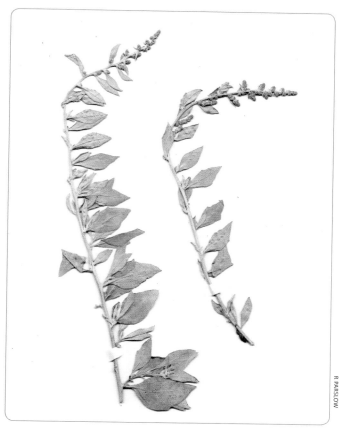

Neophyte. The plants recorded from Tolman Point, St Mary's and near the Turk's Head, St Agnes were almost certainly originally planted.

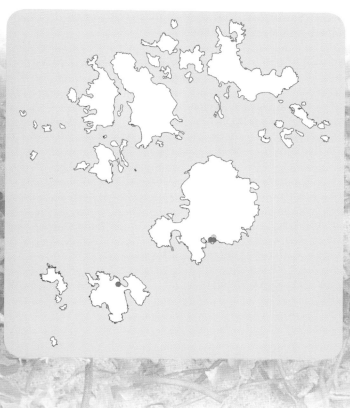

AMARANTHACEAE – Goosefoot family
Beta vulgaris ssp. *maritima* (L.) Arcang.
Sea Beet
Inhabited islands, most uninhabited islands

Native. Abundant on all the inhabited and almost all the uninhabited islands where there is any vegetation. These include islets in the Norrard Rocks, all the Eastern Isles, Rosevear in the Western Rocks, some of the islets off Teän, St Helen's, Northwethel and Round Island. Can occasionally be found inland. Sea beet is eaten locally as an alternative to spinach, so is sometimes cultivated in gardens. All shores and waste ground.

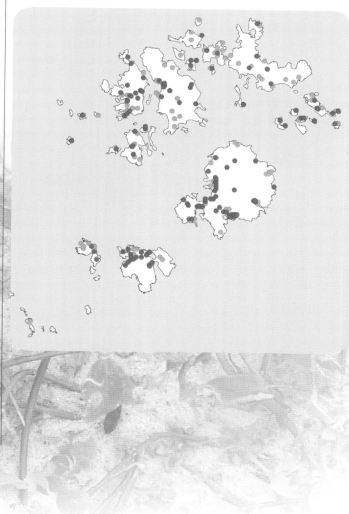

EU-DICOTS True Dicotyledons

AMARANTHACEAE – Goosefoot family
Beta vulgaris ssp. *vulgaris*
Root Beet
St Mary's and Tresco

Neophyte. An occasional escape from cultivation.

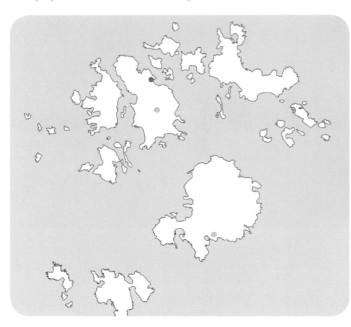

Suaeda maritima (L.) Dumort.
Annual Sea-blite
Bryher

Native. There has only been one record, from Shipman Head, Bryher by Mrs B. Graham in 1984. This unusual occurrence of just one plant in a very atypical habitat was presumed to have been bird sown. Identification was confirmed by a voucher specimen.

AMARANTHACEAE – Goosefoot family
Salsola kali ssp. *kali* L.
Prickly Saltwort
Tresco, St Martin's, St Mary's, Northwethel, Teän, Samson and Great Ganilly

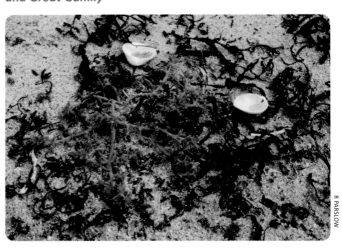

Native. This plant now has an extremely contracted distribution: Lousley recorded it from a number of localities including Bryher, St Mary's, Gugh, Annet, White Island (off Samson), Northwethel and St Martin's. The plant was last recorded from Lower Town Bay, St Martin's in 1993, from Porthcressa, St Mary's in 1983 (a small plant growing in a sandy crack in a pavement in the street), Northwethel in 1981, two places on Great Ganilly in 1994 and from Pentle Bay area on Tresco in 1993 (SV901148 and SV898138). Being very prickly it is possible the plant was not tolerated on more populated beaches on the inhabited islands, or like many other strandline species it has become a victim of more frequent storms. Recent records are from just two sites; a sandy beach at West Porth, Teän in 2009 and the landing beach, Samson in 2004, both islands with earlier records. Sandy and shingly beaches.

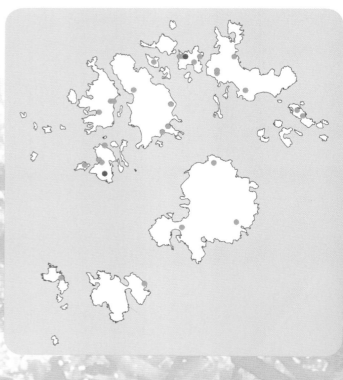

AMARANTHACEAE – Goosefoot family
Amaranthus retroflexus L.
Common Amaranth
St Mary's and Tresco

Neophyte. Casual. Recorded occasionally in fields on Tresco and St Mary's, and as a birdseed alien on St Mary's. Arable fields and waste ground.

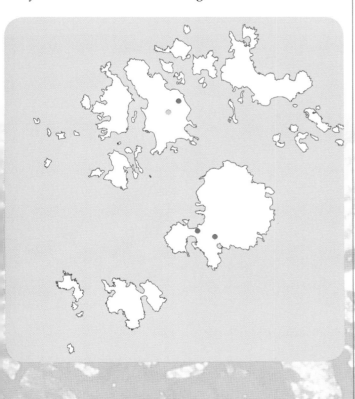

AMARANTHACEAE – Goosefoot family
Amaranthus hybridus L.
Green Amaranth
Tresco

Neophyte. Casual. Found occasionally in a few places on Tresco including around Pool Road, Borough Farm and on a rubbish dump on Appletree Banks. Cultivated and disturbed ground.

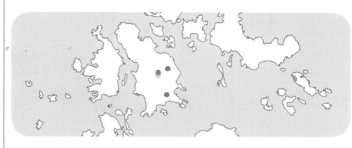

Amaranthus caudatus L.
Love-lies-bleeding
St Agnes

Neophyte. A garden escape near the school in 1992.

Amaranthus albus L.
White Pigweed
St Mary's

Neophyte. Found growing with other birdseed aliens in Ram's Valley, St Mary's in 2000.

EU-DICOTS True Dicotyledons

AIZOACEAE – Dewplant family

Many species of 'mesembs' or succulents are grown on walls and in gardens around the inhabited islands. Most are plants that originated in South Africa and places with similar climates. They were introduced to the Isles of Scilly as garden plants, especially to the Abbey Gardens on Tresco. Outside gardens it is not always clear which plants have been deliberately planted and which have become naturalised, as plants are easily propagated either from seed or from bits of plant. Several species of succulents have also been found on uninhabited islands where they have been taken by gulls, apparently as nesting material. Most of these plants do not persist, possibly due to exposure to salt spray in winter, but on larger islands such as St Helen's, some have become established. Even tiny islets such as Plumb Island off New Grimsby, Tresco has a dense topping of *Carpobrotus edulis*. Some genera, for example – *Drosanthemum* and *Disphyma* have become established in places around the coast.

Drosanthemum at Morning Point

Disphyma crassifolium by the coast

AIZOACEAE – Dewplant family
Aptenia cordifolia (L. f.) Schwantes
Heart-leaf Iceplant
Inhabited islands, also Round Island

Neophyte. Garden escape. There are scattered records from around the islands, some of which appear to be naturalised plants although it is often planted on walls. *Aptenia* is originally a South African species that has also become established in Cornwall and the Channel Islands. Usually the small flowers are deep reddish-purple, but occasionally can be dark blue. Walls and roadsides.

AIZOACEAE – Dewplant family
Drosanthemum floribundum (Haw.) Schwantes
Pale Dewplant
Inhabited islands, Round Island, St Helen's and Great Ganinick

Neophyte. Long established on St Mary's, St Martin's and Tresco. In the past it has also been recorded on St Helen's, Round Island and Great Ganinick. In some places, such as Tolman Point and Morning Point, St Mary's the plant has spread over rocks and walls completely smothering the native vegetation. The leaves are small and greyish (due to their covering of small, white papillae) and the prolific small mauve flowers are at their best in spring to early summer. It is very popular with bees. Lousley refers to a herbarium specimen from Old Town churchyard, St Mary's dated 1875. The species is also found in gardens on St Agnes and Bryher. In 2009 and 2015 the plant was photographed on Round Island.

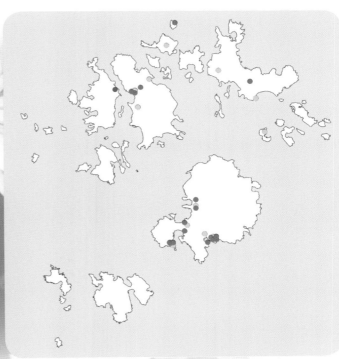

EU-DICOTS True Dicotyledons

AIZOACEAE – Dewplant family
Oscularia deltoides (L.) Schwantes
Deltoid-leaved Dewplant
Inhabited islands

Neophyte. Also originating from South Africa. This is a distinctive plant with curious triangular, keeled leaves and small mauve flowers in spring. It is frequently planted on walls around gardens from where it appears to have occasionally escaped to become established on the inhabited islands. Walls and rocky places.

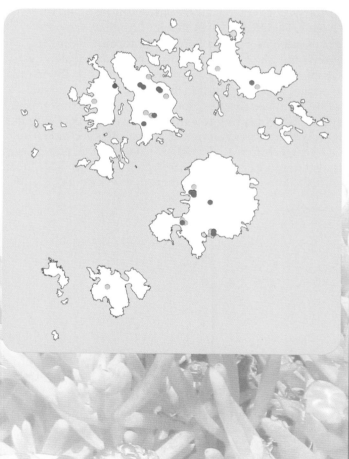

AIZOACEAE – Dewplant family
Ruschia caroli (L. Bolus)
Shrubby Dewplant
Tresco and St Mary's

Neophyte. This is another plant of South African origin. Found as a garden escape on both Tresco and St Mary's. This is a bushy shrub with deep purple-red flowers, but can be other shades. Walls and roadsides.

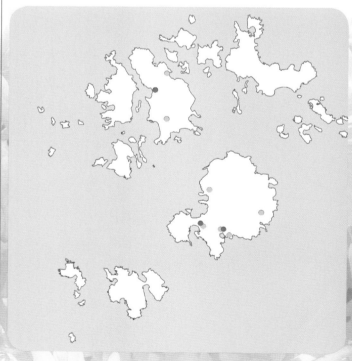

AIZOACEAE – Dewplant family
Lampranthus falciformis (Haw.) N.E. Br.
Sickle-leaved Dewplant
St Mary's, St Martin's and Tresco

Neophyte. Although grown in gardens on all the inhabited islands, the species is only apparently established on St Mary's, St Martin's and Tresco. Many colourful, large-flowered garden cultivars are grown in the islands and may occasionally be found as garden escapes. Walls and rocks.

AIZOACEAE – Dewplant family
Lampranthus conspicuus (Haw.) N.E. Br.
a dewplant
St Agnes and St Mary's

Neophyte. This species has been recorded from just three places; a wall on St Agnes (SV880081), and St Mary's (SV909104) from Alan Underhill in 1993 to 1994. A plant found on a wall at Old Town Bay, St Mary's at SV913102 in 2011 may have been this species.

EU-DICOTS True Dicotyledons

AIZOACEAE – Dewplant family
Lampranthus roseus (Willd.) Schwantes
Rosy Dewplant
St Mary's, Tresco and St Helen's

Neophyte. Also originally from South Africa. Apparently established in several places on St Mary's and Tresco. Distinguished from the other *Lampranthus* species by slight differences in the leaves which are also less obviously dotted than the previous species, but the colours are variable so a pink-flowered plant is not necessarily this plant. A plant growing on a south-facing cliff on the south side of St Helen's in 2002 was thought to be this.

AIZOACEAE – Dewplant family
Disphyma crassifolium (L.) L. Bolus
Purple Dewplant
Tresco, St Mary's, St Martin's, Bryher, Round Island

Neophyte. According to Lousley this plant was known at New Grimsby, Tresco well before 1936. The strange hanging strings of 'jelly bean'- like leaves hang down from rocks and walls and detached pieces are easily spread to make new plants. The leaves frequently turn red, increasing the resemblance to jelly beans; the sessile flowers are purple, only opening in sunshine. The plant originates from South Africa (where it also grows on the coast). It is abundant on Round Island. Walls, verges, rocks and coastal sites.

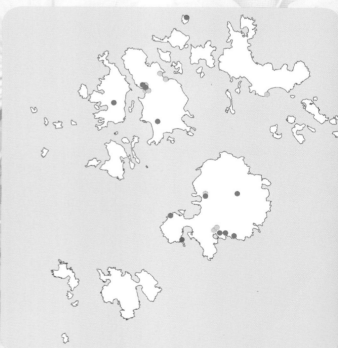

AIZOACEAE – Dewplant family
Erepsia heteropetala N.E. Br.
Lesser Sea-fig
St Mary's (and Bryher)

Neophyte. Another South African species that had been growing on the walls of the quarry on Buzza Hill, St Mary's since long before it was recorded by Lousley in 1967. The pink flowers are small and half-hidden by the calyx, they appear in June. Although the plant appears to spread by seed within the quarry it does not seem to have spread any further. A plant growing on a garden wall on Bryher had been planted there. Quarry walls.

AIZOACEAE – Dewplant family
Carpobrotus acinaciformis (L.) L. Bolus
Sally-my-handsome
All inhabited islands

Neophyte. Another plant that originates from South Africa. This species can be most easily identified by the leaves which are very curved and thickest at the ends – 'scimitar-shaped', unlike *Carpobrotus edulis* where the leaves are about the same thickness their whole length. The flowers are pinkish-purple, much like *Carpobrotus edulis* var. *rubescens*. In Preston and Sell (1988) this plant is included with *C. edulis*, but the leaf shape seems sufficiently different to recognise it as distinctive. It is much less common than *C. edulis* but is most frequently recorded from St Martin's where there are large stands at Lawrence's Bay and above Great Bay. Around the coasts, on cliffs and in dunes.

AIZOACEAE – Dewplant family
Carpobrotus edulis (L.) N.E. Br.

Hottentot Fig

All inhabited islands, also St Helen's, Round Island, Teän. Puffin Island (off Samson) Plumb Island (off Tresco), Toll's Island, as well as Great Ganinick and Menawethan in the Eastern Isles

Neophyte. The first plants were introduced by Augustus Smith to the Abbey Gardens in the middle of the nineteenth century and he also planted it around the coast as well as Marram Grass to stabilise the dunes and prevent the sand blows that had been such a nuisance to the islanders. Despite this there are some dune sites where it does not seem to have established, for example Higher Town Bay, St Martin's. By the 1920s the plant was common on most of the islands, almost certainly from being planted. It was introduced by the lighthouse keepers to Round Island. The plant was further spread by gulls until it is now found on many of the uninhabited islands, often well away from the shore. Now there is a concern that the plant has become a bit of a thug, crowding out the native vegetation and competing with some of the rare coastal plants.

Carpobrotus edulis usually appears in a yellow-flowered form that often fades to pink as it goes over. This variety is called var. *edulis* in Stace (2010). This is the common plant in Scilly. Other varities include var. *rubescens* which is pink/purple and var. *chrysophthalmus* which is also pink but with yellow bases to the petals (the petals may also be slightly more pointed). The varieties have not been mapped separately. Another plant that originates from South Africa.

Where Hottentot Fig is having an impact on the native vegetation the Wildlife Trust has targeted these in their management. It has been observed that cattle on occasion eat the plant (D. Mawer, pers. comm.). Grows on coasts, inland on rocks and in dunes.

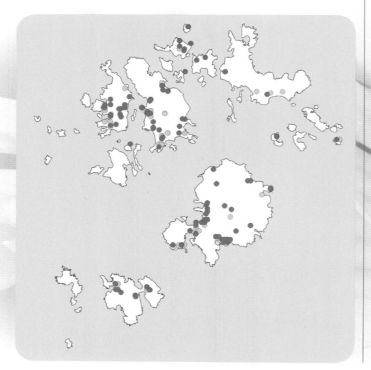

AIZOACEAE – Dewplant family
Carpobrotus glaucescens (Haw.) Schwantes
Angular Sea-fig
Tresco and St Mary's

Neophyte. So far only recorded from a few places on Tresco and from Rocky Hill (SV91511118), St Mary's. But as the plant is smaller and may be less invasive it is likely to cultivated and spread to other sites. This species has a similar habit to the other *Carpobrotus* but is distinctly smaller in all its parts. The flowers are deep purple-pink with white or pinkish bases to the petals, the leaves are triangular in section, as in *C. edulis* but smaller. Sometimes the leaves may have a slightly wider 'base' section than the other two sides, and also may be slightly curved and thicker towards the ends, although never as strongly as in *acinaciformis*. Whether this is a 'good' character or due to some crossing with *acinaciformis* needs investigation. In Australia (from where the species originates) the species is called 'pigface', the fruit is considered edible and the leaves used to sooth insect stings. Grows on walls and on the ground in dunes.

AIZOACEAE – Dewplant family
Tetragonia tetragonioides (Pall.) Kuntze
New Zealand Spinach
Tresco, St Martin's and Teän

Neophyte. Casual. Occasional records from near Great Pool, Tresco and East Porth, Teän in 1995 and 1996. Plants found on cultivated land on St Martin's (SV9215) in 2012 are thought to have been grown as a crop. Found on sandy beaches or disturbed ground.

EU-DICOTS True Dicotyledons

NYCTAGINACEAE – Marvel-of-Peru family
Mirabilis jalapa L.
Marvel-of-Peru
St Mary's

Neophyte. Occasional garden escape. Has been found in several places on St Mary's, including rubbish tips, Holy Vale and growing near the shore between Porth Mellon and Porthloo.

MONTIACEAE – Blinks family
Claytonia perfoliata Donn ex Willd.
Springbeauty
Inhabited islands

Neophyte. Locally frequent to occasionally dominant within the crop. Originally from North America it is believed to have been first introduced to Scilly as a salad crop. An arable weed in bulbfields and other cultivated ground.

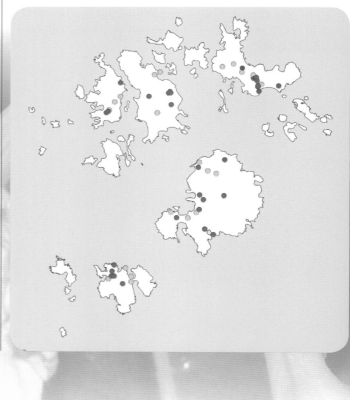

MONTIACEAE – Blinks family
Claytonia sibirica L.
Pink Purslane
St Mary's and St Agnes

Neophyte. Garden escape or casual. There is a record from Hugh Town in 1982, also from Higher Town, St Agnes in 2000.

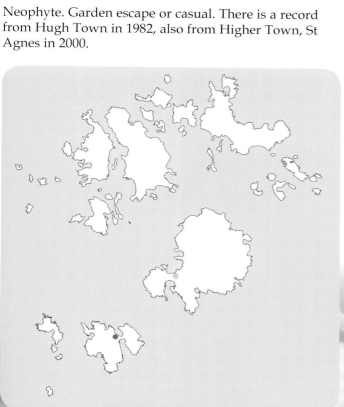

MONTIACEAE – Blinks family
Montia fontana L.
Blinks
Inhabited islands, also Samson

Native. There are scattered records from all the inhabited islands and also from Samson. On Samson it was recorded near the ruined buildings on South Hill in 1980 and then near the well in 2007. The subspecies generally recorded on Scilly is *amporitana* although ssp. *variabilis* and ssp. *chondrosperma* have been recorded. Found on damp ground on heath, grassland and occasionally on cultivated ground.

EU-DICOTS True Dicotyledons

PORTULACACEAE – Purslane family
Portulaca oleracea L.
Common Purslane
Inhabited islands

Neophyte. A persistent weed of cultivation that occurs in a few places including Troy Town on St Agnes in 2002, and several places on Tresco, including around New Grimsby, Pool Road and in the vegetable garden in the Abbey Gardens. Formerly know from fields on west of St Mary's in 1994, in 2013 it was discovered in a bulbfield at Rocky Hill. In 2011 plants were found in fields at Higher Town, St Martin's and on Bryher. As the plant grows close to the ground it can be easily overlooked. It is not known how the plant reached Scilly, various theories suggest it may have been grown as a salad or that it was introduced with seaweed mulches used to top-dress the bulbs. Cultivated and disturbed ground.

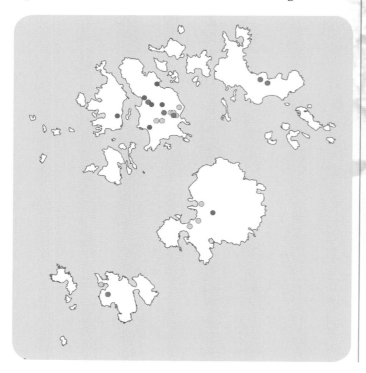

HYDRANGEACEAE – Mock-orange family
Hydrangea macrophylla (Thunb.) Ser.
Hydrangea
Tresco andary's

Neophyte. Occasional garden escape.

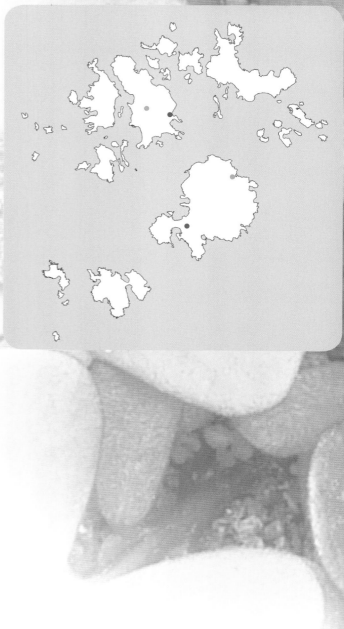

POLEMONIACEAE – Jacob's-ladder family
Gilia capitata Sims
Blue Thimble-flower
St Mary's

Neophyte. Garden escape. One record from Church Street, St Mary's (SV905105) in 2000.

PRIMULACEAE – Primrose family
Myrsine africana L.
African Boxwood
Tresco and St Mary's

Neophyte. Garden escape or planted shrub, may also be bird sown. In several places on Tresco, mostly close to the Abbey Gardens and at Borough Farm, St Mary's.

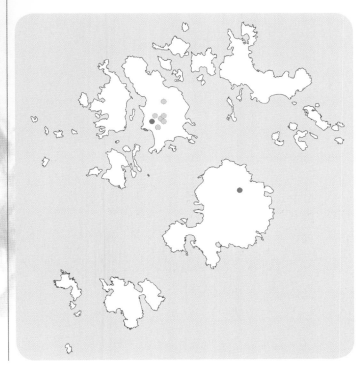

EU-DICOTS True Dicotyledons

PRIMULACEAE – Primrose family
Primula vulgaris Huds.
Primrose
St Mary's, Bryher, Tresco and Samson

Native in the British Isles, but introduced to Scilly. Restricted to just a few places where it was almost certainly originally planted or has escaped from cultivation. Found in Old Town churchyard and Watermill Lane, St Mary's; Bryher churchyard; also Appletree Banks, Tresco and on Samson where it is thought to have been introduced by the former inhabitants. Churchyards, hedgebanks and dunes.

PRIMULACEAE – Primrose family
Lysimachia nemorum L.
Yellow Pimpernel
Tresco

Native. Recorded from just three places on Tresco; SV894143 (weed in Abbey Gardens) in 1984, SV89211416 in 2010 and from Abbey Hill in 2013. It is not known whether it is native to the area or it had originated from the Abbey Gardens.

PRIMULACEAE – Primrose family
Lysimachia nummularia L.
Creeping-Jenny
Tresco and St Mary's

Neophyte. Recorded from near Great Pool and Abbey Wood, Tresco from 1953 until 1995 (may still be there), also from Higher Moors, St Mary's where it was seen near the Boardwalk (SV923109) in 1997. Found in damp places.

PRIMULACEAE – Primrose family
Lysimachia vulgaris L.
Yellow Loosestrife
St Mary's

Native. Only known from Lower Moors, St Mary's (SV913105) where it grows in a very wet area under willows, well away from the track. It was last seen in 2002. This is the wild species, not the cultivated plant *L. punctata* as previously thought. Wetland.

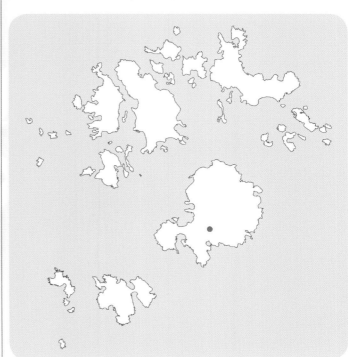

EU-DICOTS True Dicotyledons

PRIMULACEAE – Primrose family
Glaux maritima L.
Sea-milkwort
Inhabited islands, Northwethel, Great Ganilly and Teän

Native. Found at scattered sites around the coast and beside pools on the inhabited islands, although it has not been seen on St Martin's (SV917160) since 1988. It has not been recorded on Teän or Great Ganilly recently but is likely to still be present. Found by brackish pools, marshy places and under coastal rocks in more exposed places.

PRIMULACEAE – Primrose family
Anagallis tenella (L.) L.
Bog Pimpernel
Tresco, Bryher, St Agnes and St Mary's

Native. Known from Badplace Hill, Bryher (SV875162); Lower Moors (SV913105) and around Shooter's Pool, St Mary's; on St Agnes from several places on Wingletang Down and on Tresco near the Abbey Pool, Great Pool and at Merchant Point (where last seen 1995). Found in wet places on heaths, in marshes and by pools, sometimes in seasonally wet flushes.

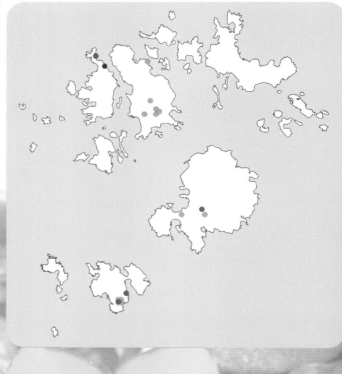

PRIMULACEAE – Primrose family
Anagallis arvensis subsp. *arvensis* L.
Scarlet Pimpernel
Inhabited islands, Samson, St Helen's, Teän, White Island (St Martin's), Nornour, Little Arthur and Great Ganilly

PRIMULACEAE – Primrose family
Centunculus minimus L.
Chaffweed
Inhabited islands

Native. A very rare plant on Scilly. Recorded from near Big Pool, St Agnes (SV878086) in 1996 and from Chapel Down, St Martin's (SV927158) in 1989, (SV942156) in 1990 and (SV94091576) in 2009. Found on Helvear Down, St Mary's (SV921126) in 2006. In 2014 a thriving group of plants was found growing along a seepage line on Shipman Head, Bryher (SV87731562). Former sites include near Lower Town, St Martin's (SV917160 - SV918163); Middle Down and near Abbey Pool, Tresco (SV896141). Typically grows in damp, sandy places such as cart ruts and tracks on heathland.

Native. A widespread and common plant found on all the larger islands. The colour forms *azurea* (deep blue) and *carnea* (flesh pink) of var. *arvensis* may be found occasionally. Grows on waste and cultivated ground, cliffs and dunes.

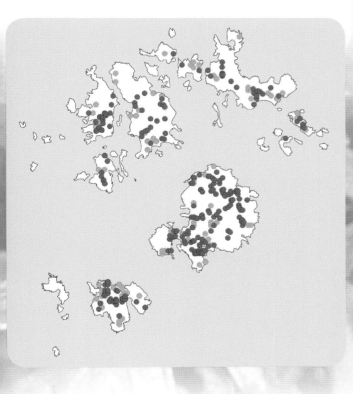

EU-DICOTS True Dicotyledons

PRIMULACEAE – Primrose family
Samolus valerandi L.
Brookweed
Tresco, St Mary's, St Agnes, Annet and Menawethan

Native. Nowhere common and easily overlooked. In 2009 it appeared in quantity around the drawdown area on Porthellick Pool when the mud was exposed. Grows around pools and wetlands, but also as on Annet and Menawethan in freshwater seepages near the shore.

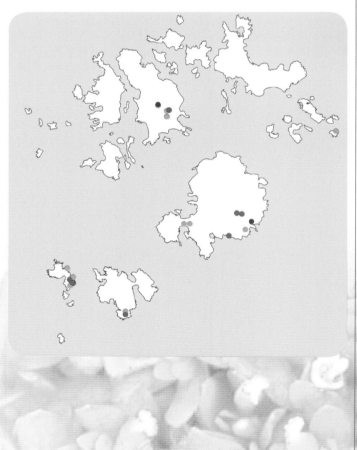

CLETHRACEAE - Lily-of-the-valley-tree family
Clethra arborea Aiton
Lily-of-the-valley-tree
Tresco

Neophyte. Garden escape on Tresco; Abbey Hill area, doubtfully established.

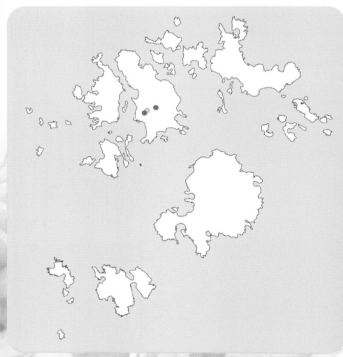

ERICACEAE – Heather family
Rhododendron ponticum L.
Rhododendron
Tresco, St Mary's and Bryher

Neophyte. There are large plantations of rhododendrons on Tresco; around Abbey Wood and Monument Hill and on the southern edge of Castle Down. In places seedlings are spreading into heathland and dunes. Recently Tresco Estate have been removing some plants due to concerns over 'Sudden Oak death'. A group of shrubs on Bryher on the side of Samson Hill may have been removed. Occasional escapes from cultivation have been recorded on St Mary's.

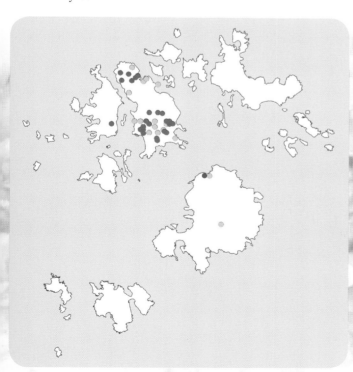

ERICACEAE – Heather family
Calluna vulgaris (L.) Hull
Heather (or Ling)
Inhabited islands, St Helen's, Samson, Teän, Nornour, Great Ganilly and the Arthur's

Native. The distribution of *Calluna* is almost identical to that of *Erica cinerea*, usually they are found together, although *Calluna* is often the dominant species. On just a few sites one or the other species occurs in isolation. *Calluna* usually flowers slightly later than *Erica cinerea* but when they are both in flower at the same time the heathland can be spectacular. Formerly heather 'turf' was cut and used for fires, thatching, even building walls. Heathlands, coasts and dune heath.

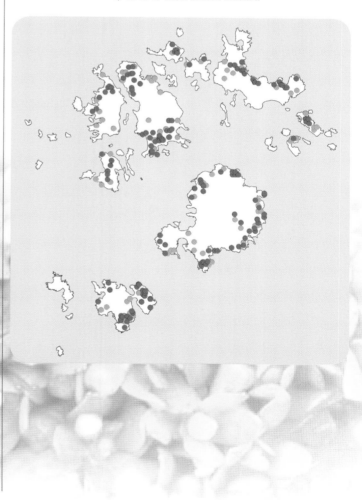

EU-DICOTS True Dicotyledons

ERICACEAE – Heather family
Erica cinerea L.
Bell Heather

Inhabited islands, St Helen's, Samson, Teän, Northwethel, Nornour, Great and Little Ganilly and the Arthur's

Native. *Erica cinerea* is co-dominant with *Calluna vulgaris* on many heathlands in Scilly. It occurs in a few places including several small islands, for example Little Ganilly and Northwethel where *Calluna* has not been recorded. Heathland, dunes and coastal sites.

ERICACEAE – Heather family
Erica arborea L.
Tree Heather

Tresco

Neophyte. Garden escape, established in woodland area outside Tresco Abbey Gardens, south of Old Grimsby and on Appletree Banks.

ERICACEAE – Heather family
Erica erigena R. Ross
Irish Heath
Tresco

Native in western Ireland, but a neophyte in Scilly. Recorded in Abbey Wood in 1995, probably planted there.

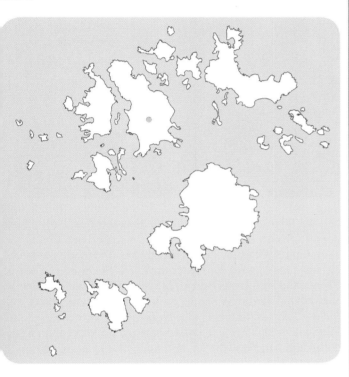

Erica vagans L.
Cornish Heath
Tresco

Native in the British Isles, with main populations in Cornwall, but not native to Scilly. Garden escape recorded on Appletree Banks in 1998.

ERICACEAE – Heather family
Gaultheria shallon Pursh.
Shallon
Tresco

Neophyte. Garden escape or planted Abbey Wood area. Possibly established Abbey Hill.

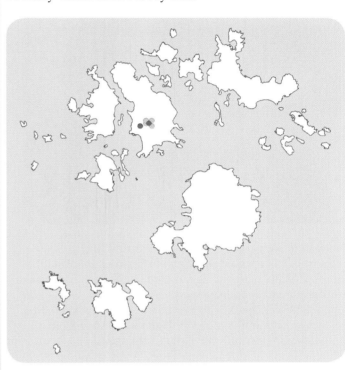

Gaultheria procumbens L.
Checkerberry
Tresco

Neophyte. Garden escape or planted near the Monument (SV892142).

EU-DICOTS True Dicotyledons

RUBIACEAE – Bedstraw family
Coprosma repens A. Rich
Tree Bedstraw
Bryher, Tresco, St Mary's and St Agnes, also Samson and Northwethel

Neophyte. Appears to be absent from St Martin's, although it is planted widely on the other inhabited islands. Mainly grown as a hedge or shelterbelt because its shiny evergreen leaves are unaffected by salt winds. The sticky yellow berries are relished by birds; leading to *Coprosma* being frequently birdsown away from cultivation, including on some of the uninhabited islands such as Samson and Northwethel. It did grow on Annet in the 1960s, but has not been seen there recently. Grows almost anywhere where the seeds can germinate; near buildings, on rock outcrops, heaths and on the coast.

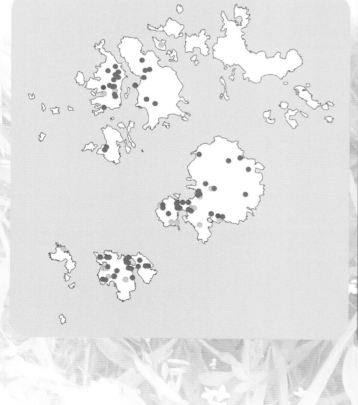

RUBIACEAE – Bedstraw family
Sherardia arvensis L.
Field Madder
Inhabited islands

Native. Scattered sites on the inhabited islands. Nowhere very common although it is found regularly in the same fields. On Bryher it has only been recorded from Hillside Farm, and on Tresco from fields on both sides of Great Pool and from Borough Farm. The suite of fields above The Bar on St Agnes have been a constant site for the species, and it is also found in fields at Lower Town. On St Mary's it is widespread on several farms including Lunnon, Rocky Hill and other places. Mainly a plant of cultivated fields, but also occasionally found in grassland and sandy places near the coast.

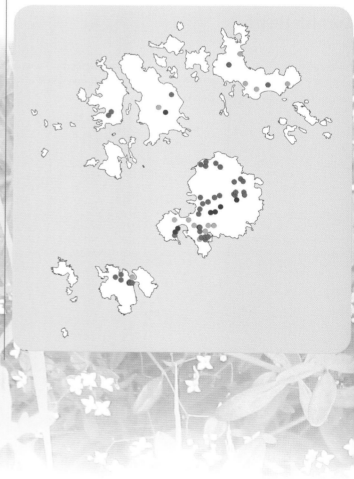

RUBIACEAE – Bedstraw family
Galium palustre L.
Marsh Bedstraw
Tresco, Bryher and St Mary's

RUBIACEAE – Bedstraw family
Galium verum L.
Lady's Bedstraw
Bryher, Tresco, St Mary's and St Martin's, also Gugh, Samson and Teän

Native. Found in wetlands around Tresco Great Pool and in both Lower and Higher Moors, St Mary's. The record from Bryher was on Samson Hill (SV877144) in 1991. Both ssp. *palustre* and ssp. *elongatum* have been recorded on Lower Moors. Ditches and marshy places.

Native. Locally frequent. Apparently absent from St Agnes although found on Gugh. Lousley recorded the species from the Eastern Isles but does not specify which island. Typically grows in dune grassland and heathy grassland.

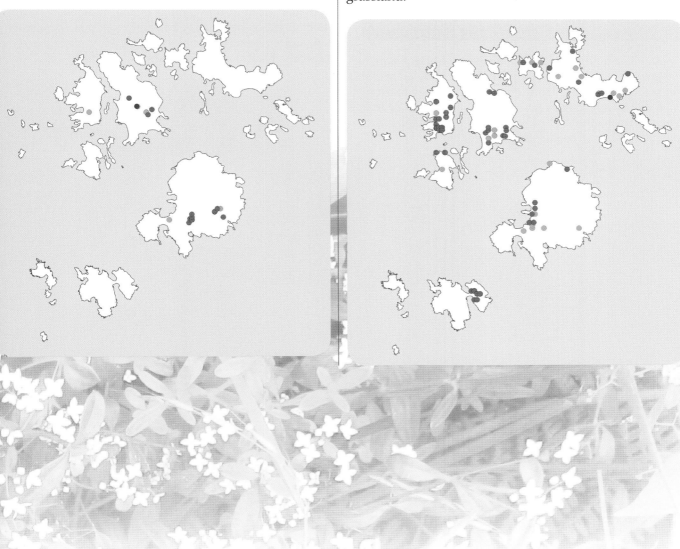

EU-DICOTS True Dicotyledons

RUBIACEAE – Bedstraw family
Galium album Mill.
Hedge Bedstraw
St Mary's, Tresco and St Martin's

Native. In a field on St Martin's 1994 and hedgebank on St Mary's 1988, possibly accidentally introduced and did not persist. Recorded from Tresco near Abbey Farm in 2014 and in a garden at Hilltop in 2015 are also probably recent introductions. A record from Higher Moors in May 2012 has not been confirmed.

RUBIACEAE – Bedstraw family
Galium saxatile L.
Heath Bedstraw
Bryher, Tresco, St Martin's and St Mary's, also Gugh and Samson

Native. Far less common on Scilly than might be expected. Although a typical heathland species, on Scilly it only occurs in scattered sites and is apparently absent from some heathland areas. The only records from Gugh are from near the Bar, last in 1995. On Bryher it was recorded from Heathy Hill in 1991 and Samson Hill in 1995. It is possible it may be found again at these sites since more management has been taking place there recently. Heathland and dunes.

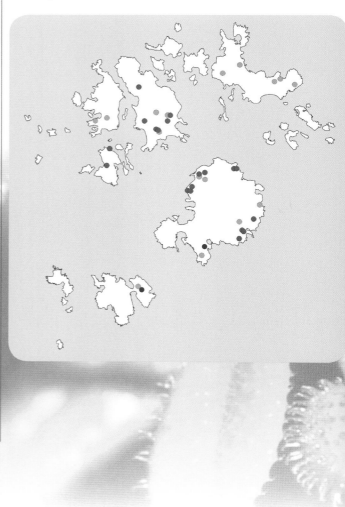

RUBIACEAE – Bedstraw family
Galium aparine L.
Cleavers
Inhabited islands and many uninhabited islands

Native. Locally abundant. It is easily spread by birds and people, which may be why it is so widespread, even being found on tiny islets such as Hedge Rock (off St Martin's), Puffin Island (off Samson), Foreman's Island and nearby Northwethel. In the Eastern Isles it has been recorded from Great and Little Ganinick, Little Arthur, Little Ganilly and Nornour. There are also records from Teän and Annet (2012). It is only absent from the Western Rocks and Norrards. Cleavers grows in hedgebanks, cultivated fields and on the coast.

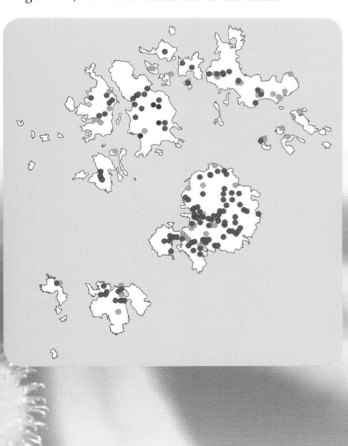

RUBIACEAE – Bedstraw family
Rubia peregrina L.
Wild Madder
Inhabited islands, also Teän, Samson, St Helen's, Little Arthur, Toll's Island, and Great Ganilly

Native. Absent from the Western Rocks and Norrards, it has also not been found on Annet. Although typically growing on the coast among brambles and taller vegetation, in Scilly it occurs at some inland sites such as on the 'Moors' on St Mary's and along hedgebanks beside roads. Scrambling in hedges, among brambles and tall vegetation on cliffs, dunes and field edges.

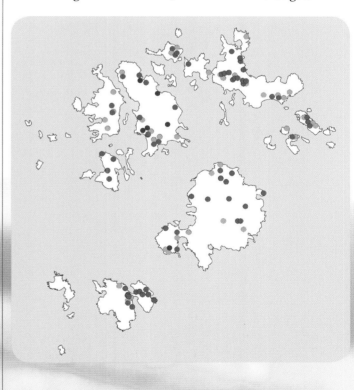

EU-DICOTS True Dicotyledons

GENTIANACEAE – Gentian family
Centaurium erythraea Rafn
Common Centaury
Inhabited islands, also Samson, St Helen's, Teän, Nornour, Great Ganilly and Middle Arthur

Native. A very variable and widespread species; on dunes or exposed sites plants can be tiny, stemless and stunted. White-flowered plants also occur occasionally. Dune grassland, cliffs, wall tops and heathlands.

APOCYNACEAE – Periwinkle family
Vinca minor L.
Lesser Periwinkle
St Mary's and Tresco

Archaeophyte. Garden escape near Longstone, St Mary's and on Appletree Banks, Tresco.

APOCYNACEAE – Periwinkle family
Vinca major L.
Greater Periwinkle
Inhabited islands

Neophyte. Well-established garden escape on the inhabited islands. The var. *oxyloba* with much narrower petals has been recorded on The Garrison, St Mary's. Shady places on roadsides, dunes, hedgebanks and quarries.

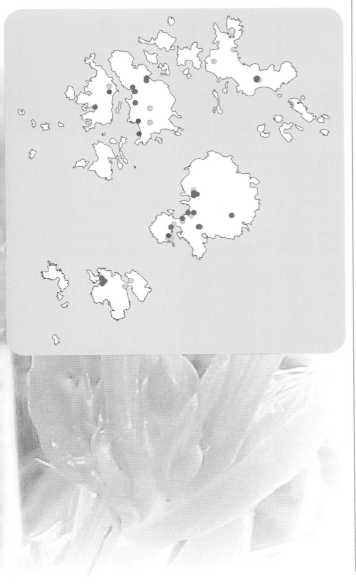

BORAGINACEAE - Borage family
Echium vulgare L
Viper's-bugloss
Bryher, Tresco, St Agnes and Gugh

Native. A rare plant in Scilly with just a few records; Bryher in 1975 in the south-west of the island (SV8714), on Tresco a plant was found on Appletree Banks (SV889142) in 2009. The only consistent population has been known from Gugh since at least 1933. Plants were still there in approximately the same area in 2009 (SV888084 to SV889083), this includes the field below the house, and the 'sand-pit' nearby. Latest records are just one or two plants by the 'sand-pit'. It was recorded from the margin of a bulbfield just above the Bar on St Agnes (SV885082) in 1995 and 2000. Sandy places.

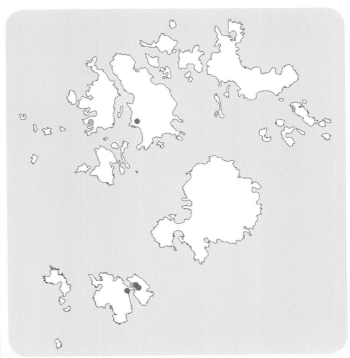

EU-DICOTS True Dicotyledons

BORAGINACEAE - Borage family
Echium plantagineum L.
Purple Viper's-bugloss
Tresco, St Martin's and St Mary's

Archaeophyte. Formerly recorded as an arable weed on Tresco, St Martin's and St Mary's, but had apparently been lost as a wild plant by the end of the 1990s (garden escapes occurred occasionally as with plants found in Church Road, St Mary's in 2000). However plants found on Tresco (SV88941521, SV89731450 and SV89751449) in 2007 and 2010 by Mr and Mrs Schofield appeared to be genuine. In 2011 plants were found in fields north of Great Pool where they were associated with a field planted with Quinoa *Chenopodium quinoa* and it is now believed to have been accidentally reintroduced as a seed contaminent. Arable fields & waste places.

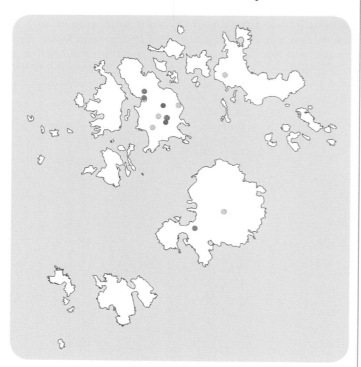

BORAGINACEAE - Borage family
Echium pininana Webb & Berthel
Giant Viper's-bugloss
Inhabited islands

An occasional garden escape. This and other garden *Echium* species (including *E.* x *scilloniensis*) occasionally occur as self-seeded plants on the inhabited islands, but tend not to persist. It is frequently impossible to determine which plants have been planted and which have derived from blown seeds (these can be many metres from the parent plant). Disturbed ground.

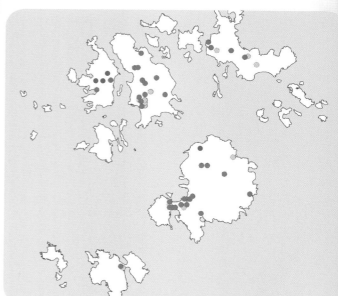

BORAGINACEAE - Borage family
Symphytum x uplandicum Nyman
Russian Comfrey
St Mary's and Tresco

Neophyte. Occasional garden escape. On St Mary's there are still plants near Watermill where they were recorded by Lousley and near Telegraph (SV91351199). Most Tresco records are from near the Abbey Gardens, or near Great Pool (places mentioned by Lousley). There is a 1975 record from the Blockhouse area. Hedgebanks and waste ground.

BORAGINACEAE - Borage family
Anchusa arvensis (L.) M. Bieb.
Bugloss
Inhabited islands

Neophyte. Locally frequent on Tresco and St Mary's with scattered records elsewhere. There is a reference to the plant growing on the 'neck' of Samson in 1980s, probably when the area was more open and sandy – it has since become much overgrown (Coulcher, 1999). A plant of cultivated fields, disturbed ground and sandy places near the coast.

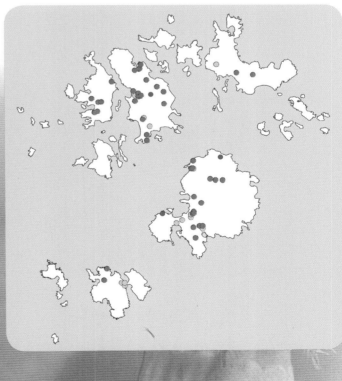

Symphytum officinale L.
Common Comfrey

Recorded in error but probably was *Symphytum* x *uplandicum*.

EU-DICOTS True Dicotyledons

BORAGINACEAE - Borage family
Anchusa azurea Mill.
Garden Anchusa
St Mary's

Neophyte. Occasional garden escape; Old Town 1966 and Porth Minick 1967.

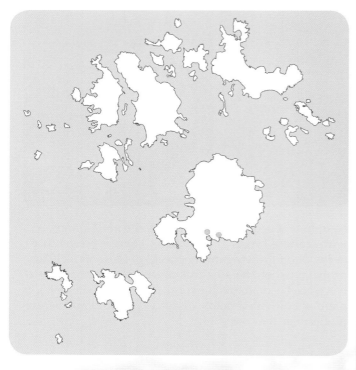

BORAGINACEAE - Borage family
Pentaglottis sempervirens (L.) Tausch ex L.H. Bailey
Green Alkanet
Tresco

Neophyte. Established on Appletree Banks since 1939, where it probably originated from a garden rubbish dump, still there in 1998. Found near the Blockhouse Cottages in 2009. Dune area and rubbish dump.

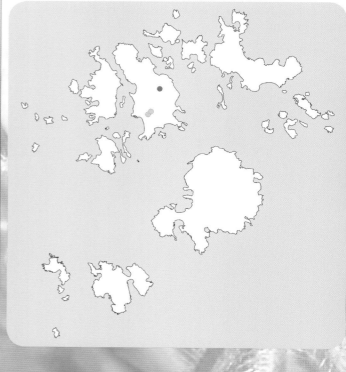

BORAGINACEAE - Borage family
Borago officinalis L.
Borage
Inhabited islands

Neophyte. A garden escape now established on all the inhabited islands, most frequently on St Mary's. Unusually a plant was found on the small, uninhabited Ragged Island, Eastern Isles in 1997 where it was presumed to have been bird-sown. Roadside verges, rubbish dumps and places near the coast where garden waste is thrown.

BORAGINACEAE - Borage family
Borago pygmaea (DC) Chater & Greuter
Slender Borage
St Mary's

Neophyte. Presumed garden escape recorded from The Garrison, St Mary's (SV901101) by Alan Underhill in 1992.

EU-DICOTS True Dicotyledons

BORAGINACEAE - Borage family
Myosotis scorpioides L.
Water Forget-me-not
Tresco

Native. Although the species has been known from Tresco since 1893, there are very few confirmed records. It was recorded from Great Pool in 1939 and occasionally there since, most recently in 2007. In 1995 it was found by Abbey Pool. Sometimes plants recorded as this species have been found to have been one of the other wetland forget-me-nots. Pool edges.

BORAGINACEAE - Borage family
Myosotis secunda Al. Murray
Creeping Forget-me-not
Tresco and St Mary's

Native. The range of this species has contracted since Lousley's account. It has only been recorded recently from near Tresco, Great Pool; St Mary's from Higher and Lower Moors and in 2014 from ponds in a field near Maypole (SV921114). Last recorded from a former site at Watermill Cove in 1984. Wetland areas around pools and in reed beds.

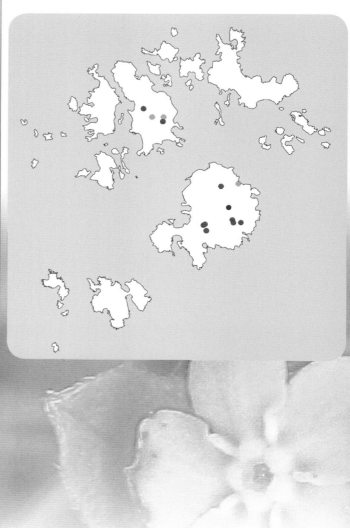

BORAGINACEAE - Borage family
Myosotis laxa Lehm
Tufted Forget-me-not
St Mary's

BORAGINACEAE - Borage family
Myosotis arvensis (L.) Hill
Field Forget-me-not
inhabited islands and Teän

Archaeophyte. Now only found on the inhabited islands. There is a record from Teän (SV911165) in 1988. Lousley had recorded it from the 'neck' on Samson (an area that has since become very overgrown). It seems it is no longer as common as in Lousley's day. Cultivated fields and waste areas.

Native. This appears to be a new species for Scilly. Our plant is subsp. *caespitosa* (Shultz) Hyl. Ex Nordh. Several plants were found in excavations beside Shooter's Pool in 2007, possibly introduced by waterfowl. Wetland.

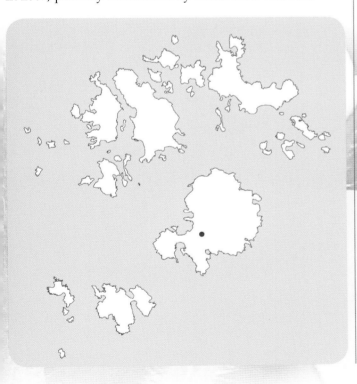

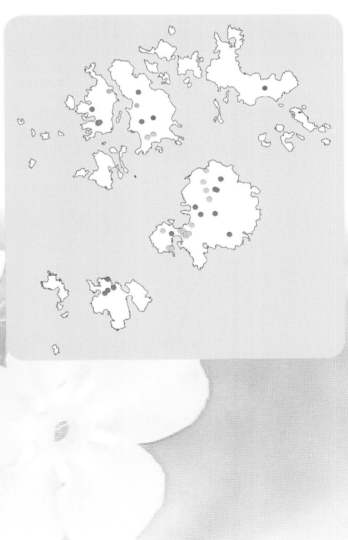

EU-DICOTS True Dicotyledons

BORAGINACEAE - Borage family
Myosotis ramosissima Rochel

Early Forget-me-not

Inhabited islands, also Gugh, Great Ganilly, Samson and Teän

Native. Found in small discrete populations, mainly near the coast. Now seems to also be absent, or not recorded recently, from Annet and St Helen's (Lousley, 1971). This could be partly due to the very few visits that are made to most uninhabited islands early in the year when the plant flowers. Dune grassland, pathways and sandy places, occasionally also in sandy fields or wall tops.

BORAGINACEAE - Borage family
Myosotis discolor Pers.

Changing Forget-me-not

Inhabited islands, also St Helen's, Annet, White Island (off St Martin's), Teän, Samson, and Great Ganilly

Native. Widespread, locally frequent in some places. It can be variable in size with tall plants in bulbfields and very tiny ones in places with shallow turf, which may be confused with *Myosotis ramosissima*. Sometimes plants are very pale green in colour with white flowers; Lousley recorded these from Appletree Banks (and Samson Hill, Bryher); similar pale plants were found in several places on Appletree Banks in 2012 and 2013. Cutivated fields, heaths and cliffs.

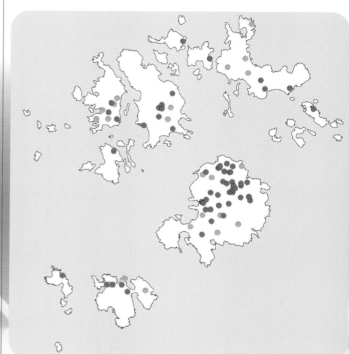

BORAGINACEAE - Borage family
Phacelia tanacetifolia Benth.
Phacelia
St Martin's and St Mary's

Neophyte. Planted as a crop usually in 'conservation' seed mixes. Also found as a garden escape or left from plantings. The fields on St Martin's attracted a lot of interest from visitors.

CONVOLVULACEAE – Bindweed family
Convolvulus arvensis L.
Field Bindweed
Inhabited islands

Native. Distribution patchy, although it appears to be frequent in some places. It seems to be missing from other apparently suitable areas. Cultivated fields, wasteland and roadsides.

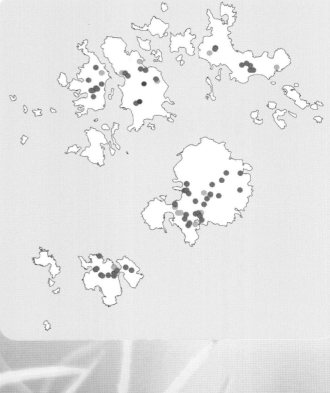

EU-DICOTS True Dicotyledons

CONVOLVULACEAE – Bindweed family
Calystegia soldanella (L.) R. Br.
Sea Bindweed
Inhabited islands, also Samson, Northwethel, Teän and Great Ganilly

Native. Almost exclusively coastal, although can extend somewhat inland on very sandy ground. Dunes and above HWM on sandy beaches.

CONVOLVULACEAE – Bindweed family
Calystegia sepium (L.) R. Br.
Hedge Bindweed
Inhabited islands and Teän

Native. Locally frequent on the inhabited islands. Ssp. *sepium* was found on Teän in 2009. Two subspecies occur; white-flowered plants have been recorded as ssp. *sepium* and pink-and-white-striped flowered plants as ssp. *roseata*. In a few places on St Mary's, for example near the pond on Porthloo Road are some pink-flowered bindweeds with inflated, overlapping bracteoles that appear to be *C.* x *lucana*, the hybrid between *C. sepium* ssp. *roseata* and *C. silvatica*. On St Mary's *C. sepium* ssp. *roseata* is common. It is also found on St Agnes (it is apparently uncommon there), Tresco, Bryher and St Martin's. A scrambler through hedges, in reedbeds, quarries and copses.

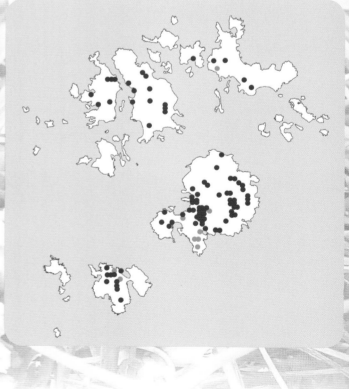

CONVOLVULACEAE – Bindweed family
Calystegia silvatica (Kit.) Griseb.
Large Bindweed
Inhabited islands

Neophyte. Occurs in similar places to *C. sepium* but is much less common. Subspecies *silvatica* has been identified on St Mary's. Some plants have pink stripes on the outside of the flower (see reference to *C.* x *lucana* under *C. sepium*). In Lousley's time it was only known from a few places on St Mary's, since when it has increased in frequency and distribution on all the inhabited islands. Hedgebanks, rubbish dumps and scrambling up trees and taller vegetation.

CONVOLVULACEAE – Bindweed family
Cuscuta epithymum (L.) L.
Dodder
St Martin's

Native. Now only known from near Top Rock, St Martin's (SV92221649). Just one or two plants in a large gorse thicket. Parasitic on *Ulex europaeus*.

EU-DICOTS True Dicotyledons

SOLANACEAE – Nightshade family
Hyoscyamus niger L.
Henbane
Tresco

Archaeophyte. A rare plant recorded most recently from Appletree Banks (SV892141) in 2007 and before that at Carn Near in 1971. There are previous records in Lousley's Flora from St Mary's, St Agnes and Teän. Found on dunes, shores and cultivated ground.

SOLANACEAE – Nightshade family
Nicandra physalodes (L.) Gaertn.
Apple-of-Peru
St Mary's, Bryher and Tresco

Neophyte. Casual. Naturalised on scattered sites on St Mary's and Tresco, where it sometimes persists. From a field on Bryher in 2014. Found on both cultivated and wasteland including beaches.

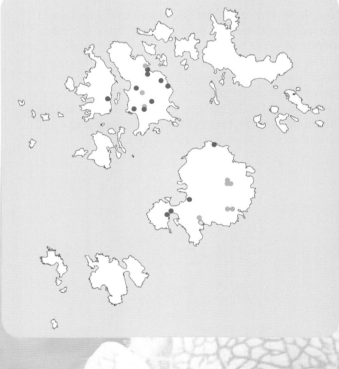

SOLANACEAE – Nightshade family
Datura stramonium L.
Thorn-apple
Tresco, St Mary's, also Teän

Neophyte, occasional sporadic appearances. This plant is very distinctive and this means that it is often reported. But because it is known to be poisonous the plants are frequently destroyed. There are recent records from St Mary's between 2002 and 2005 from Salakee Lane, Little Porth beach, a field at Old Town and on the allotments. On Tresco it has occurred at several places around the Abbey Gardens and nearby paths, also by the Heliport, Crow Point and Borough Farm. On Teän, where there are records of the plant growing on beaches they were last seen there in 1990. Cultivated fields, waste ground and sandy beaches.

SOLANACEAE – Nightshade family
Physalis peruviana L.
Cape-gooseberry
St Mary's

Neophyte. It has turned up a few times from 1996 around Porthcressa beach, St Mary's, including Little Porth (SV902104) in 2005 and 2008. Beaches above HWM.

EU-DICOTS True Dicotyledons

SOLANACEAE – Nightshade family
Solanum nigrum L.
Black Nightshade
Inhabited islands, also Samson

Native. Frequent on the inhabited islands and also found near the abandoned dwellings on Samson. The plants on Samson appeared after vegetation was cleared from around the ruined buildings and could have been associated with the fields and gardens of the former inhabitants. This is the plant, reported as 'deadly nightshade' that was apparently blamed for the death of a child on Tresco in 1966. Occurs on cultivated land, beaches and waste ground.

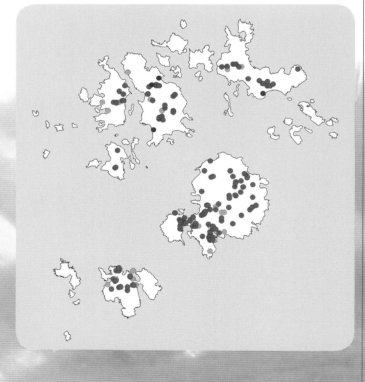

SOLANACEAE – Nightshade family
Solanum physalifolium Rusby
Green Nightshade
St Mary's, Tresco and St Martin's

Neophyte. Casual. A few scattered records from St Mary's, Tresco and St Martin's. Plants have been recorded on St Mary's; Old Town (SV911100), Parting Carn (SV915108), Normandy (SV92781115 and SV92601137) and near Bar Point (SV91151271). The plants were common in the fields at Normandy until 2009. The site on Tresco was at Borough Farm (SV89911509) and those on St Martin's at Middle Town (SV919162) and (SV923153). It had not been recorded in Scilly until 1989 (it is not in Lousley's Flora), but the plant is very similar to, and often grows with *S. nigrum* so may have been overlooked. According to Stace (2010) our plant is var. *nitidibaccatum*. A weed of cultivated fields and waste land.

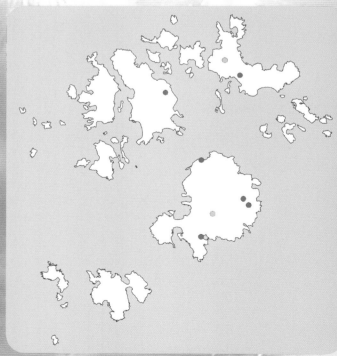

SOLANACEAE – Nightshade family
Solanum sarachoides Sendtn.
Leafy-fruited Nightshade
St Martin's

Neophyte. Casual. Formerly known from St Martin's where at one time it was abundant in bulbfields between Middle and Lower Town (Lousley, 1971). Last recorded near Lower Town (SV917160) in 1978. Weed in arable fields.

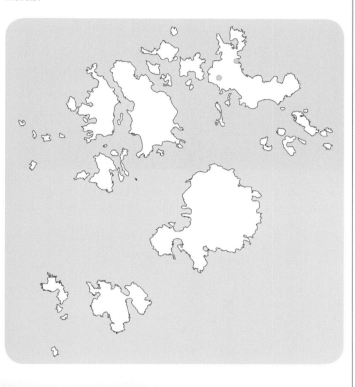

SOLANACEAE – Nightshade family
Solanum dulcamara L.
Bittersweet
Inhabited islands, also most uninhabited islands except those in the Norrads and Western Rocks

Native. Widespread and common. The fleshy, procumbent variety *marinum* occurs commonly on shingle and rocky beaches. Grows on cultivated land, waste ground and shores.

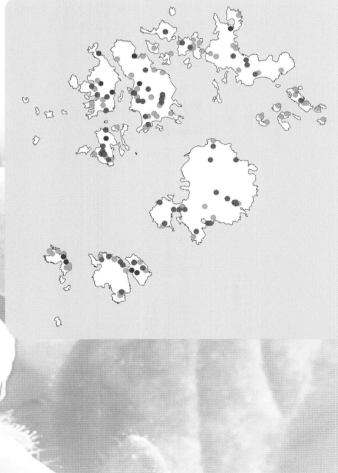

EU-DICOTS True Dicotyledons

SOLANACEAE – Nightshade family
Solanum tuberosum L.
Potato
St Mary's, Bryher, Tresco and St Martin's

Neophyte. An escape from cultivation that is occasionally found on rubbish dumps, beaches and waste land.

SOLANACEAE – Nightshade family
Solanum lycopersicum L.
Tomato
St Mary's, St Agnes, Tresco, also Samson

Neophyte. Garden escape or casual often found on beaches, especially near sewage outfalls and waste land.

SOLANACEAE – Nightshade family
Solanum laciniatum Aiton
Kangaroo-apple
St Mary's and Tresco

Neophyte. Garden escape on St Mary's and Tresco. Apparently spreading, Lousley only recorded the plant from Tresco. It produces large yellow fruits and may possibly be bird-sown. On Tresco it is established from right around the Abbey Gardens to Appletree Banks and also near Tommy's Hill (SV89701526). On St Mary's it is found in the area to the north of Trenoweth and around Jac-A-Bar, also in Hugh Town and on the track from Old Town to Peninnis. Sometimes pulled up by people concerned about its toxicity. Beaches, field edges, in shelterbelts and on waste land.

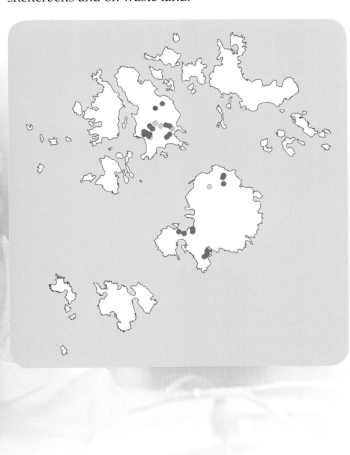

SOLANACEAE – Nightshade family
Nicotiana alata Link & Otto
Sweet Tobacco
Tresco

Neophyte. Garden escape on Tresco SV 89211409 in 2004.

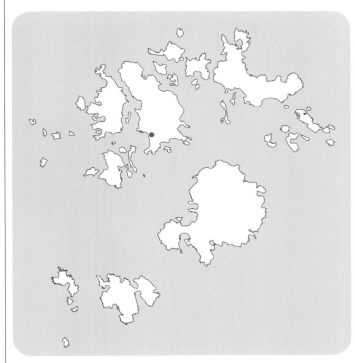

Nicotiana x *sanderae* W. Watson
(*Nicotiana alata* x *Nicotiana forgetiana*)
St Mary's and Tresco

Neophyte. Garden escape Tresco and St Mary's, sometimes in quarries or rubbish dumps.

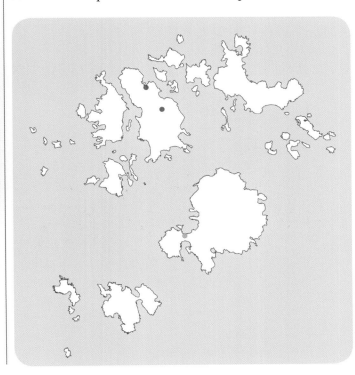

EU-DICOTS True Dicotyledons

SOLANACEAE – Nightshade family
Nicotiana forgetiana Hemsl.
Red Tobacco
St Mary's

Neophyte. Garden escape St Martin's and St Mary. Not recorded since 1997.

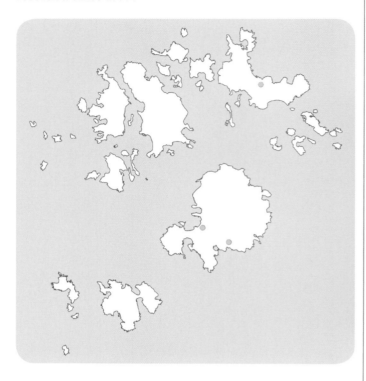

OLEACEAE – Ash family
Jasminum officinale L.
Summer Jasmine
St Mary's

Neophyte. In Hugh Town, occasional garden escape or planted.

Fraxinus excelsior L.
Ash
St Mary's and Tresco

Although native in the British Isles, this tree was introduced to Scilly. Ash was planted in Holy Vale in c.1650 (Turner [1695] in *The Scillonian*, 1964), but does not seem to have become established long term. There is a mature tree near Newford Farm, St Mary's.

OLEACEAE – Ash family
Ligustrum vulgare L.
Wild Privet
Tresco, St Martin's, Bryher and St Mary's, also Teän and St Helen's, Samson, Ganilly and Little Arthur

Native. Locally frequent in the northern part of the islands, but missing from St Agnes. Lousley records it from Samson – before 1971, but it has not been relocated. The plants in Scilly are low-growing, forming semi-prostrate shrub banks, the var. *coombei* P.D. Sell (Sell and Murrell, 2006). Dunes and coastal slopes.

OLEACEAE – Ash family
Ligustrum ovalifolium Hassk.
Garden Privet
St Mary's (including Toll's Island) and St Martin's

Neophyte. Planted shrub used as hedging but apparently also occasionally self-seeding.

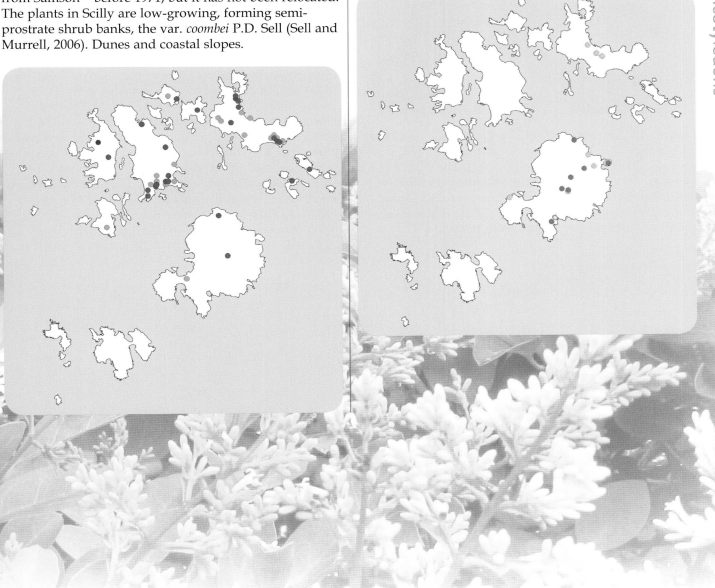

EU-DICOTS True Dicotyledons

VERONICACEAE – Speedwell family
Digitalis purpurea L.

Foxglove

Inhabited and most uninhabited islands other than the Western and Norrad Rocks

Native. An abundant and widespread species on almost all islands where there is any vegetation. This is one of the most charactistic flowering plants in the islands in summer. Foxglove is one of the first plants to colonise disturbed ground on heathlands after burning. Found along roadsides, on walls, in hedgebanks and among rough vegetation on cliff slopes and heaths.

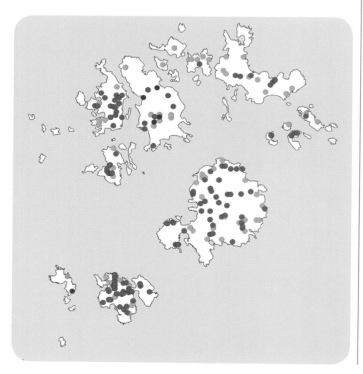

VERONICACEAE – Speedwell family
Veronica officinalis L.

Heath Speedwell

Tresco, St Mary's, (including Toll's Island) also Gugh and Samson

Native. Except on the southern half of Tresco this is an uncommon plant in Scilly. It was recorded on Gugh (SV890085) in 1995 and on the west edge of the Golf Course near Carn Morval, St Mary's in 2002 and near by in 2014, in an abandoned bulbfield (SV91051224) in 2009 and on the Garrison (where there had been recent grazing) in 2011. On Tresco it is locally common on grass paths in woodland on Monument Hill, by the side of Abbey Road and Appletree Banks. The only record from St Martin's (SV924161) was made by Owen Mountford in 2011. In 2016 David Mawer found many plants in the dunes on Samson (SV878133) where they had previously been overlooked. Grassy areas on heaths and in woodland.

VERONICACEAE – Speedwell family
Veronica montana L.
Wood Speedwell
Tresco

Native. Only known from Abbey Wood. It is apparently restricted to the small area in the woodland and around the Abbey gardens (SV893144-SV894145). The most recent records were between 2011 and 2015. Lousley considered this to be a relict woodland species which appears to fit with the distribution of other woodland plants in the same area. Woodland.

VERONICACEAE – Speedwell family
Veronica beccabunga L.
Brooklime
St Mary's

Native in the British Isles but introduced to Scilly. Found at Longstone Heritage Centre in 2005, introduced with pond plants.

EU-DICOTS True Dicotyledons

VERONICACEAE – Speedwell family
Veronica serpyllifolia ssp. *serpyllifolia* L.
Thyme-leaved Speedwell
Inhabited islands, also Samson

Native. Lousley includes this species in his Flora as 'rare', he had not seen the plant himself. D.E. Allen visited Tresco in 1975 and recorded the species in several places. Since then there have been records from all the inhabited islands. It is most frequent on Tresco and St Mary's. The only records from St Martin's are from Burnt Hill (SV936159) in 1998 and from two areas near English Island Point (SV93901544 and SV938153) in 2003 and 2006. On Bryher there are records from Green Bay (SV880149) in 1999 and Rushy Bay (SV87651415) in 2003. On St Agnes the species was found in two areas on St Agnes, the sloping path from the behind the Coastguard Cottages in 1988, since overgrown, and the path to Covean (SV884083) in 2002 and a nearby bulbfield (SV885082) in 2007. The only record from Samson was from near the ruins on South Hill in 2002. Field boundaries, paths and damp grassland.

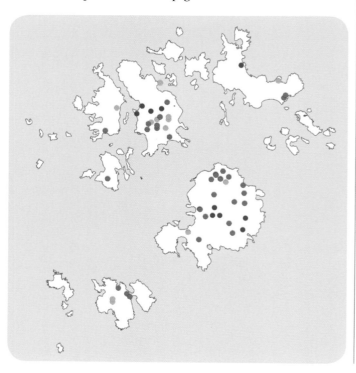

VERONICACEAE – Speedwell family
Veronica hederifolia L.
Ivy-leaved Speedwell
Inhabited islands

Archaeophyte. There are only a few scattered records of this species from the inhabited islands. The following are recent records:

Bryher - Hillside Farm (SV878148)

Tresco - Abbey Gardens 1995, Pool Road 1997, Gimbel Porth (SV891160) 2009

St Martin's - Middle Town 1997, and Lower Town (SV91591617) in 2007

St Agnes - bulbfield above Bar (SV885082)

St Mary's - fields near Carn Leh (SV913099) in 2002, Carrick Dhu (SV91751105) 2003, several records from SV9010 and Old Town (SV915105) in 2009

The subspecies *lucorum* was found in the Abbey Gardens by D.E. Allen in 1975 (as *V. sublobata*). Found in bulbfields and cultivated ground.

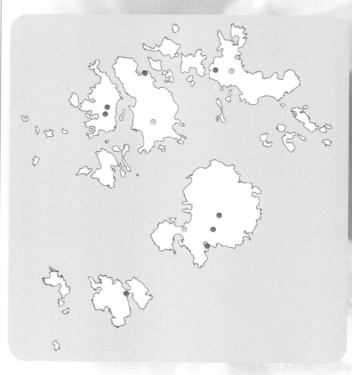

VERONICACEAE – Speedwell family
Veronica agrestis L.
Green field-Speedwell
Inhabited islands and Samson

Archaeophyte. Similar distribution and places to the previous species. The plant found on Samson in 2008 may have been associated with former cultivations or is a recent accidental introduction. Cultivated fields and disturbed, especially sandy, ground.

VERONICACEAE – Speedwell family
Veronica polita Fr.
Grey Field-speedwell
St Mary's, St Agnes and Tresco

Neophyte. A rare plant; the only recent records are from fields at Old Town (SV91140996), St Mary's in 2005 and Trenoweth (SV91891250) in 2011 and from Lower Town Farm, St Agnes in 2003. It had formerly been recorded as a weed in Tresco Abbey gardens (1956). Cultivated ground.

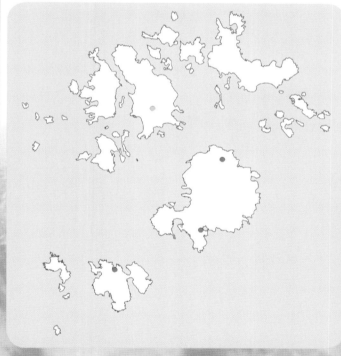

EU-DICOTS True Dicotyledons

VERONICACEAE – Speedwell family
Veronica persica Poir.

Common Field-speedwell

Inhabited islands

Neophyte. Locally frequent on St Mary's and Tresco, with just scattered records from St Agnes, Bryher and St Martin's. Bulb and other cultivated fields and waste places.

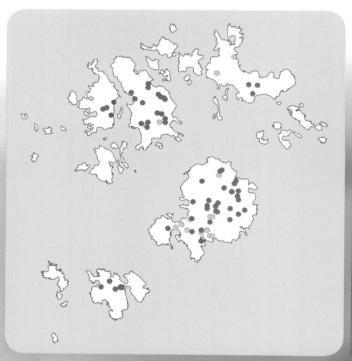

VERONICACEAE – Speedwell family
Veronica chamaedrys L.

Germander Speedwell

Inhabited islands, also Samson, St Helen's and Great Ganilly

Native. Although not found there since Lousley's Flora recorded on Teän, it is widespread and locally frequent on most of the larger islands. Grows along roadsides, field boundaries and on coastal sites.

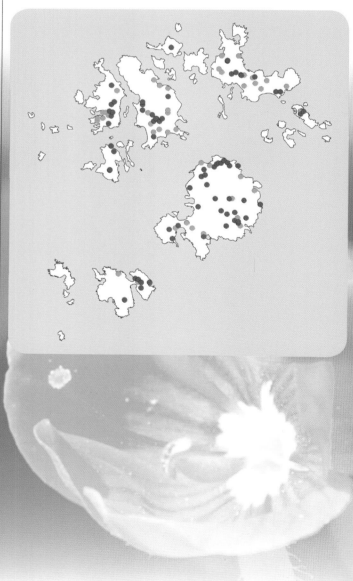

VERONICACEAE – Speedwell family
Veronica arvensis L.
Wall Speedwell
Inhabited islands, also Teän

Native. Only frequent on St Mary's otherwise from scattered localities on the other islands. Although Lousley had recorded it from Samson and the 'Eastern Isles' without details, there are no recent records from there. Cultivated and disturbed ground, wall tops and dunes.

VERONICACEAE – Speedwell family
Veronica x *lewisii* J.B. Armstr.
(*V. salicifolia* x *V. elliptica*)
Lewis's Hebe
Inhabited islands

The identification of this species was queried by Green (1973) when he suggested the name *lewisii* was a misidentification and that all the plants naturalised in Britain identified as this were *V.* x *franciscana*. Some other hedge veronicas (formerly *Hebe*) are grown in Scilly, but none has been confirmed as truly naturalised. Other species recorded include *V. salicifolia*, *V. brachysiphon* and *V. dieffenbachii*.

EU-DICOTS True Dicotyledons

VERONICACEAE – Speedwell family
Veronica x *franciscana* Eastw.
(*V. elliptica* x *V. speciosa*)
Hedge Veronica
Inhabited islands

Neophyte. Planted as hedging on the inhabited islands (see note above). The variety 'Blue Gem' is the one most often used as bulbfield hedges. As the plant freely self-seeds there are occasional escapes from cultivation.

VERONICACEAE – Speedwell family
Sibthorpia europaea L.
Cornish Moneywort
St Mary's

Native. Now only found on one place on St Mary's, on the wall side of a ditch between Newford Ponds and Watermill stream. Due to work on rebuilding the wall in 2013 most of the plants had to be relocated. Although the relocated plants failed, there was still a small amount growing on an undisturbed section of wall nearby in September 2016. At one time it was known from the Salakee ditch and a very wet field nearby, but was lost some time after 1990 when the stream was diverted and the ditch and field dried out. Always a rare plant in Scilly; several former sites known to Lousley were apparently also lost due to the resurfacing of roadside drains and changes to other wetland sites. Ditch-side.

VERONICACEAE – Speedwell family
Misopates orontium (L.) Raf.
Weasel's-snout
Inhabited islands

Archaeophyte. Scattered records, but never very common. Appearances can be sporadic, for example on St Martin's in 1988 many dozens of plants appeared from buried seed when the ground was cleared to build the foundations for the hotel. White-flowered plants occur at times. Bulb and cultivated fields, disturbed ground.

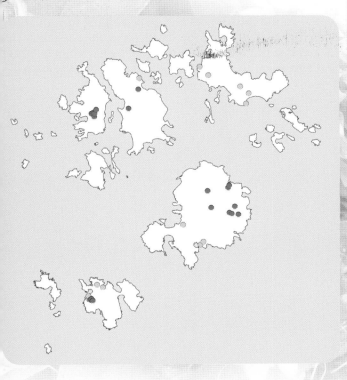

VERONICACEAE – Speedwell family
Cymbalaria muralis P. Gaertn., B. Mey. & Scherb.
Ivy-leaved Toadflax
Tresco, St Mary's and St Martin's

Archaeophyte. Established around or near habitations. The white form *alba* is found in several places on St Mary's and St Martin's. Our plant is subspecies *muralis* which grows on walls and buildings.

EU-DICOTS True Dicotyledons

VERONICACEAE – Speedwell family
Kickxia elatine (L.) Dumort.
Sharp-leaved Fluellen
Bryher, Tresco, St Agnes and St Mary's

Archaeophyte. Only scattered records from St Agnes, Bryher and Tresco, more frequent on St Mary's and absent from St Martin's. In some bulbfields the plant may be locally quite common forming low mats across the soil surface, elsewhere there may be just one or two individual plants. On Scilly this species has much rounder leaves than usual and can be mistaken for *K. spuria* if not examined closely. Bulb and other cultivated fields.

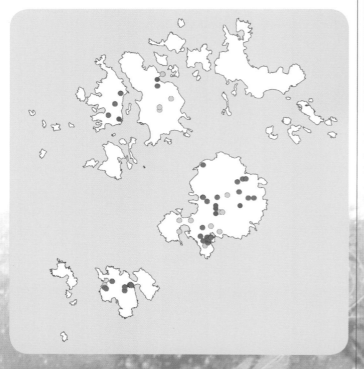

VERONICACEAE – Speedwell family
Kickxia spuria (L.) Dumort.
Round-leaved Fluellen
St Agnes

Archaeophyte. The only recent record is from a bulbfield at Higher Town, St Agnes, 1995. Not found there since and possibly may have been an error for *K. elatine*.

Linaria vulgaris Mill.
Common Toadflax
St Mary's and St Agnes

Native in the British Isles but an occasional introduction on Scilly, last known from St Mary's in the road near the incinerator in 1992, then at nearby Porth Mellon in 1993 and from near the Coastguard houses on St Agnes in 1995.

PLANTAGINACEAE – Plantain family
Plantago coronopus L.
Buck's-horn Plantain
Inhabited islands, also the uninhabited islands other than the Western and Norrard Rocks

Native. An abundant and widespread plant, predominantly coastal in distribution, but does also grow inland. Very variable in size, with some tiny plants with linear leaves lacking the characteristic lobed edges (see plants in Allseed photo on page 194). These latter plants may be mistaken for *P. maritima*. Grows on short turf, bare ground, heathland, cliffs, paths and around buildings.

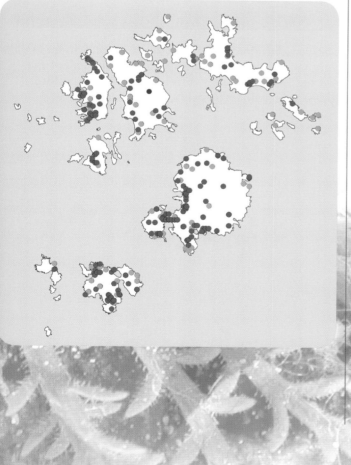

PLANTAGINACEAE – Plantain family
Plantago maritima L.
Sea Plantain
St Mary's and Tresco

Native. This is a surprisingly rare plant on Scilly. The species was not found until 1969 when plants were located between Pelistry Bay and Deep Point. Searches by Mike and Anne Gurr have extended the number of locations along the coast between Giant's Castle north to Pelistry with the best population along the cliffs on Mount Todden Down. It was found at Normandy Down (SV93131121) in 2014. There were a few records from Bar Point (SV916127/9) in 1993, but these have not been refound, neither have plants been found at Porth Mellon (SV908107) since 1996. On Tresco it has been recorded at Old Grimsby in 1996 and New Grimsby (SV889151) in 2007 and 2011. The latter records need confirmation. Coastal edge, grassland and cliff top.

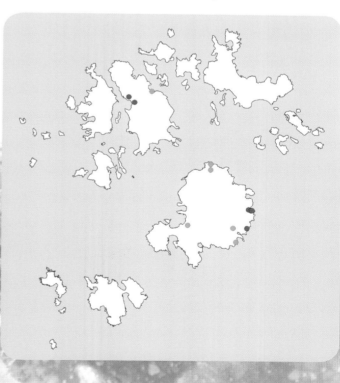

EU-DICOTS True Dicotyledons

PLANTAGINACEAE – Plantain family
Plantago major L.
Greater Plantain
Inhabited islands, also Samson, Northwethel and Teän

Native. Abundant on the inhabited islands, except St Martin's where there are just a few records. Although the subspecies are not usually recorded, in most cases the plants appear to be ssp. *major*, but ssp. *intermedia* also occurs occasionally. Grassland, cultivated land, roadsides, paths and waste ground.

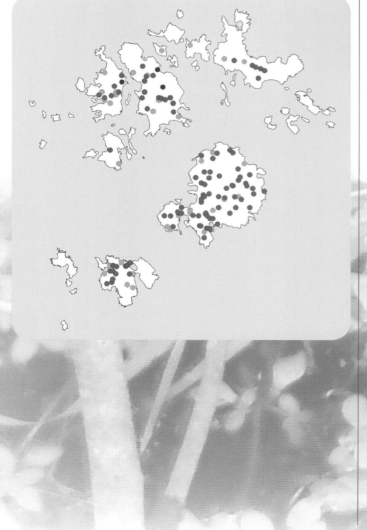

PLANTAGINACEAE – Plantain family
Plantago lanceolata L.
Ribwort Plantain
Inhabited islands, also Samson, Teän, Northwethel, St Helen's and Middle Arthur

Native. Abundant on the inhabited islands. In bulbfields very tall plants can occur especially at the end of the season. Found in many different habitats including grasslands, cultivated fields, dunes, roadsides and wasteland.

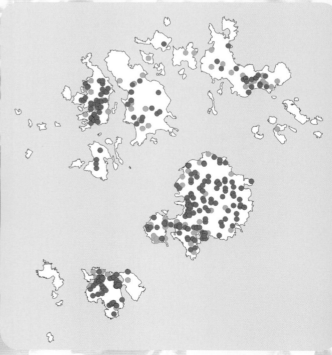

PLANTAGINACEAE – Plantain family
Littorella uniflora (L.) Asch.
Shoreweed
Tresco

Native. Only known from Abbey Pool. Although it has not been recorded since 1997. The Pool has been far less accessible due to the Heliport fencing or high water levels, so it may well still be present. On sandy mud, both on the drawdown zone as well as under water.

CALLITRICHACEAE – Water-starworts

Due to the difficulties of identifying water-starworts, some older records, although included, may now be impossible to verify. Those recently recorded are believed to have been correctly named, usually by specimens (sometimes grown on to see fruit). There is also some fluctuation in species composition from year to year even in the same water bodies.

Callitriche stagnalis Scop.
Common Water-starwort
Tresco, St Mary's, also Samson

Native. Occasional records from Abbey and Great Pool, Tresco; Higher Moors, Lower Moors and Watermill Stream, St Mary's: from Samson near Southward Well in 1992. Many recorders do not attempt to identify *Callitriche* species, but lump them together as *C. stagnalis*. Ditches and pools.

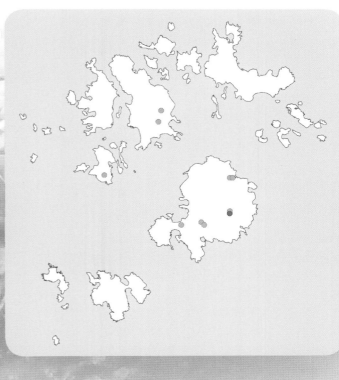

EU-DICOTS True Dicotyledons

CALLITRICHACEAE – Water-starwort family
Callitriche platycarpa Kütz.
Various-leaved Water-starwort
Bryher

Native. Recorded from Bryher in 1977 (SV8814) without further details.

Callitriche obtusangula Le Gall
Blunt-fruited water-starwort
St Mary's

Native. Has been recorded from several places, on the Watermill Stream, the ditch through Lower Moors, a pool on the Holy Vale stream and a wet field near Lower Moors. Most of these records have been verified by specimens.

CALLITRICHACEAE – Water-starwort family
Callitriche brutia Petagna
Pedunculate Water-starwort
St Mary's, Bryher and Tresco

Native. Formerly recorded from Great Pool, Tresco in 1975. There are recent records from the stream through Holy Vale and Higher Moors in 2008, 2009 and 2011, confirmed by specimens. In Stace (2010) this is treated as ssp. *brutia*.

CALLITRICHACEAE – Water-starwort family
Callitriche hamulata (Kütz. Ex W. D. J. Koch)
Intermediate Water-starwort

(This has been downgraded by Stace to a subspecies of *C. brutia*. As it has been recorded from a number of sites it is therefore mapped separately).

Native. Recorded from Bryher Little Pool in 1990 and 1991; Tresco Great Pool in 1987 and Abbey Pool 1986 and 1987; St Mary's Newford Duck Ponds in 1985, Watermill Stream in 1999 and Higher Moors in 2002 (confirmed by specimen).

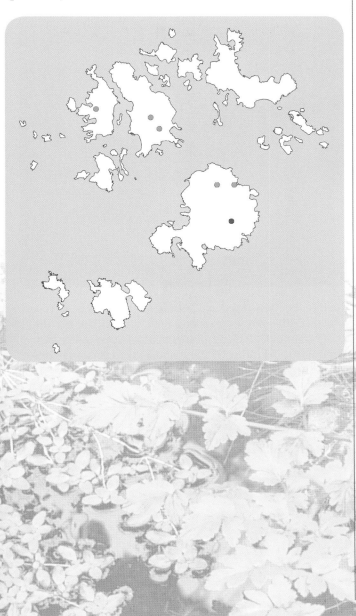

SCROPHULARIACEAE – Figwort family
Verbascum blattaria L.
Moth Mullein
Tresco and St Agnes

Neophyte. Garden escape or casual. Recorded in a quarry on Tresco (SV899148) in 1992 and at Lower Town Farm, St Agnes in 2014.

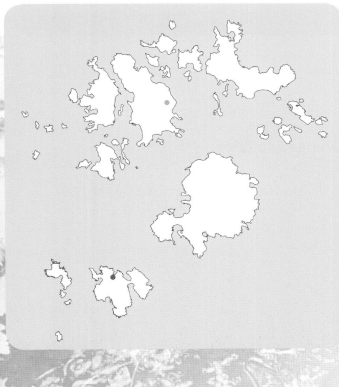

SCROPHULARIACEAE – Figwort family
Verbascum virgatum Stokes
Twiggy Mullein
Bryher and Tresco

Neophyte. Casual. Appletree Banks, Tresco in 1956 (Lousley), also from (SV874141) Bryher in 1998.

Verbascum phlomoides L.
Orange Mullein
St Mary's

Neophyte. Near Star Castle, garden escape 1993. Previously recorded by Lousley.

SCROPHULARIACEAE – Figwort family
Verbascum thapsus L.
Great Mullein
Inhabited islands, also St Helen's and Teän

Native. Scattered records, often seems to persist in some places with seedlings arising near the mature plants. As it is biennial flowering plants are not seen every year. Frequently found near the coast as well on roadsides, dunes, quarries and walls.

SCROPHULARIACEAE – Figwort family
Verbascum nigrum L.
Dark Mullein
St Mary's and Tresco

Native in the British Isles but possibly only a garden escape in Scilly. Known from a site near Telegraph (SV91251215) where a group of flowering plants were found on the roadside in 2010, still there 2012. A plant was found near the dump in 2015 (SV9098310482). There have been occasional records from near Tresco Abbey gardens without details. The Telegraph plants may have originated from a nearby garden.

SCROPHULARIACEAE – Figwort family
Scrophularia nodosa L.
Common Figwort
St Mary's and St Martin's

Native. Formerly recorded from Borough, Tresco and Trenoweth, St Mary's by Lousley and believed extinct until a plant was found in flower at the side of Garrison Walk (SV90101028) by Judith Cox in 2010. As *S. scorodonia* grows in the same area it is possible it had previously been overlooked. Unfortunately the area was herbicided later, there was no sign of the plant again until May 2015 a group of nine plants was found nearby at SV90071026. Also in 2015 Bob Dawson found a plant on St Martin's (SV9372715311).

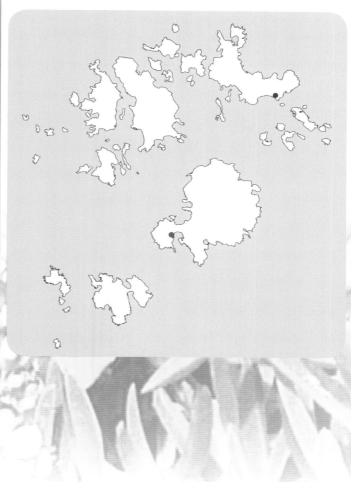

SCROPHULARIACEAE – Figwort family
Scrophularia auriculata L.
Water Figwort
St Mary's and St Martin's

Native. Lousley rejected earlier records of this species on the grounds they usually turned out to be errors for *S. scorodonia*. It was not until 1995 that a plant in Hugh Town (SV902106) was confirmed as *S. auriculata*. In 2003 a plant was found in Rocky Hill Lane (SV91431109) which had increased to 35 by 2004, with five or six plants still there in 2013. There have been several recent records from places in Hugh Town (SV902107 and SV90131044), and a 2005 record from the grounds of the Longstone Centre. In 2010 a plant was found growing near the bakery on St Martin's. Lane-sides and verges.

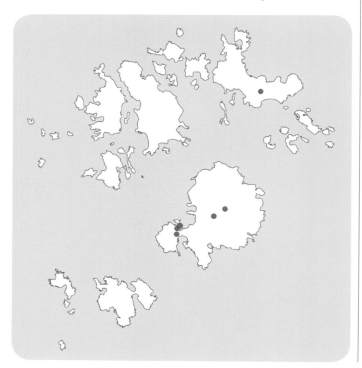

SCROPHULARIACEAE – Figwort family
Scrophularia scorodonia L.
Balm-leaved Figwort
Inhabited islands, also Samson, Annet, St Helen's, Teän, Nornour and Great Ganilly

Neophyte. Widespread and often abundant in places, for example on Tresco where it appears to be ubiquitous. At one time was absent from St Agnes although it still grew on Gugh. Since 2002 it has been refound in two places on St Agnes. Because the plant is so common it may have masked the identification of the above two species which had previously been assumed to be extinct. The larvae of a small and very local moth *Nothris congressariella* (Bruand) feed on spun leaves of this species of figwort. The plant is found in a range of habitats from among dense bracken and bramble, dunes, marshes, hedgebanks, cultivated land, ditches, cliffs and rocky shores.

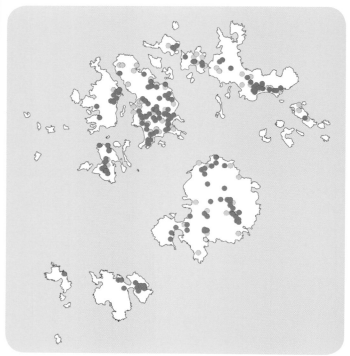

SCROPHULARIACEAE – Figwort family
Buddleja davidii Franch.
Buddleja
St Mary's and Tresco

Neophyte. An occasional garden escape. On the mainland *Buddleja* can be invasive; fortunately it does not seem to be becoming established in the wild on Scilly.

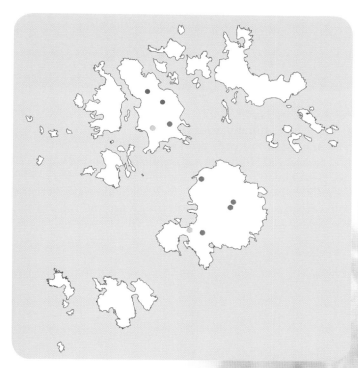

SCROPHULARIACEAE – Figwort family
Buddleja globosa Hope
Orange-ball-tree
St Mary's and Bryher

Neophyte. Garden escape; found in a shelterbelt at Trenoweth and in the Buzza quarry. Also on Bryher in 2011.

Limosella aquatica L.
Mudwort
Extinct or error

Only known from the Abbey Pool, Tresco, where it was first recorded by A.E. Gibbs from 'lake in front of Abbey' in 1883 and last recorded there by H. Downes in 1921. A record from 1923 was found to be *Elatine hexandra* (see Lousley, 1971). It is now considered these earlier records could have been misidentifications; the tiny form of *Ranunculus flammula* with spoon-shaped leaves and long pedicels found at Abbey Pool looks remarkably like *Limosella* when not in flower.

EU-DICOTS True Dicotyledons

LAMIACEAE – Dead-nettle family
Stachys sylvatica L.
Hedge Woundwort
St Martin's, Tresco and St Mary's

Native. A rare plant in Scilly. Formerly recorded from a withy bed near Tresco Abbey by Lousley. More recent records are from St Martin's rubbish dump (SV928158) in 1998 and from two sites on St Mary's, a garden at Old Town in 1993, and from McFarlands Down (SV913124) in 2008. Hedgebank and a rubbish dump.

LAMIACEAE – Dead-nettle family
Stachys palustris L.
Marsh Woundwort
Tresco and St Mary's

Native. Now very rare on Scilly. Formerly recorded from Bryher churchyard in 1953 and St Martin's 1939. The only recent records are Tresco (SV8915) in 1995 and St Mary's, Higher Moors in 1978, and beside the hedge at Halangy Down Ancient Village (SV90991233 to SV910911) in 1995, still there in 2015. Scilly records seem to be mostly on cultivated and rough ground rather than wetlands.

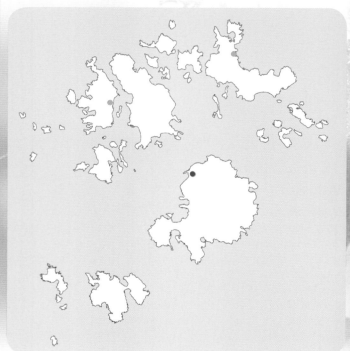

LAMIACEAE – Dead-nettle family
Stachys arvensis (L.) L.
Field Woundwort
Inhabited islands

Archaeophyte. Widespread and locally abundant. An important constituent of the arable plant flora. White-flowered plants occur occasionally. Bulb and other arable fields and disturbed ground.

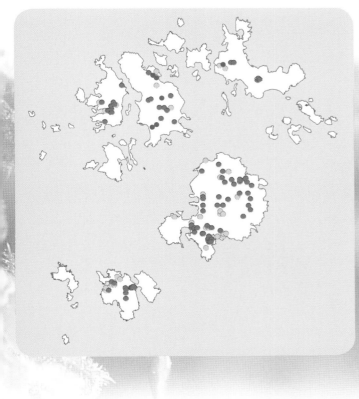

LAMIACEAE – Dead-nettle family
Betonica officinalis L.
Betony
St Martins, and Great Ganilly

Native. Only known from two places in Scilly, Great Ganilly (SV947144 –SV946146) and from Burnt Hill, Chapel Down, St Martin's (SV937159) where it was last seen in 1998. A recent record from St Mary's still needs confirmation. This is a common plant on mainland Cornwall so its rarity in Scilly is surprising. Where the plant grows on Great Ganilly is an area of heathland where *Thymus polytrichus* and *Ornithopus pinnatus* have also been found. Heathland.

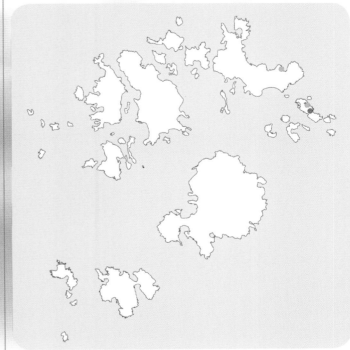

EU-DICOTS True Dicotyledons

LAMIACEAE – Dead-nettle family
Ballota nigra ssp. *meridionalis* (Bég.) Bég.
Black Horehound
St Mary's and St Martin's

Archaeophyte. Very rare. Lousley describes the plant as common around Hugh Town but it is no longer found there. It was recorded from Old Town (SV912102) in 1999, (SV91101035) in 2003 and near the school (SV911104) in 2015. Records from St Martin's include Lower Town (SV9116) in 1997 and from the quarry (SV928158) between 1992 and 1997. Waste ground, a quarry and roadsides.

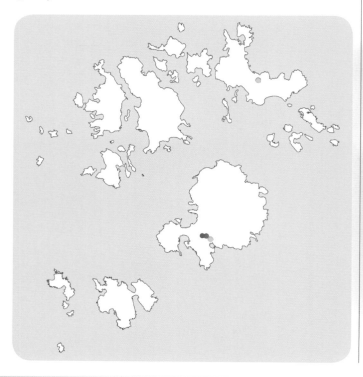

LAMIACEAE – Dead-nettle family
Lamiastrum galeobdolon ssp. *argentatum* (Smejkal) Stace
Garden Yellow Archangel
St Mary's

Garden escape at Higher Trenoweth, last seen in 1995.

Lamium purpureum L.
Red Dead-nettle
Inhabited islands

Archaeophyte. Scattered records, but nowhere common. Cultivated ground.

Lamium album L.
White Dead-nettle
Extinct

The only record was from St Mary's in 1879.

LAMIACEAE – Dead-nettle family
Lamium hybridum Vill.
Cut-leaved Dead-nettle
Inhabited islands

Archaeophyte. Similar distribution to *L. purpureum*, but nowhere very common. Cultivated fields.

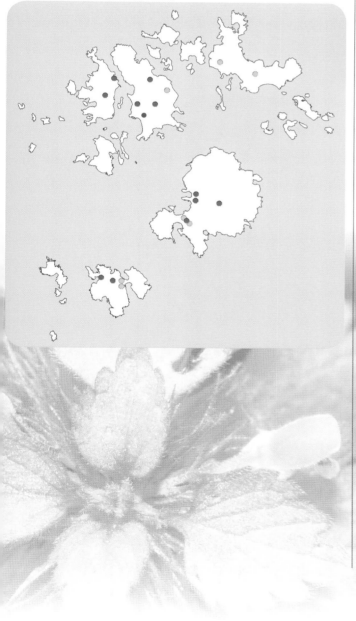

LAMIACEAE – Dead-nettle family
Lamium amplexicaule L.
Henbit Dead-nettle
St Agnes, St Mary's, Bryher and St Martin's

Archaeophyte. A rare plant in Scilly. Just a few records; St Agnes, Higher Town (SV884082) in 2005; Bryher, Hillside Farm (SV87681471) in 2004 and from St Martin's there are records from Middle and Higher Town in 1960, and most recently from Higher Town (SV935155) in 1986 and (SV922162) in 1989. In 2015 it was found growing in flower beds on Porthcressa, St Mary's. Arable fields.

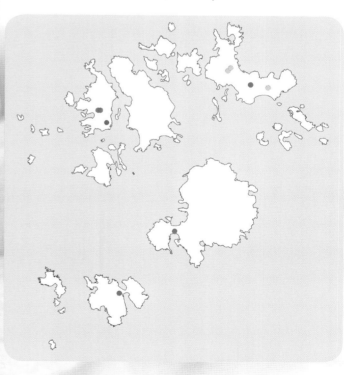

EU-DICOTS True Dicotyledons

LAMIACEAE – Dead-nettle family
Scutellaria galericulata L.
Skullcap
Tresco, also Samson

Native. Skullcap has been known from Samson since 1864. It has been found in seepages all round that island, but especially on the south-east shore around Southward Well. It was not until 1999 that it was found away from Samson when Will Wagstaff found it by Abbey Pool, Tresco (SV897142), presumably carried there by waterfowl, it has not been recorded there since. Freshwater seepages and marshy places.

LAMIACEAE – Dead-nettle family
Teucrium scorodonia L.
Wood Sage
Inhabited islands, also Annet, Samson, St Helen's, Teän, Nornour, Little Ganilly, Great Ganilly and Little Arthur

Native. Locally abundant on most of the larger islands where it is strongly associated with heathland sites and coastal slopes. Absent from farmed areas on the inhabited islands. An unusual record is one from the south end of Annet in 1990, not obviously a heathy area. Bracken slopes and heathland.

LAMIACEAE – Dead-nettle family
Ajuga reptans L.
Bugle

St Mary's

Native on the mainland but neophyte on Scilly. Garden escape. The plant with coloured leaves found at Porth Minick, St Mary's in 1999 was clearly a garden cultivar.

LAMIACEAE – Dead-nettle family
Glechoma hederacea L.
Ground-ivy

Inhabited islands, also Samson (and Puffin Island), Northwethel, St Helen's, Teän, White Island (St Martin's), Nornour, Great Ganilly and the Arthurs

Native. Widespread, but apparently less common on St Martin's and the arable areas of southern half of St Mary's, parts of Tresco and St Agnes. It has not been recorded from Teän since 1987 but could still be there. Found in a range of habitats from grass fields, hedgebanks and under bracken on cliffs and heaths.

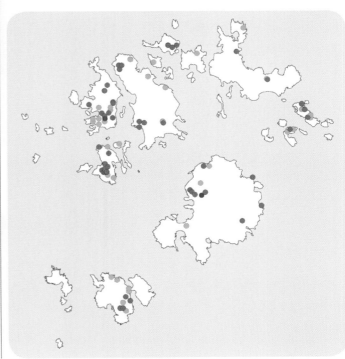

EU-DICOTS True Dicotyledons

LAMIACEAE – Dead-nettle family
Prunella vulgaris L.
Selfheal
Inhabited islands, also Samson and Great Ganilly

Native. Scattered sites and sometimes locally frequent. It is more common in the central area of Tresco and parts of St Mary's where it follows the roads. Road verges, grasslands and dunes.

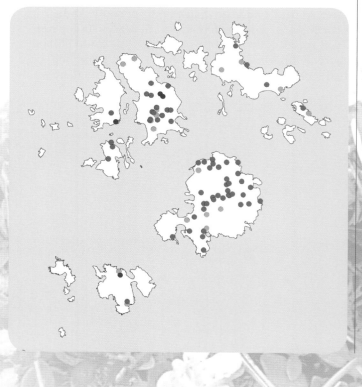

LAMIACEAE – Dead-nettle family
Melissa officinalis L.
Balm
St Mary's

Neophyte. There had been a 1956 record from St Agnes by Lousley. Now known only as a garden escape around Holy Vale (SV920114-6).

LAMIACEAE – Dead-nettle family
Clinopodium ascendens (Jord.) Samp.
Common Calamint
St Mary's and St Agnes

Native. Still found in some of the places known to Lousley. On St Mary's, Trenoweth (SV918123), Porthloo (SV909115). On St Agnes from 1978-1999 it was recorded in several places near the Lighthouse (SV878082) to Higher Town (SV884081-3) and the path to Covean. Roadsides and hedgebanks.

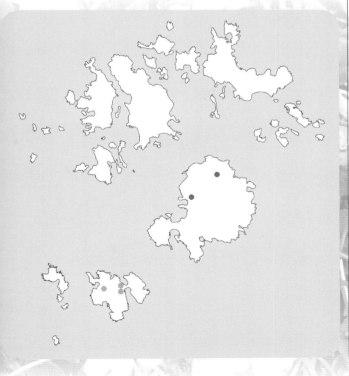

LAMIACEAE – Dead-nettle family
Thymus polytrichus A. Kern. ex Borbás
Wild Thyme
St Mary's, also Gugh and Great Ganilly

Native. Always a rare plant in Scilly. The largest population is on Gugh where it has spread from the patch of coastal dune grassland into the 'sandpit' (in c.2000) and has now spread along the paths across the neck of the island. This may be a recent re-invasion of the neck area after some cutting where it had become overgrown. A site on the St Agnes side of The Bar has long gone. The small site on Peninnis near the lighthouse is now the only one on St Mary's (SV910095). The Great Ganilly site (SV946146) found in 1966 by P. Z. MacKenzie was still there in 2002. On sandy ground, dune grassland and short turf on heathland.

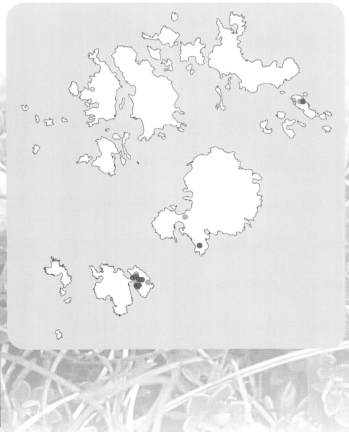

EU-DICOTS True Dicotyledons

LAMIACEAE – Dead-nettle family
Thymus x *citriodorus* (Pers.) Schreb.
Lemon Thyme
St Mary's

Neophyte. Garden escape on The Garrison 2003.

LAMIACEAE – Dead-nettle family
Lycopus europaeus L.
Gypsywort
Tresco, St Mary's and St Martin's

Native. Only found on St Mary's in Lousley's time. It is still found at Higher Moors and Porth Hellick Pool on St Mary's but is now also found on Tresco, at Abbey Pool since at least 1984 and Great Pool where it was first noted in 2007. The records from St Martin's are unusual as they are growing away from wetland, in dunes (SV917160) in 1988 and behind English Island Point (SV938153) in 1994. Pools and wetlands, but apparently also sometimes drier places.

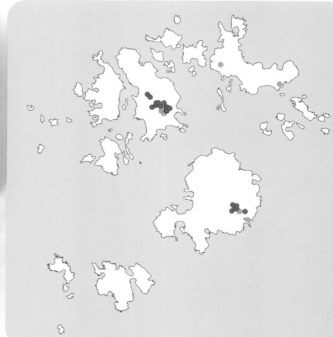

LAMIACEAE – Dead-nettle family
Mentha aquatica L.
Water Mint
St Mary's, Tresco and St Martin's

Native. A rare plant that is found only on St Mary's at both Lower and Higher Moors and also in Holy Vale. On Tresco it is only known from a boggy area at the north-eastern edge of Great Pool. It had also been recorded in 1989 from above Great Bay, St Martin's (SV923166). Wetland sites and damp ground.

LAMIACEAE – Dead-nettle family
Mentha x *suavis* Guss.
(*Mentha aquatica* x *Mentha suaveolens*)
Hybrid Mint
St Mary's

Neophyte. One record of this hybrid has been recorded near from Old Town, St Mary's (SV912102) in 1997.

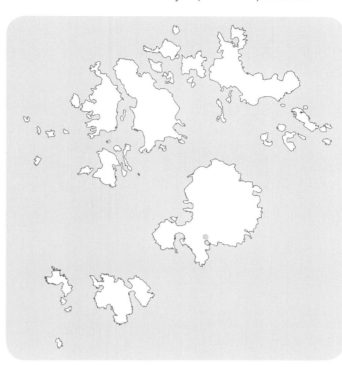

Mentha x *piperita* L.
(*Mentha aquatica* x *Mentha spicata*)
Peppermint
St Mary's

Native species that is a garden escape St Mary's, Old Town (SV912103) in 1994 and the Garrison (SV8909) in 1995.

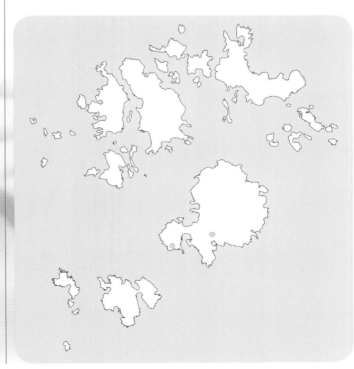

EU-DICOTS True Dicotyledons

LAMIACEAE – Dead-nettle family
Mentha spicata L.
Spear Mint
Tresco, St Martin's, St Agnes and St Mary's

Archaeophyte but an occasional garden escape in Scilly, not recorded from Bryher.

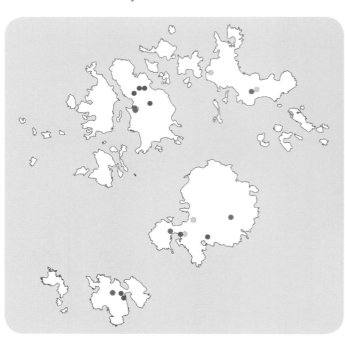

Mentha x *villosa* Huds.
(*Mentha spicata* x *Mentha suaveolens*)
Apple-mint
St Agnes, St Mary's and St Martin's

Archaeophyte but an occasional garden escape in Scilly recorded from a few scattered sites. The variety *alopecuroides* has been recorded from St Mary's and St Martin's.

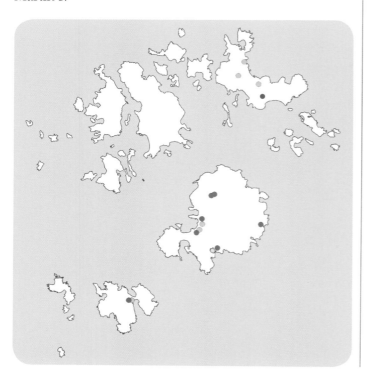

LAMIACEAE – Dead-nettle family
Mentha x *rotundifolia* (L.) Huds.
(*Mentha longifolia* x *Mentha suaveolens*)
False Apple-mint
St Mary's, St Agnes and Tresco

Neophyte. Occasional garden escape Higher Town, St Agnes; Old Town area St Mary's and Appletree Banks, Tresco.

LAMIACEAE – Dead-nettle family
Mentha suaveolens Ehrh.
Round-leaved Mint
St Mary's, Tresco and St Martin's

Native in the British Isles but an occasional garden escape in Scilly with most records from St Mary's.

LAMIACEAE – Dead-nettle family
Mentha requienii Benth.
Corsican Mint
St Mary's

Neophyte. The only known site was in the gravel outside Hugh House, on The Garrison. Although the area had been treated with weedkiller the plants had managed to survive until 2014. In 2015 no plants could be found and the usual area had been invaded by *Soleirolia soleirolii*.

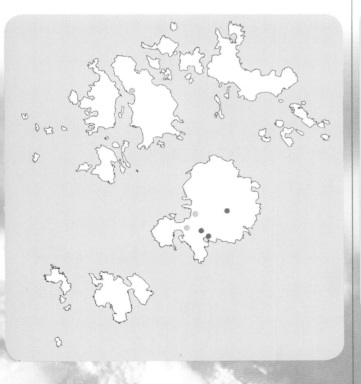

Mentha pulegium L.
Pennyroyal
Extinct

Last known from Holy Vale in 1915, 'probably near small ponds' according to Lousley.

EU-DICOTS True Dicotyledons

LAMIACEAE – Dead-nettle family
Rosmarinus officinalis L.

Rosemary

St Mary's and St Martin's

Neophyte. Occasional garden escape.

LAMIACEAE – Dead-nettle family
Salvia reflexa Hornem.

Mintweed

St Mary's

Neophyte. Casual. A few plants found in Ram's Valley where birdseed had been scattered in 2000.

OROBANCHACEAE – Broomrape family

Eyebrights

This difficult group is widespread on all the inhabited (and larger uninhabited islands according to Lousley). Some earlier identifications were made by H.W. Pugsley and P. F. Yeo. These included the following:

Euphrasia micrantha, Euphrasia curta, Euphrasia confusa, Euphrasia brevipila, Euphrasia anglica var. *gracilescens*

Since then it has not been possible to confirm these identifications. Some recent material has been confirmed by Alan Silverside.

Euphrasia L.
Eyebrights
Inhabited islands

Map showing all records.

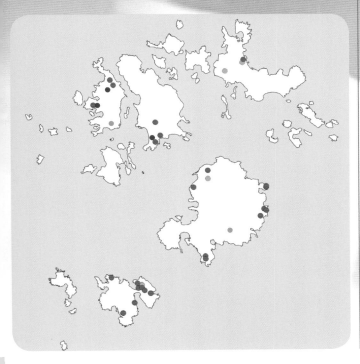

OROBANCHACEAE – Broomrape family
Euphrasia tetraquetra (Bréb.) Arrond.
Eyebright
Inhabited islands

Native. This is the species most often recorded in Scilly, often appears as small, stunted plants – but this is not diagnostic. Some records have been confirmed. Grassy places on cliffs, heathland and dunes.

Euphrasia nemorosa x *Euphrasia confusa*
Eyebright

Native. Confirmed records from Wingletang Down, St Agnes and Pelistry Bay, St Mary's.

EU-DICOTS True Dicotyledons

OROBANCHACEAE – Broomrape family
Parentucellia viscosa (L.) Caruel
Yellow Bartsia
St Mary's, Tresco, Bryher, also Gugh

Native. Locally frequent on Tresco and St Mary's with occasional records from Bryher and elsewhere. Lousley recorded the species from St Agnes without details but it has not been seen there since. There was a flush of the species on the Gugh in 1983 on bare ground after a heath fire. Heathland, damp fields, bulbfields and marshy ground, quite often in drier habitats than usual.

OROBANCHACEAE – Broomrape family
Pedicularis sylvatica L.
Lousewort
Inhabited islands, also Samson

Native. Locally frequent on sites around the coast and on heathland. Plants with white flowers are not infrequent. Heathland and dunes.

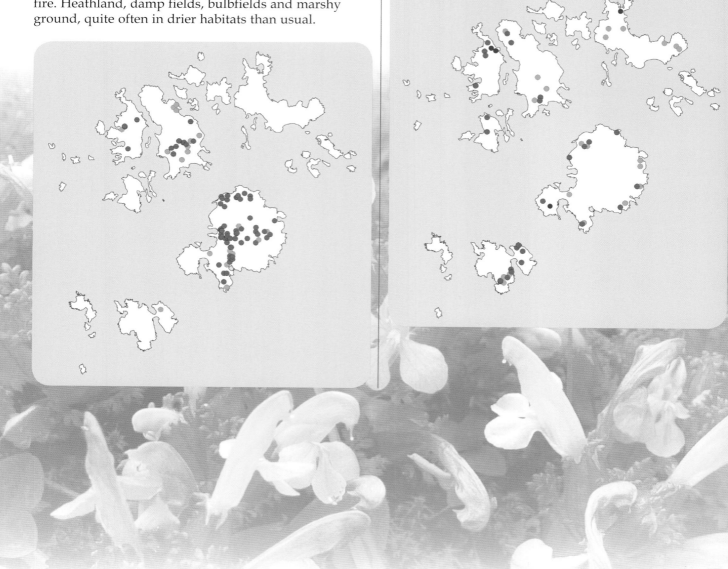

OROBANCHACEAE – Broomrape family
Orobanche minor Sm.
Lesser Broomrape
Tresco, St Mary's, Bryher, also Gugh

Native. Some of the plants on Tresco are parasitic on *Gazania* and others possibly on *Carpobrotus*. In other places the host appears to be *Trifolium repens* or *Daucus carota*. Plants from Veronica Farm, Bryher and those from the *Gazania* beds on Tresco have been confirmed as *O. minor* subsp. *maritima* by Fred Rumsey (2008). *Var. compositarum* has been recorded from Abbey Gardens, but apparently from the same place as subsp. *maritima*. Grows in gardens, under a hedge, sandy fields and dune grassland.

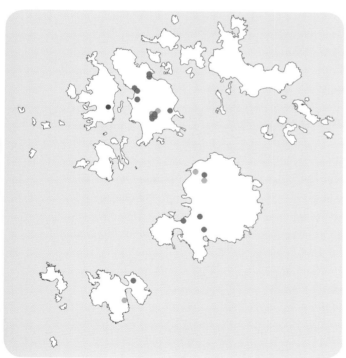

Orobanche hederae Duby
Ivy Broomrape
Extinct

Native. Recorded by Lousley in Abbey Gardens, Tresco parasitic on ivy in 1967 where it had been known for some years. It was last recorded in 1983 by Peter Clough (the head gardener at the time).

ACANTHACEAE – Bear's-breech family
Acanthus mollis L.
Bear's-breech
St Mary's, Tresco, St Martin's, St Agnes and Gugh

Neophyte. A long-established garden escape. Most common on St Mary's, with plants also on Tresco, St Martin's and St Agnes. In September 2010 a large plant was photographed on Gugh by Helen Parslow and Mark Phillips. Many of the clumps have been established for many years. Lousley mentions a colony on St Agnes that had been there since before 1800. The plants are usually found on shady roadsides, banks, shores and rubbish tips.

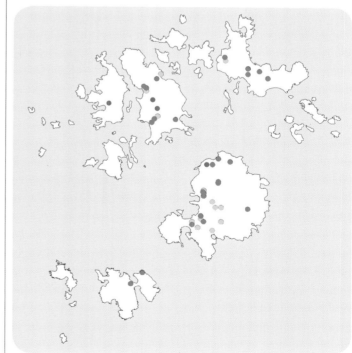

EU-DICOTS True Dicotyledons

VERBENACEAE – Vervain family
Verbena officinalis L.
Vervain
St Mary's

Archaeophyte. No longer known from Tresco, but still found in a few places on St Mary's; between Porth Minick, Tolman Point and Old Town, and in the Industrial Estate, Hugh Town. It was last seen on The Garrison in 1978. Roadsides and on waste land.

AQUIFOLIACEAE – Holly family
Ilex aquifolium L.
Holly
Tresco, St Mary's, and St Martin's, also Samson and Gugh

Native on the mainland but on Scilly a garden escape that is often bird-sown. A plant was found on Great Ganilly in 1970, but has not been recorded there since. Although there is evidence of Holly in the pollen record, and it is one of the species said to have been present in the islands in the 1600s and 1700s (Turner [1695] in *The Scillonian*, 1964; and Heath, 1750), there is no evidence that there are any native Holly trees on Scilly now.

CAMPANULACEAE – Bellflower family
Campanula rapunculus L.
Rampion Bellflower
St Mary's

Archaeophyte. Garden escape on The Garrison (SV901105) in 1995.

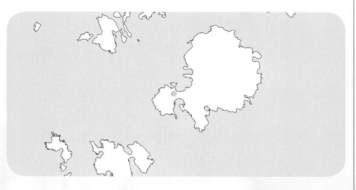

Campanula portenschlagiana Schult.
Adria Bellflower
St Mary's

Neophyte. Garden escape recorded on The Garrison (SV90111039) in 2007, there has been some confusion between this species and *C. poscharskyana*.

CAMPANULACEAE – Bellflower family
Campanula poscharskyana Degen.
Trailing Bellflower
St Agnes and St Mary's

Neophyte. Garden escape, formerly on St Agnes. Still in Hugh Town and on the Garrison. Walls. Banks.

Wahlenbergia hederacea (L.) Schrad. Ex Roth
Ivy-leaved Bellflower
Extinct

Native. Lousley records the species from wet turf by Porthellick Pool with *Hydrocotyle vulgaris* in 1963. The area is now very grown over with tall vegetation. There are no recent records.

EU-DICOTS True Dicotyledons

CAMPANULACEAE – Bellflower family
Trachelium caeruleum L.
Throatwort
Tresco

Neophyte. Escape from Abbey Gardens into woodland.

CAMPANULACEAE – Bellflower family
Jasione montana L.
Sheep's-bit
St Mary's, St Martin's and Tresco

Native. A common plant on St Mary's, but now known from only a few sites on St Martin's and one on Tresco. The Tresco site was near Old Grimsby (SV892155) in 2008. On St Martin's it is found mainly around the Lower Town, the Plains, Top Rock and Burnt Hill. It was formerly recorded from St Agnes and Teän by Lousley, but without details. On St Mary's some plants are very pale blue. Found on stone hedges, heathland, dunes and coastal grassland.

CAMPANULACEAE – Bellflower family
Lobelia erinus L.
Garden Lobelia
St Mary's, St Martin's and Tresco

Neophyte. Frequent garden escape sometimes self-seeding into pavement cracks.

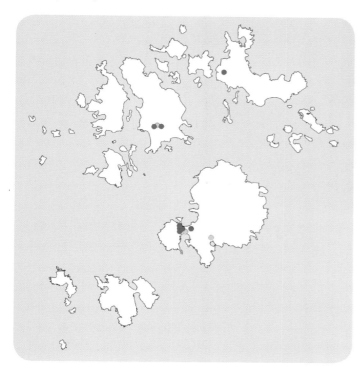

MENYANTHACEAE – Bogbean family
Menyanthes trifoliata L.
Bogbean
St Mary's

Native in the British Isles but a neophyte on Scilly. Grows in the pool at Jac-a-Bar, where it had been introduced.

ASTERACEAE – Daisy family
Arctium minus (Hill) Bernh.
Lesser Burdock
Inhabited islands, also Samson

Native. Common on the inhabited islands. On Samson there is a clump beside the ruined buildings. Often found along roadsides, field borders and woodland edges.

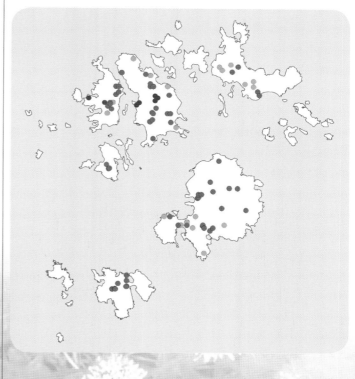

Carlina vulgaris L.
Carline Thistle
St Mary's

Extinct or error. A record from St Mary's (SV905105) by D.G. Gibbs in 1988 is unconfirmed.

EU-DICOTS True Dicotyledons

ASTERACEAE – Daisy family
Arctium nemorosum Lej.
Wood Burdock
Tresco and St Mary's

Native. A plant from Abbey Drive, Tresco collected by R. Lancaster in 1969 was confirmed by F. Perring.

ASTERACEAE – Daisy family
Carduus tenuiflorus Curtis
Slender Thistle
Bryher, Tresco, St Martin's and St Mary's, also Teän and Great Ganilly

Native. There are very few recent records, although Lousley recorded the plant as very common. Many of the appearances are sporadic, but in a few places plants may persist for several years, for example the disturbed area by Carn Near quay, Tresco; the sandy neck on Great Ganilly and on Teän. Usually coastal, on rocky and sandy shores, dunes, roadsides and rubbish dumps.

ASTERACEAE – Daisy family
Cirsium vulgare (Savi) Ten.
Spear Thistle
Inhabited islands, also Samson, St Helen's, Teän, Annet, and the Eastern Isles

Native. Locally frequent on most of the inhabited and larger uninhabited islands. In the Eastern Isles it has been recorded on Great Ganilly, Nornour, Little and Middle Arthur, Little Ganilly, Little Ganinick, and Menawethan. Occurs on field edges, roadsides, disturbed ground and rocky shores.

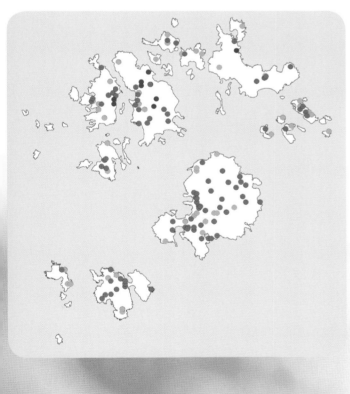

ASTERACEAE – Daisy family
Cirsium palustre (L.) Scop.
Marsh Thistle
St Mary's

Native. A rare plant in Scilly, it is only known from along the stream in Holy Vale down through Higher Moors and Porthellick Pool. Streamside and marshy places.

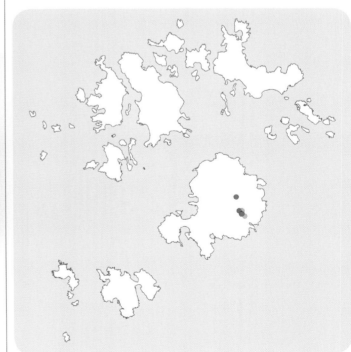

EU-DICOTS True Dicotyledons

ASTERACEAE – Daisy family
Cirsium arvense (L.) Scop.
Creeping Thistle
All inhabited islands, also Teän, Middle Arthur, Menawethan, Great Ganilly

Native. Widespread and locally frequent on the inhabited islands, less so on St Martin's. Not recorded recently from the Eastern Isles. A stand of plants without the usual prickly spines that was known from St Martin's near the cricket field pool from 1986 until 1997 was identified by David McClintock as var. *mite* (now var. *integrifolium* Wimm. & Grab) photo above right. Unfortunately after a few years the plants disappeared from the brambly corner where they grew and have not reappeared. *Cirsium arvense* grows in pastures, by roadsides and in rough grassland, including on some of the Eastern Isles.

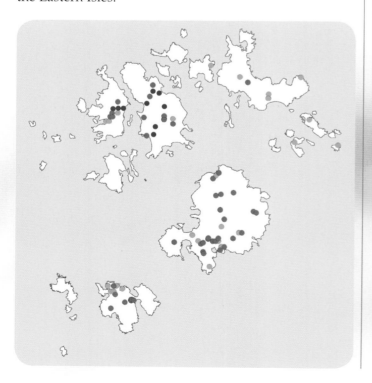

ASTERACEAE – Daisy family
Centaurea cyanus L.
Cornflower
Tresco and St Agnes

Archaeophyte. Records from near Great Pool, Tresco in 1967; Covean, St Agnes in 1970 and Merchant's Point, Tresco (SV8916) in 2014 were probably garden escapes.

ASTERACEAE – Daisy family
Centaurea nigra L.
Common Knapweed
Inhabited islands

Native. Scattered records, but nowhere very common. Knapweed is mostly found in small corners of relict, species-rich grassland. On Bryher, knapweed grows in the churchyard and at Rushy Bay. On Tresco it grows at the west end of Great Pool and along the track to the Gardens. On St Martin's there are records from Lower Town (SV917160), the churchyard, and a field near Chapel Down (SV936158). On St Agnes there are recent records from Wingletang without exact locality and on the abandoned farmland on Gugh. Found in grasslands, heaths, grass verges and churchyards.

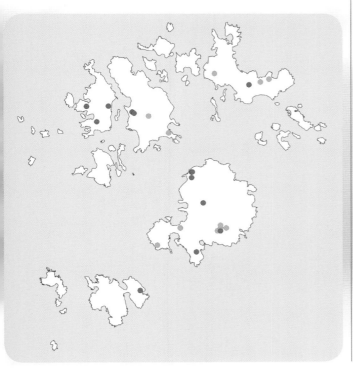

ASTERACEAE – Daisy family
Cichorium intybus L.
Chicory
St Martin's, St Agnes, Tresco and St Mary's

Archaeophyte. Recorded from St Agnes in 1966 but not since. Only known recently from St Mary's (SV898108) in 2011 and from the Lower Moors extension in 2014. On St Martin's it appears to persist in a field at Higher Town (SV927156-SV928156) and occasionally elsewhere in fields beside the road. Near Gimble Porth, Tresco in 2015. Roadside and former cultivations.

EU-DICOTS True Dicotyledons

ASTERACEAE – Daisy family
Lapsana communis L.
Nipplewort
St Mary's, Tresco and St Agnes

Native. It was formerly also found on Bryher and St Martin's (Lousley, 1971). Only now known from a few scattered sites, on St Mary's, St Agnes and Tresco. On St Agnes it was known from Covean in 1998 and was found at the Lighthouse in 2007. On Tresco most records are from around the Abbey Gardens and Great Pool, also at Old Grimsby (SV89341545). Found on roadsides, field edges, cultivated ground and near habitations.

ASTERACEAE – Daisy family
Hypochaeris radicata L.
Cat's-ear
Inhabited islands, also Samson (and Puffin Island), St Helen's, Northwethel, Teän, Little Ganilly, Great Ganilly, Little Arthur and Middle Arthur

Native. Possibly one of the most ubiquitous and abundant plants in Scilly. There are very few places where it does not grow; it appears to be missing from Annet and just a few smaller islands, and was last noted on Northwethel in 1981. Found in many habitats; grassland, heathland, dunes, roadsides and on stone hedges.

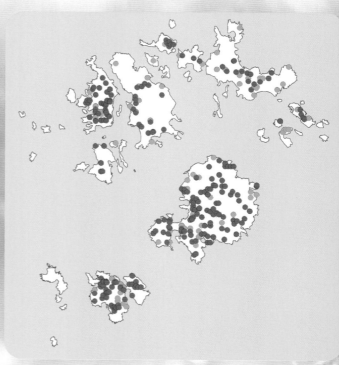

ASTERACEAE – Daisy family
Scorzoneroides autumnalis (L.) Moen
Autumn Hawkbit
Bryher, Tresco, St Mary's and St Agnes, also Northwethel, Teän and Samson

Native. There are not many records of this species, and those are all from the west of the islands. It has been recorded most frequently from St Agnes. On Bryher it has been recorded from around the Pool, Rushy Bay and Shipman Head. On Tresco it has been noted on Abbey Hill and the dump (SV892141). There are several sites on St Mary's including Peninnis, Seaways Farm (SV905118), Rose Hill and Lower Moors. Grasslands, especially damp areas and on sandy ground.

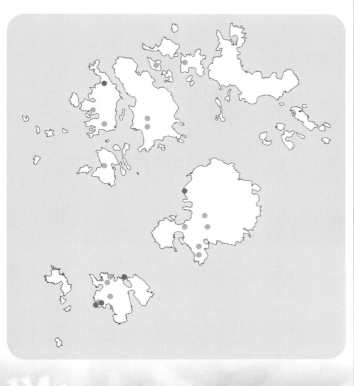

ASTERACEAE – Daisy family
Leontodon hispidus L.
Rough Hawkbit
Bryher, St Mary's, also also Gugh and Teän

Native. A very rare plant in Scilly. All the records are recent which suggests it had been overlooked previously. Bryher from Popplestone Bank in 2000 and Shipman Head 2007, Gugh sand pit (SV889083) in 2000 and 2002 and West Porth on Teän in 2007. In 2014 it was recorded on St Mary's (SV918117 and SV918118). Sandy grassland.

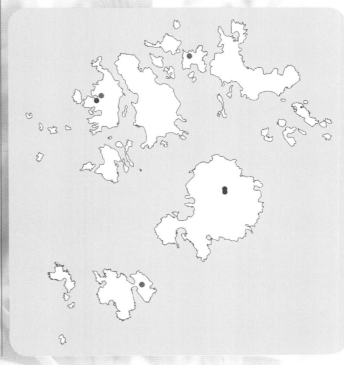

EU-DICOTS True Dicotyledons

ASTERACEAE – Daisy family
Leontodon saxatilis Lam.
Lesser Hawkbit
Inhabited islands, also St Helen's, Teän, Samson, Little Ganilly and Great Ganilly

Native. Locally frequent in discrete areas on many islands with the exception of the smaller ones. Not known from Annet and last recorded from Little Ganilly in 1983. Typically found on grassy and sandy areas, often in great profusion in short, dune grassland.

ASTERACEAE – Daisy family
Helminthotheca echioides (L.) Holub
Bristly Oxtongue
Bryher, St Martin's, St Mary's and Tresco, also Samson and Gugh

Archaeophyte. Common around Hugh Town, Old Town and the southern part of St Mary's with scattered records elsewhere. There are just a few records from St Martin's, including near Higher Town quay, the dump at Middle Town and at Lower Town. On Bryher plants are found along the road system from near the quay, south to Veronica Farm and Hell Bay hotel. The plant was found on Samson and Gugh in 1996. Often near habitations, rough ground and along roadsides.

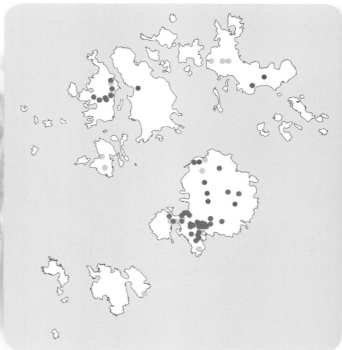

ASTERACEAE – Daisy family
Sonchus arvensis L.
Perennial Sowthistle
Inhabited islands, also Samson and St Helen's

Native. Scattered in a few places around the islands. Apparently does not persist long in one location. The St Helen's record was in 1987 and the plants on Samson were found in the dunes near the landing place in 2010. Roadsides, cultivated ground, near beaches and in wetlands.

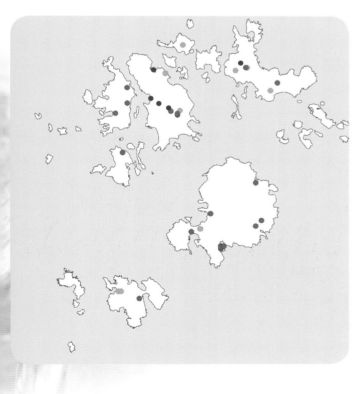

ASTERACEAE – Daisy family
Sonchus oleraceus L.
Smooth Sowthistle
Inhabited islands, also St Helen's, Northwethel, Teän, Round Island, Samson, Annet and most of the Eastern Isles

Native. Widespread and locally abundant except on rocky islets. Lousley considered this species to be more common in man-made habitats than *S. asper*, but this does not seem to hold now. May be marginally more common than *S. asper*, they grow in the same habitats roadsides, cultivated ground, dunes, rocky beaches and wetland.

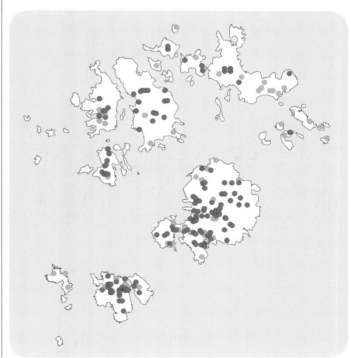

EU-DICOTS True Dicotyledons

ASTERACEAE – Daisy family
Sonchus asper (L.) Hill
Prickly Sowthistle

Inhabited islands, also St Helen's, Foreman's Island, Teän, White Island (St Martin's), Samson, Annet and most of the Eastern Isles

ASTERACEAE – Daisy family
Lactuca serriola L.
Prickly Lettuce

St Mary's

Archaeophyte. What may have been this species appeared in a field near Normandy (SV928111) in May 2015 but unfortunately the field was cut before the plants could be verified. A record from St Mary's in 2004 is without location.

Native. Widespread and locally frequent. Formerly grew on Northwethel. Grows in the same places as the *Sonchus oleraceus*; roadsides, cultivated ground, rocky beaches, dunes and wetland.

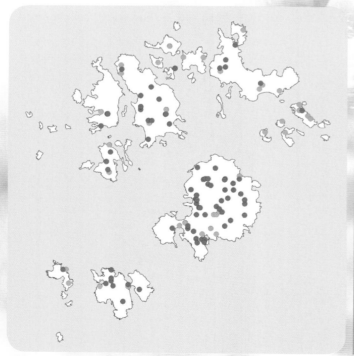

ASTERACEAE – Daisy family
Taraxacum F.H. Wigg.
Dandelions
Inhabited islands, also Teän and Samson

Native. This difficult group needs investigation in Scilly as there are clearly a number of different microspecies present. Dandelions are locally frequent in Scilly but apparently absent from parts of Tresco and St Mary's. Small, delicate, deeply pinnate-leaved dandelions that grow in dune grassland and in open sandy places appears to belong to group 'Obliqua'. Lousley called these plants *T. laevigatum*. Found on cultivated land, roadsides, dunes and near habitations.

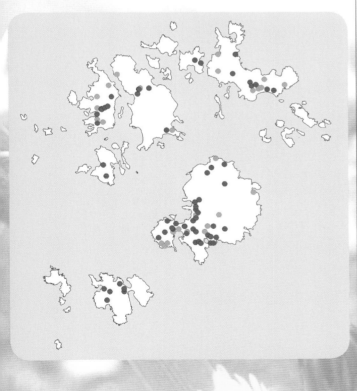

ASTERACEAE – Daisy family
Crepis biennis L.
Rough Hawk's-beard
St Mary's

Native in the British Isles but neophyte on Scilly. There are just two casual records from St Mary's, one in 2012 and one (SV92541144) from a roadside ditch in 2006.

EU-DICOTS True Dicotyledons

ASTERACEAE – Daisy family
Crepis capillaris (L.) Wallr.
Smooth Hawk's-beard
Inhabited islands, also Samson, Teän and Great Ganilly

Native. Widespread. Very tiny, stunted plants may occur in some areas of very short grassland such as at Rushy Bay, Bryher. Grassy areas on roadsides, field edges, sandy grassland and tracks on rough ground.

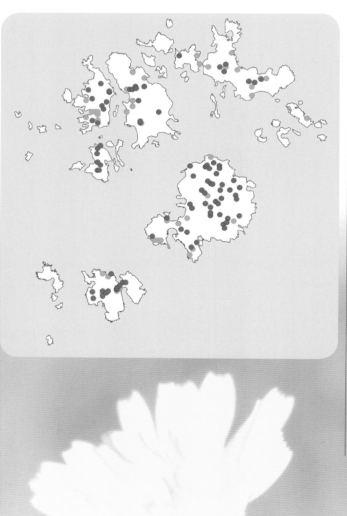

ASTERACEAE – Daisy family
Crepis vesicaria ssp. *taraxacifolia* (Thuill.) Thell. ex Schinz & R. Keller
Beaked Hawk's-beard
Inhabited islands

Neophyte. According to Lousley the plant was very local in 1939, but spread quickly over the following twenty years on St Mary's and later followed a similar course on St Agnes. From 1995 it has been found on St Martin's, and it has now been found on Bryher at Veronica Farm (SV87851480) in 2005 and Shipman Head (SV879155) in 2007. The only record from Tresco is from Borough Farm (SV89841507) in 2005. Roadsides, edges of fields and in bulbfields.

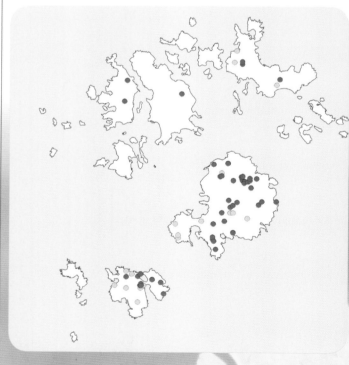

ASTERACEAE – Daisy family
Pilosella officinarum F.W. Schultz & Sch. Bip.
Mouse-ear Hawkweed
St Martin's and St Mary's

Native. Very rare. Lousley includes herbarium records from Tresco (Herb. D-S before 1940) and St Mary's (Ralfs in 1870s) although he had not seen the plant himself. For a long time the plant eluded botanists until it was found in 2002 in a small area of dune grassland above Great Bay, St Martin's (SV924163) where it was still present in 2008. The same year Colin Wild found plants in an abandoned bulbfield on St Mary's (SV91051224) that is reverting to heathland. Voucher specimens were obtained from the field in 2009. Since then the plants have apparently been overwhelmed by taller vegetation. In 2015 a small group of plants were found growing on the mound of Bant's Carn chambered tomb. Dune grassland and regenerating heath.

ASTERACEAE – Daisy family
Hieracium umbellatum ssp. *bichlorophyllum* (Druce & Zahn) P.D. Sell & C. West
Hawkweed
St Mary's and Tresco

Native. Found in approximately the same localities mapped in Lousley (1971). The main localities are between Halangy Down, Bant's Carn, Bar Point and the coast around to Innisidgen and Higher Trenoweth. There was another location between High Cross Lane and Salakee but there have been no records from there since 1993. A new find was in the north of Tresco (SV891156) in 2011. Heathy places and bracken fields, often by tracksides.

EU-DICOTS True Dicotyledons

ASTERACEAE – Daisy family
Gazania rigens L.
Treasureflower
Inhabited islands, also Round Island

Neophyte. Frequently planted and self-seeds away from gardens. It was seen on Round Island in 1987 when the lighthouse was still manned, taken there as a garden plant. Although there are many colour varieties the one often found established has yellow or orange ligules with a basal black spot, sometimes with a central white spot in the black. A plain yellow var. *uniflora* also occurs. The escape of garden-based cultivars is bound to confuse the situation. Wall-tops, verges, dumps and coastal sites.

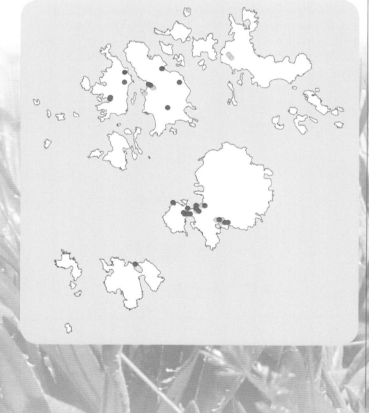

ASTERACEAE – Daisy family
Filago vulgaris Lam.
Common Cudweed
St Mary's, St Agnes, Tresco and Bryher

Native. *Filago vulgaris* has an unusual distribution pattern; although very common and widespread on St Mary's there are only a few scattered records from the other islands. On Bryher it has been found by the campsite (SV877153) in 2007, growing on a path. There are records from Old Grimsby on Tresco in 1992, from just north of Great Pool in 2007 and from the Gardens in 2009. From St Agnes it was recorded from Middle Town in 1992 and from bulbfields (SV88460823) and Barnaby Lane (SV882077) in 2007. There are no records from St Martin's. On St Mary's it can be a notable member of the weed flora of arable fields. Bulb and other arable fields, occasionally other disturbed ground.

ASTERACEAE – Daisy family
Gnaphalium sylvaticum L.
Heath Cudweed
St Mary's

Native in the British Isles but neophyte on Scilly. Only known from a pathway to Longstone (SV917109) where a plant was found in 1997 and again in 2000 but not subsequently. This rather unlikely occurrence is presumed to be an accidental introduction, although the path is partly though a shelterbelt over heathy ground.

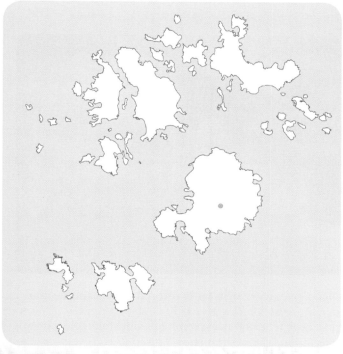

ASTERACEAE – Daisy family
Gnaphalium uliginosum L.
Marsh Cudweed
Inhabited islands

Native. The distribution of this species is similar to that of *Filago vulgaris*, common on St Mary's but rare or occasional on the other inhabited islands. As with *F. vulgaris* this is a common member of the weed flora of bulb and other arable fields, often they grow together. Gateways, disturbed ground, damp tracks and wet ground in fields and by pools.

ASTERACEAE – Daisy family
Gnaphalium luteoalbum L.
Jersey Cudweed
St Mary's and Tresco

Native or alien in the British Isles but neophyte on Scilly. This species has an interesting history. It was first noticed in the fields at Trenoweth Research Station, Higher Trenoweth (SV918123-4) in 2004. Since then it has spread to further sites on the island (SV913118), (SV920116) and (SV921116). In 2009 it was recorded in the churchyard on Tresco. In 2013 when surveying bulbfields on St Mary's with Mark Spencer we discovered seedlings were widespread in fields on Rocky Hill Farm (SV892155, SV913118, SV916111), also Lunnon Farm (SV923117). It is now known that farm machinery had driven to and from Rocky Hill Farm to Trenoweth, presumably accidentally taking Jersey cudweed seeds on the wheels. Although the origin of the first plants is not known it is possible the plant was originally an accidental introduction with horticultural material. Disturbed and cultivated ground.

ASTERACEAE – Daisy family
Helichrysum bracteatum (Vent.) Andrews
Strawflower
St Martin's and Tresco

Neophyte. A garden escape on St Martin's (SV937156) in 1994 and Racket Town Lane Tresco (SV89481510) in 2011.

Helichrysum petiolare Hilliard & B.L. Burtt
Silver-bush Everlastingflower
Tresco, Bryher, St Agnes and St Mary's, also Teän

Neophyte. Although most records of this species are closely associated with gardens, there are now a number of places where it is becoming established away from habitations. For example on Tresco it is growing in the dunes and is also spreading along the coast and on St Agnes on the cliff edge. Concern that it is spreading into natural habitats has led to IoSWT removing some plants that are becoming invasive. A plant found on Teän in 2007, was presumed to be bird-sown but seed may have blown in. Dunes, heath and waysides.

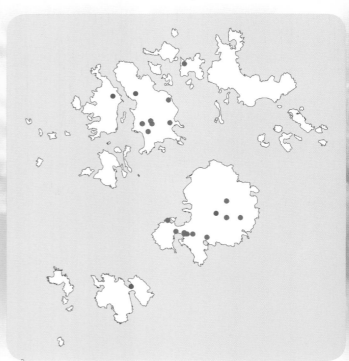

ASTERACEAE – Daisy family
Plecostachys serpyllifolia
(P.J. Bergius) Hilliard & B.L. Burtt
Cape Everlastingflower
St Mary's and St Martin's

Neophyte. Known from the wall on Newman's Battery, St Mary's for many years. Presumed originally planted there, but seedlings have spread nearby. In 2009 a plant was found growing in a hedgebank near the pub on St Martin's.

ASTERACEAE – Daisy family
Pulicaria dysenterica (L.) Bernh.
Common Fleabane
Tresco and St Mary's

Native. Although known from St Mary's since 1852 it has apparently always been rare. There are records from two places on St Mary's without exact locations, Holy Vale and a wet field near Shooter's Pool in 1972. Found at the east end of Great Pool, Tresco (SV89841449) in 2002, and both Shooter's Pool (SV912108) and Lower Moors (SV913107), St Mary's in 2007. Ditches and pool edges.

Inula helenium L.
Elecampane
Extinct

Believed last seen in 1876 near the former Castle Ennor, Old Town, St Mary's.

EU-DICOTS True Dicotyledons

ASTERACEAE – Daisy family
Solidago virgaurea L.
Goldenrod
St Mary's, St Martin's and Tresco, also St Helen's, Nornour, Great Ganilly, Great Arthur and Little Arthur

Native. Still found in approximately the same places as mapped by Lousley (1971). The slight changes probably just reflect more recording rather than a new extension of range. It appears to be mainly found along the coastal heathlands along the west of Tresco, the north and east coasts of St Mary's, heathland on St Martin's (where it is a real feature around Chapel Down), on Teän and also several of the Eastern Isles. Heathy areas and cliff edges.

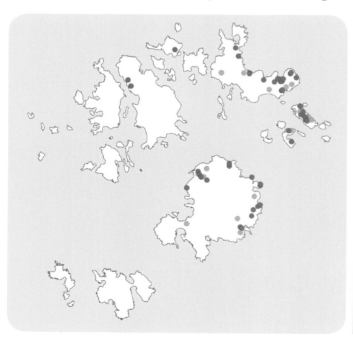

ASTERACEAE – Daisy family
Solidago canadensis L.
Canadian Goldenrod
St Mary's

Neophyte. Garden escape around Hugh Town.

ASTERACEAE – Daisy family
Aster x *versicolor* Willd.
Late Michaelmas-daisy
St Mary's and St Martin's

Neophyte. Occasional garden escape.

ASTERACEAE – Daisy family
Aster novi-belgii L.
Confused Michaelmas-daisy
St Agnes and Tresco

Neophyte. Garden escape on St Agnes last recorded in 2014 and Old Grimsby, Tresco 2015. Rough ground.

EU-DICOTS True Dicotyledons

ASTERACEAE – Daisy family
Aster lanceolatus Willd.
Narrow-leaved Michaelmas-daisy
St Agnes and Tresco

Neophyte. Garden escape on St Agnes last recorded in 1993 and Old Grimsby, Tresco 2015. Rough ground.

ASTERACEAE – Daisy family
Chrysocoma coma-aurea L.
Shrub Goldilocks
Tresco, St Martin's and St Mary's

Neophyte. A garden escape that has become established in several places; these include several sites on Tresco, notably Appletree Banks and Pentle Bay, also near the Monument on Abbey Hill. Also in Higher and Middle Town, St Martin's. There is a record from Porth Seal, St Martin's (Clement & Foster, 1994) but it has not been recorded since. Most of the sites on St Mary's and St Martin's are not far from gardens. Dunes, disturbed ground and verges.

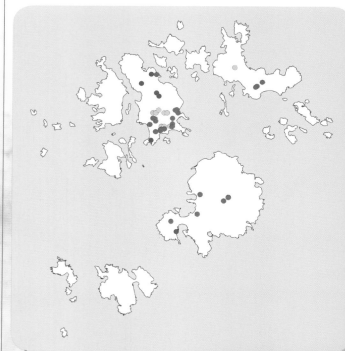

ASTERACEAE – Daisy family
Erigeron glaucus Ker Gawl.
Seaside Daisy
St Martin's, Tresco and St Mary's

Neophyte. A garden escape; Hugh Town, St Mary's; formerly at Lower Town (1996) and near Great Bay, St Martin's (1989); also a wall at Old Grimsby, Tresco. Most records are close to habitations and have originated from garden throw-outs.

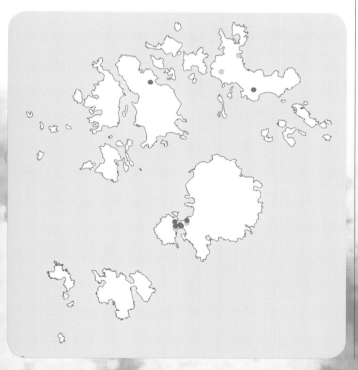

ASTERACEAE – Daisy family
Erigeron karvinskianus DC.
Mexican Fleabane
Bryher, Tresco, St Martin's and St Mary's

Neophyte. Garden escape often abundant on walls and around habitations.

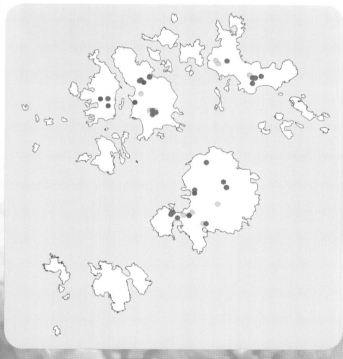

EU-DICOTS True Dicotyledons

ASTERACEAE – Daisy family
Conyza canadensis (L.) Cronquist
Canadian Fleabane
St Mary's, Bryher and Tresco

Neophyte. Possibly a recent arrival in Scilly. There are very few records since it was recorded on Tresco in 1996. This is in contrast to the other *Conyza* species which have spread rapidly since their arrival. Plants have been recorded from The Garrison, Trenoweth, Telegraph hill and the Industrial Estate on St Mary's. On Tresco there are records from near Great Pool and from the middle of the island (SV89481503). There are records from Church Quay on Bryher. All are since 2000. Disturbed and waste ground, often near habitations.

ASTERACEAE – Daisy family
Conyza floribunda Kunth (J. Rémy)
Bilbao Fleabane
St Mary's

Neophyte. Not recorded on Scilly until 2015 when the large (150 cm or more), branched and robust plants were already spreading along Telegraph Road, Pendrathen Quarry and Porthcressa. All *Conyza* plants should be checked carefully in future to identify the species and also the possibility of hybrids arising. Verges and waste ground.

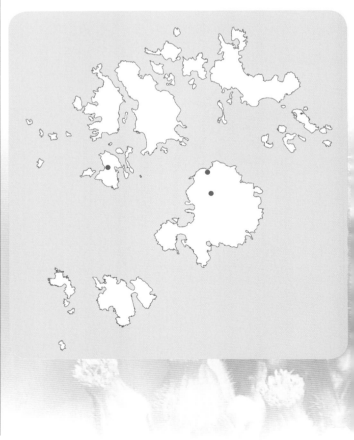

ASTERACEAE – Daisy family
Conyza sumatrensis (Retz.) E. Walker
Guernsey Fleabane
St Mary's, Tresco, St Martin's, Bryher and St Agnes, also Samson

Neophyte. Another recent arrival on Scilly which is spreading rapidly, although it was not recorded until 2007 probably due to confusion between this species and *C. canadensis*. So it is possible it may have already been present for several years before being recognised. Now plants are found on Tresco in the dunes at Appletree Banks and beside Great Pool and on Bryher along the roadside near Watch Hill and further south (SV87881493). In 2010 six plants were found in the dunes near the landing site on Samson, although these were removed, some were in the same place in 2015. Two plants were found on St Martin's (SV932157) in 2011 and two were also found by the toilets on the quay at St Agnes. In 2014 it was found in quantity in fields near the Industrial Estate and Porth Mellon, also in Holy Vale on St Mary's. The light seeds are easily parachuted to new sites so it seems likely that it will quickly become established all over the islands. Roadsides, cultivated fields and dunes.

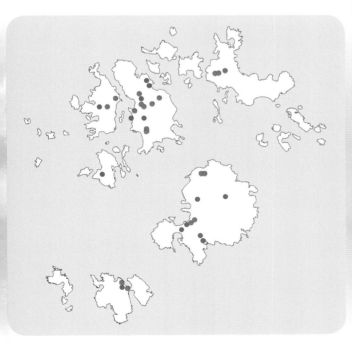

ASTERACEAE – Daisy family
Conyza bonariensis (L.) Cronquist
Argentine Fleabane
St Mary's and St Martin's

Neophyte. First recorded from the Green Farm and Watermill area of St Mary's in 2003 by E. Sears. Since then it has spread rapidly in the same area and also spread to Trenoweth, The Garrison, Pelistry, Porth Mellon and the Industrial Estate in Hugh Town. In 2012 Martin Goodey found that a field at Seaways Farm had been invaded by over 1,000 plants. Some of the sites where it occurs are some of the same places where *C. canadensis* and other *Conyza* species are also found. In 2009 three plants had been found by the roadside at Lower Town, St Martin's. The fluffy pappus hairs carry the seeds easily over a distance and it is likely that like *C. sumatrensis* and *C. floribunda*, it is going to become a very invasive weed. The distinctive 'jizz' of this species with its long side branches sometimes enables it to be picked out from the other *Conyza* species at a distance. Arable fields, gardens and waste ground.

EU-DICOTS True Dicotyledons

ASTERACEAE – Daisy family
Olearia paniculata (J.R. & G. Forst.) Druce
Akiraho
Tresco

Neophyte. Garden escape near Tresco Abbey and surrounding area.

ASTERACEAE – Daisy family
Olearia avicenniifolia (Raoul) Hook
Mangrove-leaved Daisy-bush
Tresco

Neophyte. Garden escape or planted.

ASTERACEAE – Daisy family
Olearia x *haastii* Hook. f.
Daisy-bush
Tresco and St Mary's

Neophyte. Garden escape or planted.

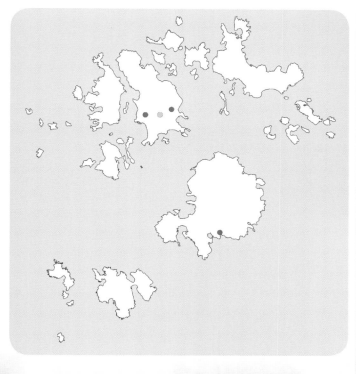

ASTERACEAE – Daisy family
Olearia macrodonta Baker
New Zealand Holly
Tresco and St Mary's

Neophyte. Garden escape or planted.

EU-DICOTS True Dicotyledons

ASTERACEAE – Daisy family
Olearia traversii (F. Muell) Hook. f.
Ake-ake
Inhabited islands

Neophyte. Frequently planted as hedging around bulbfields and in gardens. Some individual trees may have originated as escapes from cultivation although this is not confirmed. This *Olearia* had a period of popularity as an alternative to *Pittosporum* when that species proved susceptible to the hard frost during January1987. Since then it seems to have fallen out of favour, possibly because it was found to have brittle branches that snapped in strong winds.

ASTERACEAE – Daisy family
Olearia solandri Hook. f.
Coastal Daisy-bush
St Mary's

Neophyte. Tolman Point, probably planted there.

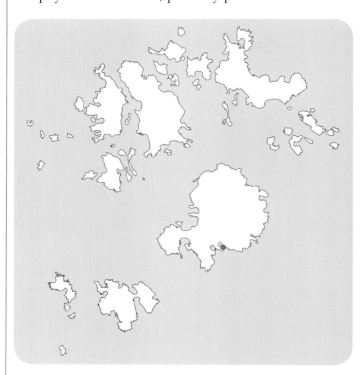

Bellis perennis L.
Daisy
Inhabited islands, also Samson, St Helen's, Teän and Nornour

Native. Abundant in most grassy habitats over the islands, but less common on the rocky coastal heathland. Where it grows in bulbfields plants can be larger than usual. Patches of plants without ray florets or with double flowers occur occasionally. At one time the 'white' lines on the St Agnes grass tennis court was marked by daises – presumably due to the lime content. Grassland, pathways and trodden ground.

ASTERACEAE – Daisy family
Tanacetum parthenium (L.) Sch. Bip.
Feverfew
St Mary's

Archaeophyte. Occasional garden escape, found in a few places on St Mary's.

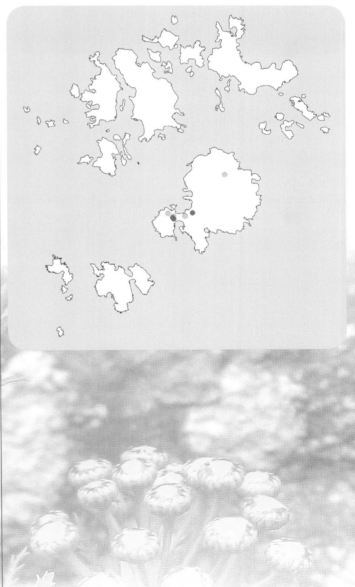

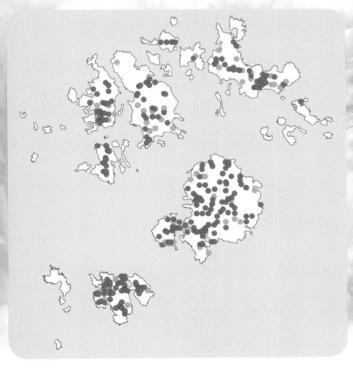

EU-DICOTS True Dicotyledons

ASTERACEAE – Daisy family
Tanacetum vulgare L.
Tansy
Bryher, St Mary's, Tresco and St Agnes

Native in the British Isles but neophyte on Scilly. A garden escape or medicinal plant that has become established in just a few places (Borlase recorded the species without locality in 1756). Still occurs on Tresco, in Back Lane (SV89131561), most recently in 2015, where it had been known since before 1852. There have been plants growing on the verge near the telephone box in Middle Town, St Agnes since at least the 1980s. It has been recorded on Hillside Farm, Bryher (SV87761471) in 2004 and St Mary's from a bulbfield (SV918112) in 1996 and Old Town (SV914103) in 2015. Roadsides, verges or near buildings.

ASTERACEAE – Daisy family
Artemisia vulgaris L.
Mugwort
Bryher and St Mary's

Archaeophyte. Found in just a few localities. On Bryher it occurs in a band from near the hotel across to the church and then up towards Hangman's Bay. On St Mary's there is a cluster of records around Hugh Town as far as Porth Mellon and King Edward's Road, and an isolated record from near Bant's Carn (SV91141248). Most plants are found on roadsides and waste places.

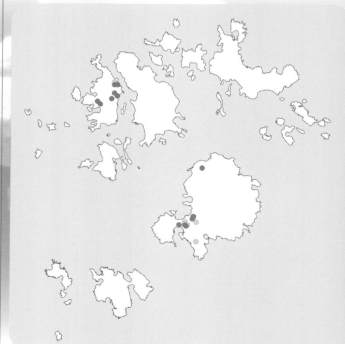

ASTERACEAE – Daisy family
Artemisia absinthium L.
Wormwood
St Agnes, Tresco, St Mary's and St Martin's

Archaeophyte. Formerly known from St Agnes where it grew near the lighthouse in 1978, from Higher Town until 1995 and from Porthcressa, St Mary's in 1988. A plant in a field near St Martin's church was last recorded in 1998. Plants recorded from Tresco SV890149 and SV892157/8 in 2015 were possible garden escapes. Garden and edge of path.

ASTERACEAE – Daisy family
Artemisia maritima L.
Sea Wormwood
St Martin's and St Agnes

Native. Known from two places on St Martin's; a field near the church (SV928157) in 1997 and Lower Town (SV916162) in 2005. On St Agnes a clump of the plant present at Higher Town (SV884083) since the early 1970s at least, was still there in 1997. Lousley failed to find the species and rejected a record from 1905 as unlikely to occur. Verges and field edges.

EU-DICOTS True Dicotyledons

ASTERACEAE – Daisy family
Santolina chamaecyparissus L.
Lavender-cotton
St Agnes, St Mary's, St Martin's and Tresco

Neophyte. Occasional garden escape.

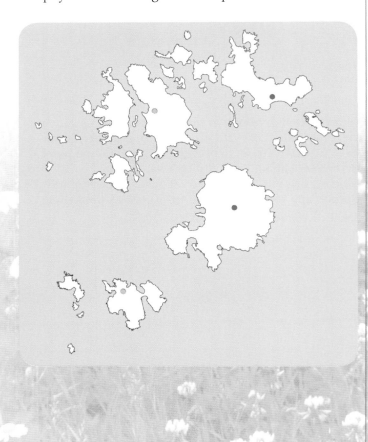

ASTERACEAE – Daisy family
Achillea millefolium L.
Yarrow
Inhabited islands, also Samson, Northwethel, Great Ganilly and Little Arthur

Native. Frequent to locally abundant on most islands, apparently less so on Tresco. Lousley also recorded it from Teän and St Helen's, where it may still be present. Although the flowers are usually white or pale pink, deeper pink-flowered plants occur on some coastal sites. Grassland and heathland, dune banks and roadsides.

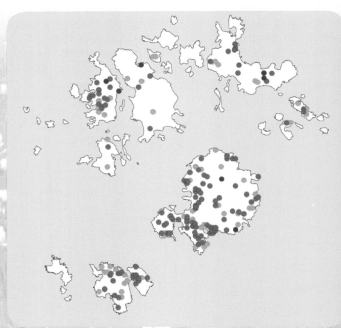

Achillea maritima (L.) Ehrhend. & Y. P. Guo
Cottonweed
Extinct

Native. Last seen by Lousley on a coastal site on St Martin's in 1936 where there had been hundreds of plants in 1922. Lousley put the extinction down to climate, possibly aided by seaweed dumping or tethering a grazing animal on the spot!

ASTERACEAE – Daisy family
Chamaemelum nobile (L.) All.
Chamomile
Inhabited islands, also Annet

Native. This is one of the great botanical delights of the Isles of Scilly. Walking over any grassland where the plant grows scents the air with its distinctive perfume. There had been no records from the uninhabited islands until 2002 when it was found on Annet near the landing site (probably accidentally introduced by visitors to the island). Double-flowered plants have been found occasionally on Bryher. Chamomile is a feature of grassy places on heathland and around the coast, mown grasslands, short turf and pathways.

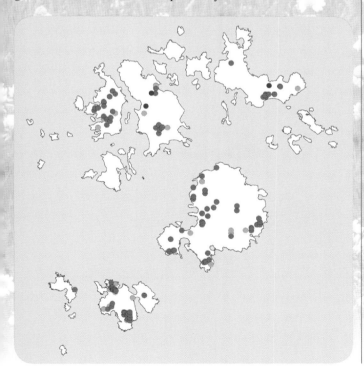

ASTERACEAE – Daisy family
Anthemis punctata Vahl
Sicilian Chamomile
St Martin's and Tresco

Neophyte. Records from St Martin's (SV925157) in 2006 and Tresco (SV898148) in 2014 on paths in woodland, are of garden origin. Our plant is subsp. *cupaniana* (Tod. Ex Nyman) R. Fern.

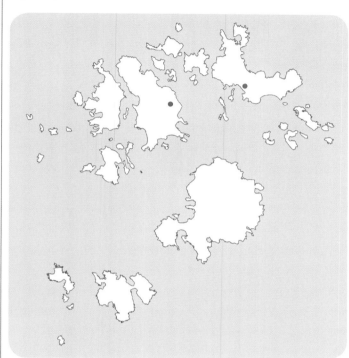

EU-DICOTS True Dicotyledons

ASTERACEAE – Daisy family
Anthemis arvensis L.
Corn Chamomile
St Mary's

Archaeophyte. No recent records, former garden escape Carn Friars area 1986 and Trenoweth 1984.

Anthemis cotula L.
Stinking Chamomile
Extinct?

Archaeophyte. Formerly St Mary's, St Agnes, Bryher in Lousley. No recent records.

ASTERACEAE – Daisy family
Glebionis segetum (L.) Fourr.
Corn Marigold
Inhabited islands

Archaeophyte. Especially abundant on St Mary's and St Agnes, scattered elsewhere. The species is known locally as 'Bothams'. In some fields it is so abundant as to dominate the crop and makes a vivid display. The bulbfield regime seems to suit the plant better than other forms of arable cultivation. Usually found in mixed weed communities in the bulb and other arable fields but sometimes almost a monoculture. Very pale-flowered plants occur occasionally. Arable fields and disturbed ground.

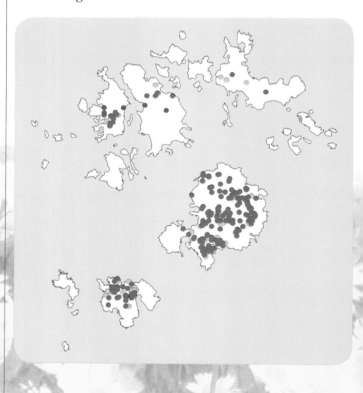

ASTERACEAE – Daisy family
Leucanthemum vulgare Lam.
Oxeye Daisy
Bryher, St Mary's, Tresco and St Agnes

Native. An uncommon plant in the Isles of Scilly. There are recent records only from Bryher, Tresco (SV894150) and St Mary's, it has not been recorded from St Martin's since Lousley. Plants on St Agnes originated from a wildflower mix sown in a field in Barnaby Lane. On Bryher it occurs in and near the church, and on St Mary's it is found in a few fields near the airfield, Harry's Walls, near Old Town church and roadsides. Some of these records are known to have originated from wild flower seed mixtures so it is difficult to be sure which populations are native. Churchyards, grasslands, walls and verges.

ASTERACEAE – Daisy family
Matricaria chamomilla L.
Scented Mayweed
St Agnes, St Mary's and St Martin's

Archaeophyte. Locally frequent in a few fields in the north of St Agnes and the Peninnis headland on St Mary's. Elsewhere known from scattered sites but apparently absent from Bryher. On Tresco it occurs by Great Pool and Borough Farm. On St Martin's it grows between Churchtown Farm and a field north of the church. Interestingly this plant appears to have been totally overlooked by Lousley who does not record it at all. Mainly cultivated fields, but also roadsides and wasteland.

EU-DICOTS True Dicotyledons

ASTERACEAE – Daisy family
Matricaria discoidea DC.
Pineappleweed
Inhabited islands, also Samson

Neophyte. Lousley considered this plant to be local and only recorded it from St Mary's, Tresco and St Martin's. It has spread rapidly since and is now common on St Agnes and Bryher as well. On Samson the plant was found near the landing and on a footpath and was almost certainly accidentally introduced by visitors to the island. Common on pathways, cultivated ground and especially in gateways, it gets carried on the wheels of vehicles and by feet.

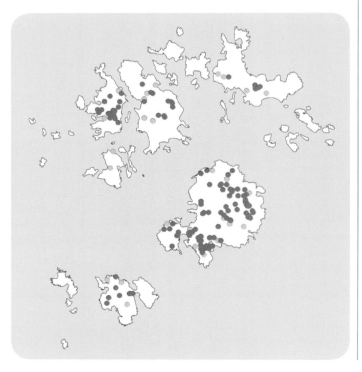

ASTERACEAE – Daisy family
Tripleurospermum maritimum (L.) W.D.J. Koch
Sea Mayweed
Inhabited islands, most uninhabited islands

Native. Widespread and locally abundant around the coasts. Absent from more-rocky islets of the Western and Norrard Rocks. Very fleshy plants are found on seabird nesting sites and double flowers also occur occasionally. Grows on cliff slopes, sandy beaches, rocky coasts and on field edges and roadsides.

ASTERACEAE – Daisy family
Tripleurospermum inodorum (L.) Sch. Bip.
Scentless Mayweed
St Agnes, Bryher, Tresco and St Mary's

Archaeophyte. Only scattered records from the inhabited islands other than St Martin's. There appears to have been some confusion between the identification of this and the previous species. Stace (2010) comments that this species might be better as a subspecies of *T. maritimum*. Generally found more inland than the previous species on cultivated ground and waste places.

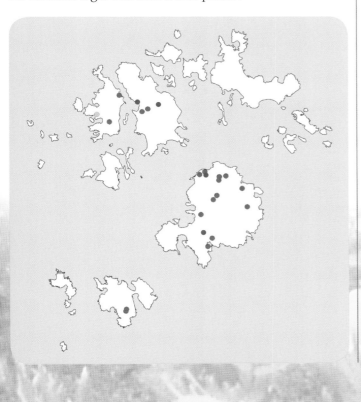

ASTERACEAE – Daisy family
Cotula australis (Sieber ex Spreng.) Hook. f.
Annual Buttonweed
Tresco, St Martin's and St Mary's

Neophyte. An uncommon garden escape established in a few places including the Abbey Gardens, Tresco; the formal area outside Hugh House on The Garrison (now lost?) and a few cultivated fields on St Mary's; and a field near the cricket field on St Martin's (SV932154).

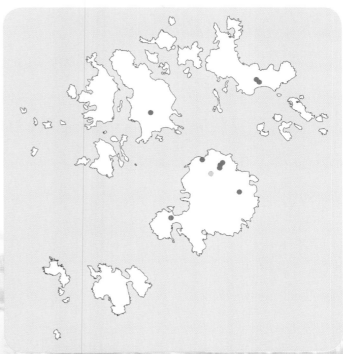

EU-DICOTS True Dicotyledons

ASTERACEAE – Daisy family
Senecio cineraria DC.
Silver Ragwort
Bryher, Tresco and St Mary's

Neophyte. Occasional garden escape. It has been on Popplestone Bank, Bryher since 1999.

ASTERACEAE – Daisy family
Senecio jacobaea L.
Common Ragwort
Inhabited islands, also Annet, Samson, Teän, White Island (off St Martin's), Nornour, Great Innisvouls and Great Ganilly

Native. Widespread and locally frequent. This would be a much more common plant in pastures or on heathland on the inhabited islands if it was not that it is regularly 'pulled' to protect grazing stock from eating it. Grows on grassland, roadsides, dune grassland and disturbed ground.

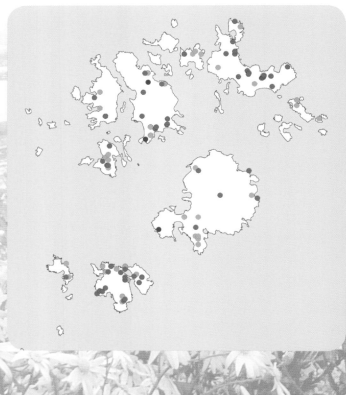

ASTERACEAE – Daisy family
Senecio glastifolius L. f.
Woad-leaved Ragwort
Tresco, St Martin's and St Mary's

Neophyte. Garden escape established in several places especially on Tresco. It is self-seeding and likely to spread further. Currently it is common around the Abbey Gardens (where it has probably been self-seeding for some years), the dune systems at both Appletree Banks and above Pentle Bay. It has also been recorded as a garden throw-out or escape including on The Garrison, St Mary's (SV898105) and near the cricket field on St Martin's (SV932154). Hedgesides, dunes and wasteland.

ASTERACEAE – Daisy family
Senecio grandiflorus P.J. Bergius
Purple Ragwort
St Mary's

Neophyte. A garden escape. At Porth Mellon, St Mary's in 1992 until 1993.

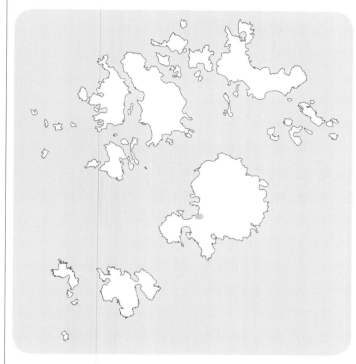

ASTERACEAE – Daisy family
Senecio squalidus L.
Oxford Ragwort
St Mary's, also Teän

Neophyte. A rare plant in Scilly with only two records; a garden at Watermill, St Mary's in 1995 and most unusually, from Teän (SV905165) found during botanical surveys by CeC in 2007.

ASTERACEAE – Daisy family
Senecio vulgaris L.
Groundsel
Inhabited islands, also Annet, Samson and Teän

Native. Locally abundant. No recent records from St Helen's or the Eastern Isles where Lousley had recorded the species. It is a common plant on cultivated ground, roadsides, waste grounds and disturbed ground including dunes.

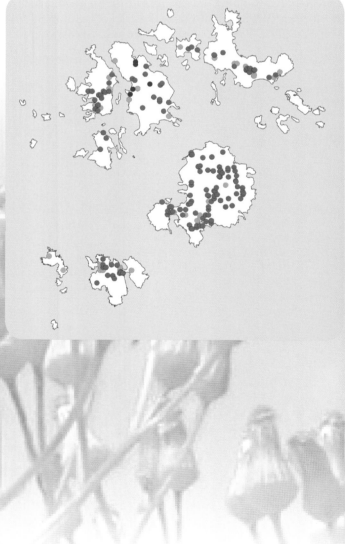

ASTERACEAE – Daisy family
Senecio sylvaticus L.
Heath Groundsel
Inhabited islands, also Annet, Teän, Menawethan, Great Ganilly and Great Innisvouls

Native. Locally frequent on most islands, but only recorded from two places on St Martin's (SV917160), the last time in 1996 and (SV93011531) in 2012. Recorded from Samson near Southward Well and the landing beach in 1984 then from the 'neck' in 2015. Lousley also knew the plant from the Eastern Isles including Nornour and Great Innisvouls, where he commented it grew under heavy salt concentrations. Widespread on heathland, in fields, near the coast and on sandy ground.

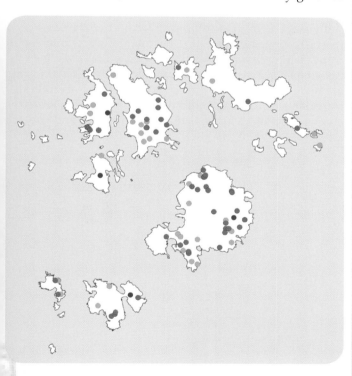

Senecio viscosus L.
Sticky Groundsel
Error?
Recorded occasionally L on St Mary's but needs confirmation. Where specimens or photographs could be checked they have turned out to be *S. sylvaticus* which is also sticky.

ASTERACEAE – Daisy family
Senecio minimus Poir.
Toothed Fireweed
Tresco, Bryher, also Samson

Neophyte. Believed accidentally introduced with garden material to Abbey gardens. The species originates from New Zealand. It first started to spread on Tresco in the 1990s; the first plants were found by Tony Butcher in October 2000 (SV89101419). In 2004 a large number of plants were found during a Wild Flower Society visit to the island led by Prof. C. Pogson, but not identified until a specimen was sent to Eric Clement by Derek Thomas (Clement, 2004). At that time about 100 plants were growing in a open area on Abbey Hill, where it was also noted by Dr G. Halliday. Since then it has spread rapidly and is now found in many places on the island. In May 2009, about 40 non-flowering plants were seen in the dunes at Rushy Bay, Bryher and later the same year eight plants were found in dunes near the landing beach on Samson, these were pulled by IoSWT staff in an effort to stop it spreading. But it is almost certainly too late to curb what is likely to become a pernicious weed in future. Dunes, woodland glades and waste ground.

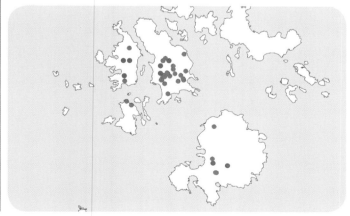

EU-DICOTS True Dicotyledons

ASTERACEAE – Daisy family
Pericallis hybrida B. Nord.
Cineraria
St Mary's, St Martin's and Tresco

Neophyte. Garden escape established and abundant in several places on St Mary's; Hugh Town, Old Town churchyard and Higher Trenoweth. It is found in several places on Tresco and St Martin's and formerly on St Agnes. Plants are found in a great range of colour varieties. Around habitations and churchyards.

ASTERACEAE – Daisy family
Delairea odorata Lem.
German-ivy
St Agnes, Tresco and St Mary's

Neophyte. Established garden escape. Lousley first recorded the species in 1939, but it had been known on Scilly since 1926. It is still found in most of the same places recorded by Lousley. The plant attracts interest from visiting botanists who are often puzzled by the ivy-shaped leaves as the yellow 'groundsel-like' flowers are not produced until winter. Mostly found scrambling over rocks or rough ground near the coast, on walls, climbing up trees, stone hedges, near houses and rubbish dumps.

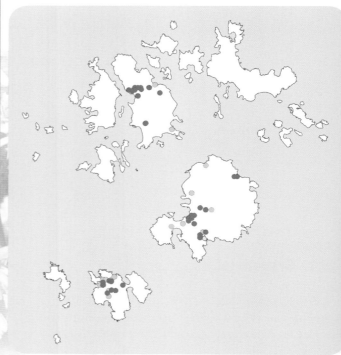

ASTERACEAE – Daisy family
Brachyglottis repanda J.R. & G. Forst.
Hedge Ragwort
Tresco, St Mary's, St Martin's and St Agnes

ASTERACEAE – Daisy family
Tussilago farfara L.
Colt's-foot
St Mary's

Native. Recently extinct. Only known from one site in a gutter in Hugh Town, St Mary's, where it had been since 2000, although it eventually disappeared in 2009 after the road was resurfaced. Lousley recorded the species from Tresco in 1939 and St Mary's in 1967, but there were no other records until the one in 2000. Other records for the species are almost certainly misidentifications for *Petasites fragrans* which also has large leaves. There is an intriguing 1981 record from Northwethel by P. Clough.

Neophyte. A planted tree or hedge on the inhabited islands. Often found away from gardens although there is no evidence it is naturalised. Two other *Brachyglottis* have also been recorded as possible garden escapes, *B.* x *jubar* (known previously as *B.* 'Sunshine') and *B. monroi* Monro's ragwort.

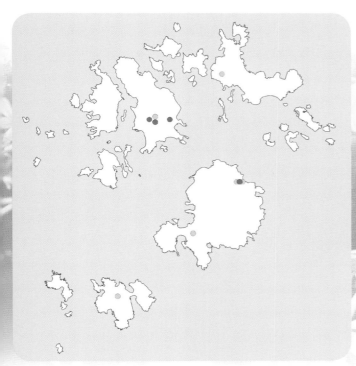

EU-DICOTS True Dicotyledons

ASTERACEAE – Daisy family
Petasites fragans (Vill.) C. Presl
Winter Heliotrope
Inhabited islands

Neophyte. Established alien on all the inhabited islands other than St Martin's where it has only been found at the dump. Found in scattered localities where it can form locally dominant patches. These sites in many cases have been known since at least Lousley's time, the St Agnes site for example since 1948. Roadsides, under trees and in a wet field.

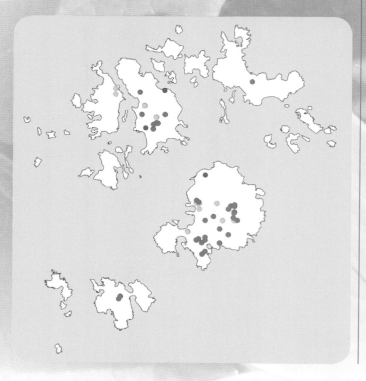

ASTERACEAE – Daisy family
Calendula officinalis L.
Pot Marigold
Bryher, Tresco, St Martin's and St Mary's, also Samson

Neophyte. Garden escape that has occurred on all the inhabited islands except St Agnes. Also found growing on the shore on Samson in 2000, but did not persist. Occurs in a few localities often close to habitation, also on the shore where garden refuse is thrown. Found on roadsides, dune banks, quarries, waste and disturbed ground.

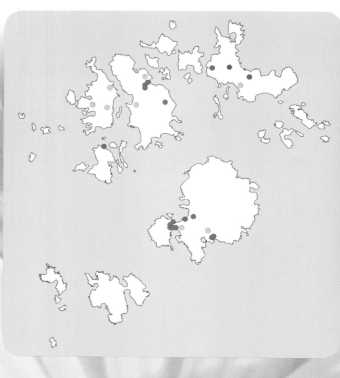

ASTERACEAE – Daisy family
Calendula arvensis L.
Field Marigold
St Mary's

Neophyte. The first record was from a bulbfield on St Mary's in 1960, in Lousley (1971) without locality. In 1972 Clare Harvey found it was abundant in two fields at Rocky Hill. Currently occurs in a few fields between Trenoweth, Rocky Hill, Sunnyside and Parting Carn. In some fields it can be a frequent member of the weed community. Bulb and other arable fields.

ASTERACEAE – Daisy family
Osteospermum jucundum (E. Phillips) Norl.
Cape Daisy
Tresco, St Agnes, St Mary's and Bryher

Neophyte. There are several different *Osteospermum* cultivars, but the one generally known as 'Tresco Purple' is the one most often found established on Scilly. Waste garden material is often dumped on roadsides or banks near the shore from where the plants can become established. As the brittle stems break and root quite easily it can spread vegetatively if not by seed. Found on banks near the sea, edges of fields and rubbish dumps.

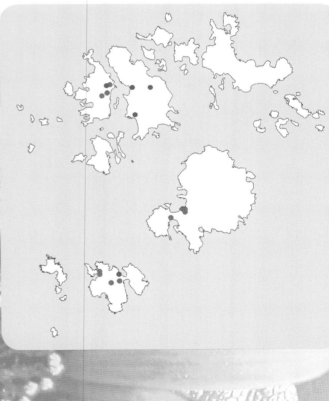

EU-DICOTS True Dicotyledons

ASTERACEAE – Daisy family
Osteospermum ecklonis (DC) Norl.
a cape daisy
Tresco, St Mary's and Bryher

Neophyte. Although there are records of this species, it is often impossible to be certain when dealing with garden cultivars.

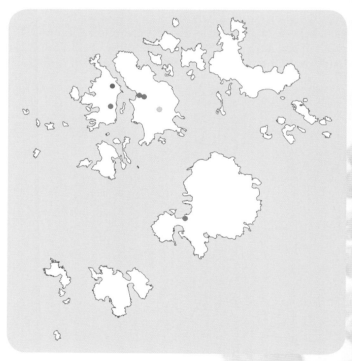

ASTERACEAE – Daisy family
Ambrosia artemisiifolia L.
Ragweed
St Mary's

Neophyte. A bird-seed annual recorded in Hugh Town in 2000 and Old Town in 2009.

ASTERACEAE – Daisy family
Helianthus annuus L.
Sunflower
St Mary's and Bryher

Neophyte. Garden escape recorded in 1988 and 2004 from St Mary's. On Bryher in 2012 among dumped soil (SV875146).

Helianthus petiolaris Nutt.
Lesser Sunflower
Tresco

Neophyte. Garden escape near the Gardens in 1996.

ASTERACEAE – Daisy family
Helianthus x *multiflorus* L.
Thin-leaved Sunflower
St Mary's

Neophyte. Rare garden escape near Green Farm, St Mary's 1993 and 1994.

Helianthus x *laetiflorus* Pers.
Perennial Sunflower
St Mary's

Neophyte. Garden escape St Mary's, 1994.

EU-DICOTS True Dicotyledons

ASTERACEAE – Daisy family
Galinsoga parviflora Cav.
Gallant-soldier
St Mary's

Neophyte. Casual. Only one record, a plant in Silver Street, Hugh Town in 2009. Probably introduced with a pot plant.

ASTERACEAE – Daisy family
Galinsoga quadriradiata Ruiz. & Pav.
Shaggy-soldier
Tresco and St Mary's

Neophyte. A rare weed recorded from Newford Farm (SV91711196) and from a garden at Watermill, St Mary's (SV919121), also from a street in Hugh Town in 2010 and 2011. On Tresco it has been recorded from fields north of Great Pool (SV89301557 and SV89641463) and around Abbey Farm Estate Office. Cultivated and disturbed ground.

ASTERACEAE – Daisy family
Bidens tripartita L.
Trifid Bur-marigold
Tresco

Native. Believed to be a recent arrival, probably introduced by waterfowl. The first record was in 1994 when a few plants were discovered growing in a muddy area beside Great Pool, Tresco. Since then the plant has now spread to other parts of the lake margin and also to Abbey Pool in 1998. Wetland areas, around pool edges.

ASTERACEAE – Daisy family
Eupatorium cannabinum L.
Hemp-agrimony
St Mary's and St Agnes

Native in the British Isles but neophyte on Scilly. Accidental introduction. A plant was found near Old Town Church in 1975, then one was found growing in a garden on St Agnes in 1981 where it had apparently been introduced with plant material from the mainland. That plant did not persist. Similarly introduced to the garden at the Longstone Centre where it may now be established.

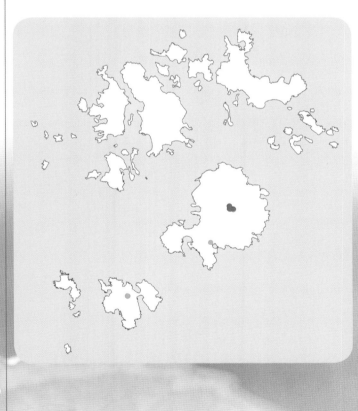

EU-DICOTS True Dicotyledons

ASTERACEAE – Daisy family
Ageratum houstonianum Mill
Flossflower
St Martin's

Neophyte. Garden escape at Higher Town, St Martin's (SV937156) in 1994.

ESCALLONIACEAE – Escallonia family
Escallonia macrantha Hook & An.
Escallonia
Inhabited islands

Neophyte. Formerly planted as hedging, but may have fallen out of favour recently. An occasional escape from cultivation or as relict in abandoned fields.

CAPRIFOLIACEAE – Honeysuckle family
Sambucus nigra L.
Elder
Inhabited islands, also Samson and St Helen's

Native. Scattered Elder bushes are found on all inhabited islands in places where it is not possible to tell whether they have been planted, are native or bird-sown. It is not clear when Elder was introduced to Scilly although it is suspected it may have been brought to Tresco by the Tavistock monks in the 11th century. That there were elders on Tresco, and that they were common enough for the island to be named Trescau ('homestead of elder-trees') in 1305 is significant. In his Itinerary of 1535-43 Leland wrote; 'There is another cawled Inisschawe [Tresco], that is to sey, the Isles of Elder, by cawse yt bereth stynkkyng elders' (Leland in Smith, 1906 to 1910). There are two Elder trees on Samson that are believed to have been planted by the former inhabitants, so these may be more than 170 years old. Another bush grows by the Oratory on St Helen's where it may also have been planted for medicinal use when the island was inhabited. Most of the trees now appear to be associated with habitations or spread by birds. Hedgebanks and near houses.

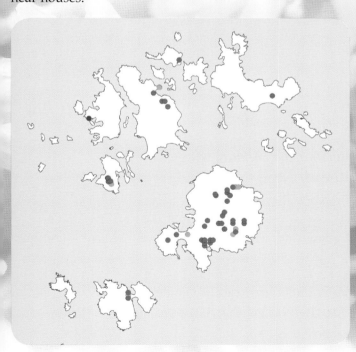

CAPRIFOLIACEAE – Honeysuckle family
Viburnum opulus L.
Guelder–rose
St Martin's

Native in the British Isles but neophyte on Scilly. A garden escape. Higher Town 2011.

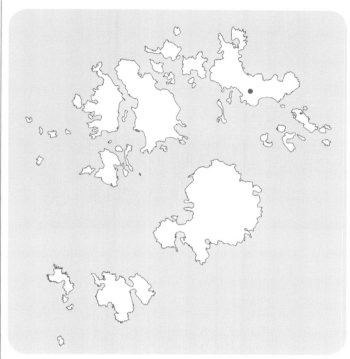

Viburnum tinus L.
Laurustinus
St Agnes

Neophyte. Garden escape or planted.

EU-DICOTS True Dicotyledons

CAPRIFOLIACEAE – Honeysuckle family
Lonicera periclymenum L.
Honeysuckle
Inhabited islands, also Teän, White Island (St Martin's), Toll's Island (St Mary's), St Helen's, Northwethel and most of the Eastern Isles

Native. Abundant even on the small islands, although not recorded on Annet. It has been found on Great and Little Ganilly, Middle and Little Arthur, Great Ganinick and Nornour in the Eastern Isles. As Lousley commented, Honeysuckle in Scilly has limited options as a climber and not only scrambles over gorse bushes and brambles but also grows along the ground in places. Plants near the shore often have very fleshy leaves. The large and highly-scented flowers can be pale yellow in some places and deep red in others. Widespread in heathland habitats, on cliff slopes, hedgebanks, on walls and in dunes.

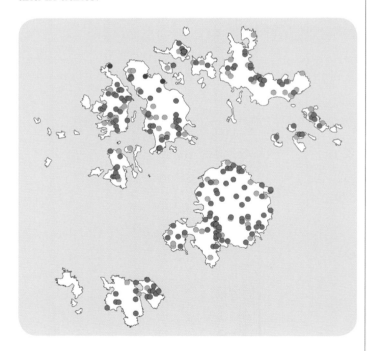

Lonicera caprifolium L.
Perfoliate Honeysuckle
Extinct

A plant outside the gardens on Tresco has not been seen since 1964.

VALERIANACEAE – Valerian family
Valerianella locusta (L.) Laterr.
Common Cornsalad
Inhabited islands, also Teän

Native. A common plant according to Lousley's Flora, but now only recorded from a few scattered sites and apparently declining. The only records from Bryher were from near the quay (SV880149) in 1995. Records from Tresco are from around Great Pool, the most recently near the Swarovski hide in 2005. The records from St Agnes are almost all from the suite of fields above the Bar (SV885082-3) and in 1994 from fields at Lower Town. The variety *dunensis* P.D. Sell was formerly recorded from West Porth, Teän by David Allen in 1961, but not since. Arable fields and disturbed ground.

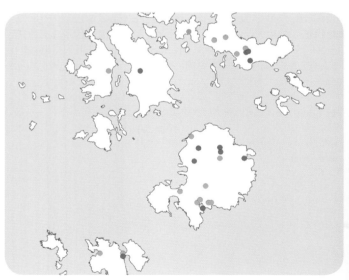

Valerianella carinata Loisel
Keel-fruited Cornsalad
Extinct?

Formerly at Higher Town, St Agnes, and bulbfields near Lunnon and near Old Town church, St Mary's (Lousley). No recent records.

Valerianella dentata (L.) Pollich
Tooth-fruited Cornsalad
Extinct?

A specimen was collected by Lousley on Tresco in 1966.

VALERIANACEAE – Valerian family
Valeriana officinalis L.
Common Valerian
St Mary's

Native on the mainland but neophyte on Scilly. Occurred as a casual by Old Town church, St Mary's, found by L. Knight in 2003 and still there until 2004. Possibly introduced with plants for the churchyard.

VALERIANACEAE – Valerian family
Valeriana dioica L.
Marsh Valerian
St Mary's

Probably native in Scilly, but absent from Cornwall. Its presence on Scilly could be explained as a natural introduction from birds - Lower Moors is a well-known site that attracts migratory or wintering waterfowl that may have introduced seeds. A plant that appeared in June 2000 in a wet area in Lower Moors (SV913105) cleared to benefit orchids, probably arose from buried seed. The area has grown over again and the plant has not been recorded since.

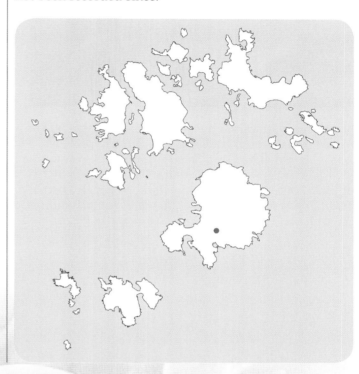

EU-DICOTS True Dicotyledons

VALERIANACEAE – Valerian family
Centranthus ruber (L.) DC.
Red Valerian
Inhabited islands, also Samson

Neophyte. Naturalised on all the inhabited islands. The record from Samson in 2013 is unusual and probably was a plant that had originated from material brought to the island by gulls or washed up. Plants are mostly red or pink-flowered but white-flowered plants also occur. Often grows near habitations, on walls, rubbish dumps, roadsides and quarries.

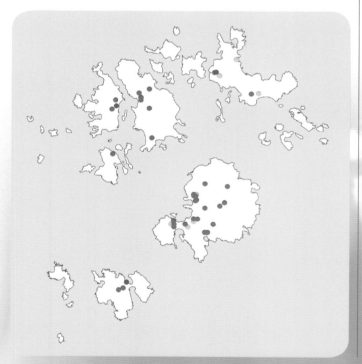

DIPSACACEAE – Teasel family
Dipsacus fullonum L.
Wild Teasel
Bryher, Tresco and St Mary's

Native or alien in the British Isles but a neophyte on Scilly. It is occasionally introduced as a garden plant, or as a component of bird seed. On at least one occasion seeds have been spread deliberately! Despite this it does not seem to have become established although a few plants turn up at different sites from time to time. Roadsides and waste ground.

GRISELINIACEAE – Broadleaf family
Griselinia littoralis (Raoul) Raoul
New Zealand Broadleaf
Tresco and St Mary's

Neophyte. Naturalised or planted in Tresco Wood and possibly on Peninnis Farm Trail, St Mary's. This is a popular hedging plant in Cornwall that self-sows readily, so may well become more common in Scilly.

PITTOSPORACEAE – Pittosporum family
Pittosporum crassifolium Banks & Sol. Ex A. Cunn.
Karo (usually called **Pittosporum**)
Inhabited islands, many uninhabited islands

Neophyte. Pittosporum is the most often planted and successful hedge species in Scilly. It was introduced from New Zealand; an illustration of a specimen from Tresco Abbey Gardens was published in 1872 (Lousley, 1971) although the date of its introduction to the Abbey Gardens is given as 1890 (King, 1985).

Pittosporum was found to be superior to other hedging species such as Tamarisk and Hedge Veronica, being both fast-growing and windproof. Additionally it was found to be palatable to cattle so hedge trimmings could be fed to them when there was not much grass. Lousley comments that the plant was only known as a planted hedge until 1952 when he noted many young plants growing on rock cairns that were already five to six feet tall. At the time although he assumed birds were spreading the seeds he was possibly unaware that it was blackbirds and thrushes that were responsible. The only disadvantage of Pittosporum as hedging is that it is not completely hardy and can be cut back or killed in cold winters. For example, in January 1987, when many hedges died. Although some of the later replanting was with *Olearia* and other shrubs much of the Pittosporum grew back from the base and as the alternative shrubs were less successful, so it has returned to favour. Pittosporum occurs both as a planted hedge and naturalised, although the map probably does not give

EU-DICOTS True Dicotyledons

a true picture of its distribution. It is now established on all the inhabited and many uninhabited islands including the larger ones such as Samson, Annet, St Helen's, White Island (St Martin's), Northwethel, Round Island and Great Ganilly, but it has also colonised quite small islets such as Plumb Island off Tresco and Toll's Island off St Mary's. As it is bird-sown it is likely to turn up anywhere. There is a concern that the shrubs are shading out native plants and lichens, for example on The Gugh, so some control measures are planned. Granite carns, heathland and on the coast.

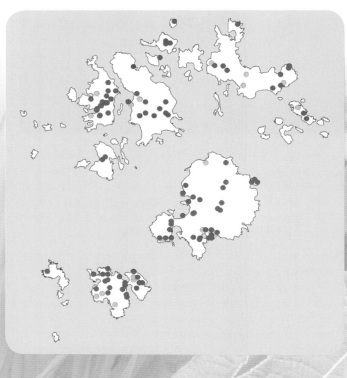

PITTOSPORACEAE – Pittosporum family
Pittosporum tenuifolium Gaertn.

Kohuhu

Tresco and St Mary's

Neophyte. Escape from cultivation or planted outside Abbey Gardens, Tresco and near Juliet's Garden, St Mary's. This includes the species with smaller, crinkly-edged leaves that is favoured by flower-arrangers.

ARALIACEAE – Ivy family
Hedera hibernica (G. Kirchn.) Bean
Atlantic Ivy
Inhabited islands, also Samson, Toll's Island (St Mary's), Teän, St Helen's, Northwethel, Great Ganinick, Little and Great Ganilly

Native. Locally abundant. The common ivy on Scilly is *Hedera hibernica*. The cultivar 'Hibernica' (Irish Ivy) recorded on St Mary's at Old Town is of garden origin. Some *Hedera helix* Common Ivy may occur in Hugh Town, St Mary's. Atlantic Ivy grows on walls, over rocks and on the ground both in woodland and open heathland including on cliffs.

ARALIACEAE – Ivy family
Fatsia japonica (Thunb.) Decne. & Planch.
Fatsia
St Agnes, Tresco, Bryher and St Mary's

Neophyte. Occasional plants found away from gardens may be self-sown.

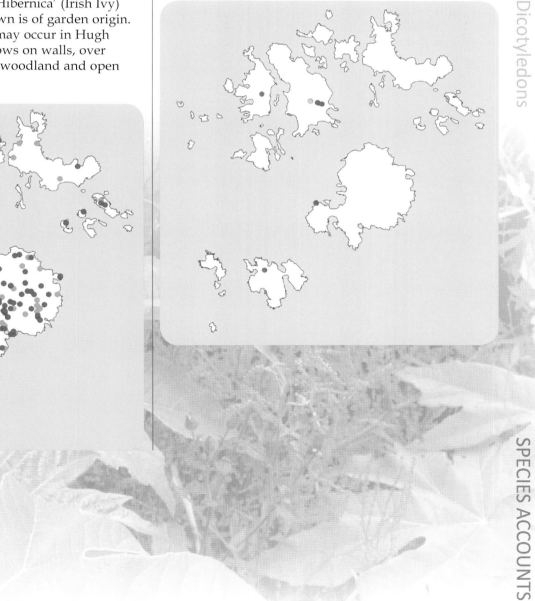

EU-DICOTS True Dicotyledons

HYDROCOTYLACEAE – Pennywort family
Hydrocotyle vulgaris L.
Marsh Pennywort
Inhabited islands

Native. Locally frequent to abundant on the inhabited islands. There are no records from the uninhabited islands perhaps because there are few wetland sites there. Grows in all manner of damp places in grassland, heathland, ditches, pathways, around pool margins and in seasonal ponds such as the craters on Salakee Down, St Mary's.

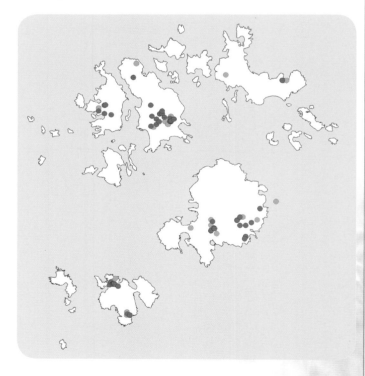

Hydrocotyle ranunculoides L. f.
Floating Pennywort
St Mary's

Neophyte. Only known in 2005-6 from a pool at the Longstone Centre, St Mary's where it had been planted.

APIACEAE – Carrot family
Eryngium maritimum L.
Sea-holly
Bryher, Tresco, St Martin's, St Agnes (also St Mary's)

Native. Locally frequent in a number of discrete sites on all the inhabited islands other than St Mary's. Populations may fluctuate from season to season, due to storms washing the tops of the plants away where they grow on sandy banks at the top of beaches. The only recent record from St Mary's is a casual record of a small plant growing at the base of a wall on The Garrison in 1991 which did not survive. There are no records from Teän since Lousley's Flora. It grows above HWM on sandy and shingle beaches and dune banks.

APIACEAE – Carrot family
Anthriscus sylvestris (L.) Hoffm.
Cow Parsley
Tresco, St Martin's, St Mary's and St Agnes

Native. Widespread in the northern half of St Mary's, apparently spreading, with just occasional records from other islands. These include Lower Town, St Martin's in 1995 and by the roadside (SV92461593) in 2007. On St Agnes it was found by Porth Killier (SV881086) and Barnaby Lane (SV883078) in 1995 and in a field above the Bar (SV885082) in 1997. On Tresco the only record was from by Great Pool (SV899144) in 2008. Lousley commented that the plant was probably a recent arrival to Scilly. Certainly it has only been recorded from St Agnes, Tresco and St Martin's comparatively recently, and does not appear to have spread far. Lousley recorded it from near the ruins on Samson, but has not been found there since. Even on St Mary's the trend seems to have been a slow colonisation of the island along road verges. Grows along roadsides, field margins and on waste ground.

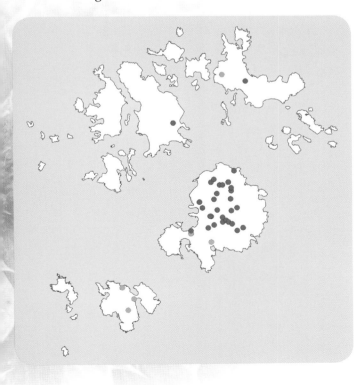

APIACEAE – Carrot family
Anthriscus caucalis M. Bieb.
Bur Chervil
St Martin's, Tresco, St Agnes, also Teän

Native. Apparently in decline. Only found in any amount on St Martin's, but there are now fewer available sites with the downturn in bulb farming. On Tresco it was recorded from Pool Road in 1993 and in a bulbfield at Town's Hill in 2001. There is only one record from St Agnes, from a bulbfield at Higher Town that has since been converted to grassland. There is a recent record from Teän in 2007 (very likely having spread from St Martin's). Found in sandy places, bulbfields and along roadsides.

Anthriscus cerefolium (L.) Hoffm.
Garden Chervil
Extinct

Neophyte. An old record from the vegetable garden at the Abbey Garden, Tresco in 1927 was clearly an escape from cultivation.

EU-DICOTS True Dicotyledons

APIACEAE – Carrot family
Scandix pecten-veneris L.
Shepherd's-needle
St Martin's formerly St Mary's

Archaeophyte. Apparently now restricted to St Martin's where its range has been contracting. Although it may be abundant in individual fields it has decreased overall. Found mostly in fields at Lower Town, scattered elsewhere on the island. There are a few casual records from St Mary's, including from Holy Vale and in a garden in Hugh Town (1996), where it was accidentally introduced with plant material. A weed in bulb and arable fields.

APIACEAE – Carrot family
Smyrnium olusatrum L.
Alexanders
Inhabited islands, also Samson and Round Island

Archaeophyte. An ancient introduction to Scilly. It is now abundant on all the inhabited islands and also occurs on Samson and is still on Round Island. In the past it has been blamed for an outbreak of blisters around the mouths of local children: they had been using hollow stems as pea-shooters and whistles. But more likely the offending plant was *Heracleum sphondylium* Hogweed. When fruiting the seeds are eaten by rats who climb into the plants making audible cracking sounds as they eat them. There has been some revival in interest in using the plant as a pot herb for flavouring stews; it tastes a bit like celery. Grows mainly on roadsides, pathways, by the coast, field edges and in waste places.

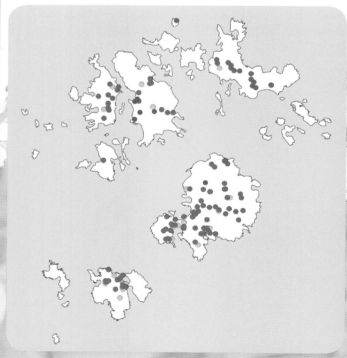

APIACEAE – Carrot family
Conopodium majus (Gouan) Loret
Pignut
St Martin's

Native. Now only found on St Martin's, although it had been recorded from St Mary's and Tresco in the 1800s. Almost all the recent records are from between SV933157 and SV93701568 and the edge of Chapel Down. In May 2014 the fields were white with thousands of the plants in flower. There is one record in 1988 from SV917160 in the west of the island. Often found under bracken on heathland and in a pasture.

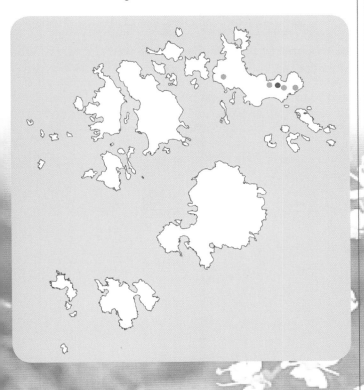

APIACEAE – Carrot family
Aegopodium podagraria L.
Ground-elder
Tresco, St Mary's and St Agnes

Archaeophyte. There are only a few scattered records from along roadsides and around habitations, only one of which is post-1995. For some reason this normally invasive plant appears to be very rare on Scilly. On Tresco it has been recorded at Borough Farm, near the church and Abbey Gardens. On St Mary's there are records from near Star Castle and on King Edward's Road (including one from there in 2008). On St Agnes there is a record from near the Parsonage. Most of these records are a while ago and it is suspected some could be attributed to confusion with emerging *Smyrnium olusatrum* which looks very similar. Roadsides and hedgebanks.

Pimpinella saxifraga L.
Burnet Saxifrage
Extinct

Recorded in 1952 and again (three plants) in 1956 between Porth Mellin and Thomas' Porth, St Mary's.

EU-DICOTS True Dicotyledons

APIACEAE – Carrot family
Berula erecta (Huds.) Coville
Lesser Water-parsnip
St Mary's and Tresco

Native. Very rare, there have been no confirmed records since 1997. On Tresco it has been recorded from Great Pool (SV894146) and on St Mary's from Lower Moors (SV913105) and Newford Duck Ponds. This appears to be another plant which suffers from confusion with a similar species, in this case *Apium nodiflorum*. Although Lousley has notes of the species in Lower Moors and on the north side of Porthellick Road we have not been able to find it recently. In ditches, pools and the 'Moors'.

Oenanthe lachenalii C.C. Gmel.
Parsley Water-dropwort
Extinct

The only record was from near Great Pool, Tresco in 1923.

APIACEAE – Carrot family
Crithmum maritimum L.
Rock Samphire
Inhabited islands, also Teän, Northwethel, St Helen's, Samson, Annet, White Island (off St Martin's), the Arthurs, Nornour, Great Ganilly and Little Ganilly

Native. Abundant around the coastal fringe where it is one of the constant species in the NVC MC1 *Crithmum maritimum – Spergularia rupicola* maritime rock-crevice community. This plant has been used pickled in the past. There is an account by Robert Heath in 1750 of the preservation of samphire for pickling by covering it in strong brine and then pickling in vinegar – when it was 'sent in small Casks to distant Parts for Presents' (Heath, 1750). In his 'Observations', Borlase (1756) comments that the 'Sampier – is the best and largest kind (far superior to the Cornish)'. There has been a recent resurgence in interest in using the plant both as a pickle and as an herb. A typical plant of the coastal fringe, growing from just above HWM on rocks, walls and cliffs, usually found not far from the sea.

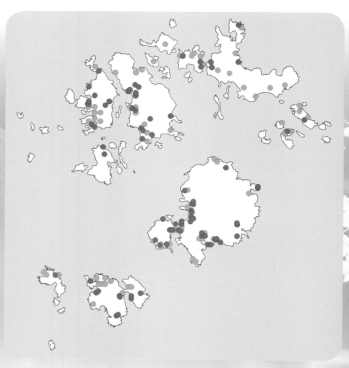

APIACEAE – Carrot family
Oenanthe fistulosa L.
Tubular Water-dropwort
Tresco and St Mary's

Native. A rare species recorded from a site among reeds beside Great Pool, Tresco (SV89461463). On St Mary's it has been found in two places on Lower Moors among marshy vegetation (SV913105) and under willow carr (SV91281058). And also in Porthloo fields (SV910111), by the boardwalk from Porthloo Lane to studios (SV90981114) and the Lower Moors extension (SV91091109). Only the Tresco site and the Porthloo Lane site were refound in 2009 during Threatened Plant Project surveys. In 2011, Martin Goodey found a few plants at a new site beside Shooter's Pool, St Mary's (SV912107), still there in 2013. There were several plants at the Lower Moors willow carr site in 2014. At some other former sites it has not been relocated, possibly due to competition with more robust or taller vegetation. Wet fields and reedbeds.

APIACEAE – Carrot family
Oenanthe crocata L.
Hemlock Water-dropwort
St Mary's, St Agnes, Bryher and St Martin's

Native. Locally abundant on wetland sites on the inhabited islands; apparently absent from Tresco. The main populations are on St Mary's where it follows the line of wetland habitats that cross St Mary's from Lower Moors to Porthloo and Holy Vale to Porthellick. Also in ditches, fields and wet flushes elsewhere. As it is very toxic, in some places attempts have been made to eradicate it to protect farm stock. There apparently has been an increase in range; Lousley only recorded it on St Mary's and Bryher. On St Agnes a few plants grow in a wet flush on Wingletang and beside a footpath behind Covean. On St Martin's plants grow near the cricket field pool and a verge at Higher Town. Some of these sites are in drier habitats than would be usual on the mainland. Found usually in wet habitats including ditches, around pools and seepages, also in fields and on verges.

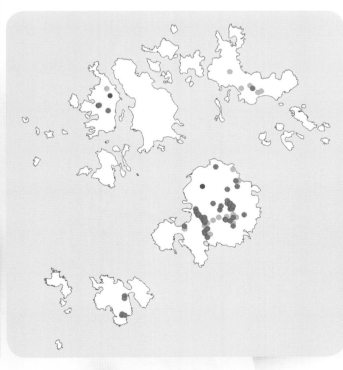

EU-DICOTS True Dicotyledons

APIACEAE – Carrot family
Foeniculum vulgare Mill.
Fennel
Inhabited islands

Archaeophyte. Garden escape, locally frequent, especially on St Mary's. Appears to have spread further in recent years, moving out from centres where the plants have been growing for years. Grows on roadsides, on walls near houses, sandy areas near beaches and on waste ground.

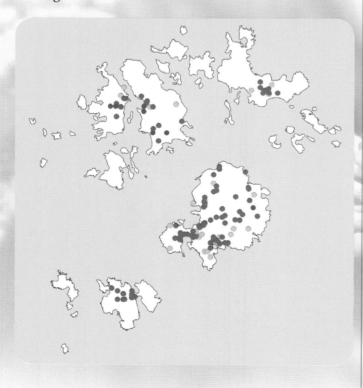

APIACEAE – Carrot family
Conium maculatum L.
Hemlock
St Mary's, also St Helen's and Samson

Archaeophyte. It is difficult to understand what has happened to this species. Lousley records it as frequent and from all the inhabited islands, but this is no longer the case. It is now a rare plant only recorded from Lower Moors, St Mary's (last in 1995), from near the Pest House on St Helen's where it gets cut down in the annual management work and may now have disappeared, and on Samson, most recently in 2007. Grows on waste ground and around buildings.

Bupleurum subovatum Link ex Spreng.
False Thorow-wax
Extinct

Neophyte. There are a few casual records, from Old Grimsby, Tresco 1968 and The Garrison, St Mary's in 1967, possibly from bird-seed.

APIACEAE – Carrot family
Apium graveolens L.
Wild Celery
St Mary's, St Martin's and Bryher

Possibly native or neophyte on Scilly. The plant recorded from St Martin's tip in 2014 was believed to be var. *dulce* – the cultivated form. The plant found on the Garrison may also be this. What may be the native plant was found on Bryher (SV880151) in 2014.

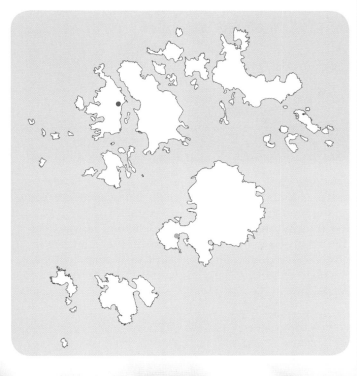

APIACEAE – Carrot family
Apium nodiflorum (L.) Lag.
Fool's-water-cress
Tresco and St Mary's

R PARSLOW

Native. Locally frequent in some wetland areas. Around Great Pool, Tresco; in Lower and Higher Moors, St Mary's and along the associated watercourses, also Salakee Lane ditch, Newford Duck ponds to Watermill stream, also pools at Longstone and Carrick Dhu. Plants in Holy Vale stream, Lower and Higher Moors, St Mary's are exceptionally large and robust. It is believed some records of *Berula* may be misidentifications for this species. Marshes, ditches, streams and pools.

EU-DICOTS True Dicotyledons

APIACEAE – Carrot family
Apium inundatum (L.) Rchb. f.
Lesser Marshwort
Bryher, Tresco and St Mary's

Native. A rare plant in Scilly restricted to a very few sites; Little Pool, Bryher; Abbey and Great Pools, Tresco; Lower and Higher Moors, St Mary's. Sometimes refound when ditches are cleaned out as on the Moors. Grows on bare mud in marshy places and beside pools.

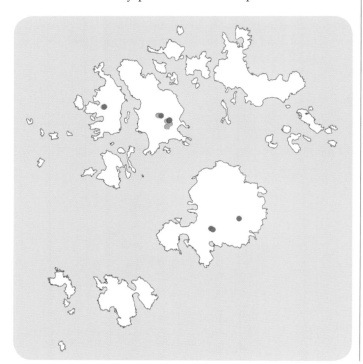

APIACEAE – Carrot family
Petroselinum crispum (Mill.) Fuss
Garden Parsley
St Mary's and St Martin's

Garden escape that persists on The Garrison, St Mary's and was found in Old Town in 2012; it was also found on the roadsides in Middle and Higher Town, St Martin's in the 1990s. There is an unusual record from South Hill, Samson by J. Bevan in 1973 which there is no reason to doubt. The mild climate appears to allow the plant to survive through winter in Scilly. Walls and roadsides.

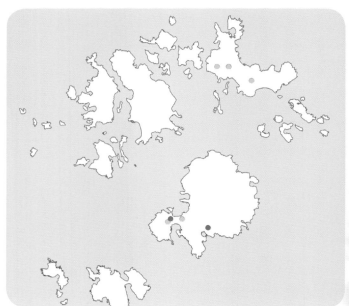

Ammi majus L.
Bullwort
Extinct

One casual record from Ram's Valley, Hugh Town, St Mary's 1987.

APIACEAE – Carrot family
Angelica sylvestris L.
Wild Angelica
St Martin's, Tresco, St Agnes and St Mary's, also Great Ganinick

Native. This species has a very patchy distribution, occurring on just four of the inhabited islands. It was last recorded on Great Ganinick in 1990, but could still be there. Where it is found it can be locally frequent, as in heathland on the north side of St Martin's where it seems to be growing in quite a dry habitat. Has increased since the map in Lousley's Flora. On St Mary's, along the edge of the reedbed on Higher Moors some plants grow exceptionally tall, up to three metres (something also mentioned in Lousley, 1971), but most are much shorter. It has also invaded the bar between Porth Hellick and Porthellick pool. Damp grassland, heathland and marshy places.

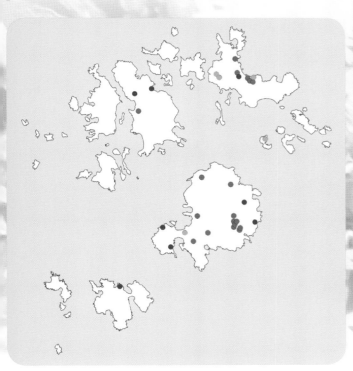

APIACEAE – Carrot family
Angelica archangelica L.
Garden Angelica
Tresco

Recorded from Pool Road, Tresco in 2007 as a garden escape. There is a possibility the plants were actually the very similar *Melanoselinum decipiens* which is spreading outside the Gardens. See page 424.

EU-DICOTS True Dicotyledons

APIACEAE – Carrot family
Melanoselinum decipiens
(Schrad. & J.C. Wendl.) Hoffm.
Madeira Giant Black Parsley
Tresco

Garden escape. This species is not included in Stace (2010) and has only recently been found away from the Abbey Gardens where it is well-established. Plants have been found along Pool Road and near Old Grimsby. It is a very robust plant, with dark leaves and pinkish flowers that grows to two metres tall or more.

APIACEAE – Carrot family
Pastinaca sativa ssp. *sylvestris*
(Mill.) Rouy & E.G.Camus
Wild Parsnip
St Mary's

Native. A rare plant in Scilly, possibly still on The Garrison where it was last recorded in 1990 and 1997, also found on the track to Salakee Farm (SV91971088) in 2007. There are no recent records from St Martin's where it was last recorded in about 1939. Roadsides and cliffs.

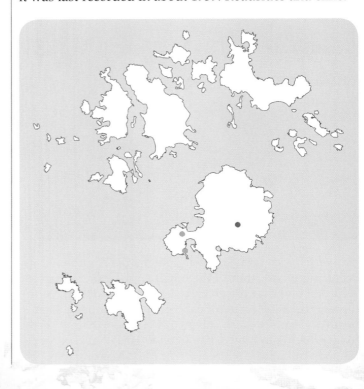

APIACEAE – Carrot family
Heracleum sphondylium L.
Hogweed

Inhabited islands, also Annet, Samson, St Helen's, Foreman's, Teän, Nornour, Great Ganinick and Great Ganilly

Native. A widespread and abundant species on the larger islands. It is very possible it was this species that caused the blisters and burns (phytophotodermatitis) suffered by local children using the hollow stems for peashooters, rather than *Smyrnium olustrum* which was blamed at the time. Grows in a range of habitats; field edges, cultivated land, roadsides and rough slopes on cliffs.

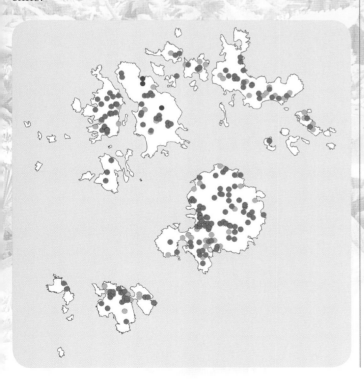

APIACEAE – Carrot family
Torilis japonica (Houtt.) DC.
Upright Hedge-parsley

St Mary's

Possibly native to Scilly. There is just one casual record of this species, possibly accidentally introduced with trees imported from the mainland. Found in one of the fields managed by the IoSWT (SV911109) in 2006. The only previous record was in 1909 about which Lousley was dubious.

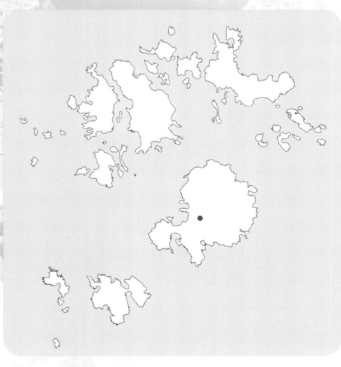

EU-DICOTS True Dicotyledons

APIACEAE – Carrot family
Torilis nodosa (L.) Gaertn.
Knotted Hedge-parsley
St Agnes and St Martin's

Native. Now a very rare plant with only two relatively recent records; near Porth Killier, St Agnes in 1999 and in fields between Middle and Higher Towns, St Martin's in 1994. Clearly something has happened to the population since Lousley described it as frequent on all the inhabited islands. Arable, mainly bulbfields.

APIACEAE – Carrot family
Daucus carota L.
Wild Carrot
Inhabited islands, also Toll's Island, Samson and Great Ganilly

Native. Apparently absent from the other uninhabited islands. It was not until 2009 that the species was recorded on Great Ganilly during a botanical survey. Where they have been separated ssp. *carota* is most frequently recorded inland, with ssp. *gummifer* on the coast. Coastal grasslands, along roadside verges and the margins of cultivated fields.

MONOCOTS - Monocotyledons

ARACEAE – Lords-and-Ladies family
Zantedeschia aethiopica (L.) Spreng.
Altar-lily
Inhabited islands

Neophyte. Most frequent on St Mary's. Relict of cultivation often in large stands along hedgebanks, field edges, in scrub and under trees.

ARACEAE – Lords-and-Ladies family
Arum italicum Mill.
Italian Lords-and-Ladies
Inhabited islands, also Samson

ssp. *neglectum* (F. Towns.) Prime
Native. This is the common arum in Scilly. It was assumed to be *A. maculatum* until identified by Lousley in 1939. Can sometimes be a pest in cultivated fields, when it invades from the hedgebanks. Field edges, hedgebanks, under trees and in dunes.

ssp. *italicum*
Plants that are clearly ssp. *italicum* with strong white veining are known from Bryher (SV87811477) and St Mary's, a few plants near Old Town churchyard, in Doiley Wood, near the rubbish dump and in Watermill Lane. Almost certainly garden escapes.

MONOCOTS - Monocotyledons

Since 2007 attempts by several botanists have been made to review the status of the arums in Scilly as some of the plants on St Mary's do not appear to be 'good' *A. italicum* ssp. *neglectum*. Often plants with white veins and ones without grow together, the leaf shapes are equally variable. There is a suspicion that some plants may be hybrids – perhaps a long-standing hybrid swarm? Only DNA investigation will be able to confirm this.

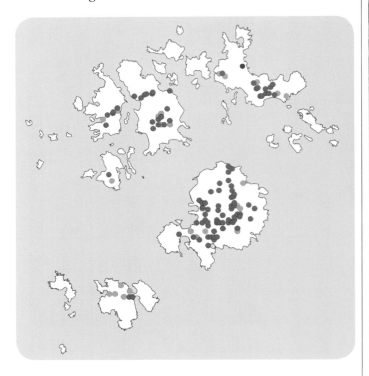

LEMNACEAE – Duckweed family
Lemna minor L.
Common Duckweed
St Mary's, St Martin's, Tresco and Bryher

Native. Restricted to a few wetland sites: Great Pool, Tresco; Little Pool, Bryher; Coldwind Pit and the cricket field pool, St Martin's; most ponds on St Mary's, the stream through Holy Vale and Higher Moors and the ditch through Lower Moors. Fluctuates in abundance especially since the arrival of *L. minuta* with which it frequently grows and which appears to be dominant. Streams and pools with still or only gently moving water. The smaller *Lemna minuta* plants are with the larger plants of *Lemna minor* in the photo above.

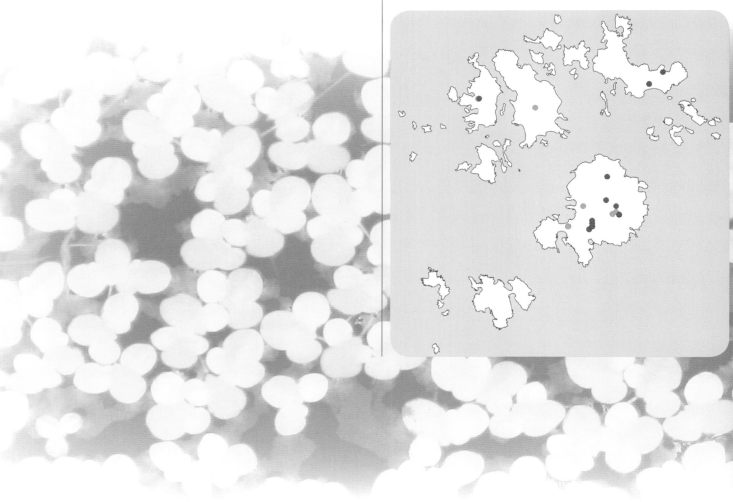

LEMNACEAE – Duckweed family
Lemna minuta Kunth
Least Duckweed
St Mary's, Tresco, also St Helen's, Teän

Neophyte. Since it was first recorded in 1994 this species has taken over from *L. minor* as the most common *Lemna* species. Recorded from a number of water-bodies including a pool in Tresco Abbey Gardens; on St Mary's in both Lower and Higher Moors, Watermill Stream, Holy Vale Stream, Newford duck ponds and Jac-A-Bar pond. Was found in a water-filled tank (see page 46) on the old pump on Teän (still there 2009) and a tiny pool near the beach on St Helen's, also in 2009. The species seems to fluctuate in frequency and appearance and apparently is able to survive, at least for a short while, in very small amounts of water. Pools and ditches as above.

ALISMATACEAE – Water-plantain family
Baldellia ranunculoides (L.) Parl.
Lesser Water-plantain
St Mary's

Native. Formerly grew in the wet fields between Porthloo and the Lower Moors extension until about 1988, but may have been lost when the fields dried out due to lowering of the water table. Possibly grew in Shooter's Pool before the recent restoration. The most recent record was from Lower Moors (SV913105) in 2000, but needs confirmation. Unless flowering, the leaves of *Ranunculus flammula* might be mistaken for this species. Wetlands.

MONOCOTS - Monocotyledons

ZOSTERACEAE – Eelgrass family
Zostera marina L.
Eelgrass

Native. There are extensive seagrass beds in Scilly especially on the sandflats between St Martin's, Tresco and St Mary's, and also between Tresco, Bryher and Samson. There are smaller populations in Cove between St Agnes and Gugh and elsewhere. After storms plants often get torn up and washed ashore or float about in the tide. Since 1992 *Zostera* populations are monitored regularly by Natural England (and predecessors), (Fowler, 1992; Bull and Kenyon, 2015). Marine.

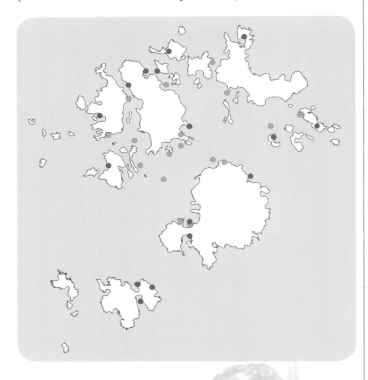

POTAMOGETONACEAE – Pondweed family
Potamogeton polygonifolius Pourr.
Bog Pondweed
St Mary's

Native. Very rare, only found in pools in Lower Moors (SV913105-SV914103) and Shooter's Pool (SV91251076). It may now only occur in the latter pool. There are earlier records from Higher Moors in 1967 and from Tresco in 1955. Pools.

Potamogeton natans L.
Broad-leaved Pondweed
Extinct
Formerly Higher Moors and Tresco (1906), from herbaria specimens. Not recorded by Lousley so probably long gone.

Potamogeton perfoliatus L.
Perfoliate Pondweed
Extinct
Formerly on Tresco (1876-7).

POTAMOGETONACEAE – Pondweed family
Potamogeton crispus L.
Curled Pondweed
St Martin's

Native. Found in the St Martin's cricket field pool in May 2011. This is the first record for Scilly. It is assumed this was introduced to the pool by waterfowl. The pool dried out later in the year, so it is not known if the plant has survived. Freshwater pool.

POTAMOGETONACEAE – Pondweed family
Potamogeton pectinatus L.
Fennel Pondweed
Inhabited islands and Northwethel

Native. Often, but not always in saline or brackish pools; including Bryher Pool, the seasonal pool on Northwethel, Big Pool on St Agnes, the Tresco Pools, St Martin's cricket field pool and formerly some pools on St Mary's. The population of this plant fluctuates markedly and it is often found in the same places as *Ruppia maritima*. Pools.

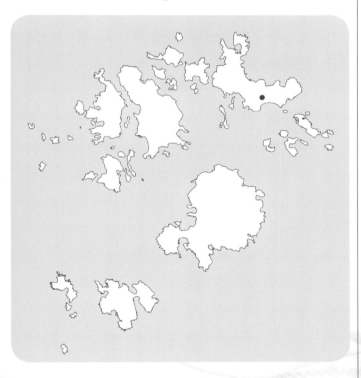

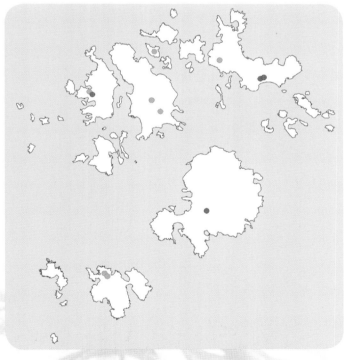

Potamogeton pusillus L.
Lesser Pondweed
Extinct
Formerly pools Tresco and St Mary's (1876 and 1890).

MONOCOTS - Monocotyledons

RUPPIACEAE – Tasselweed family
Ruppia maritima L.
Beaked Tasselweed
Inhabited islands

Native. Found in the larger brackish pools such as Bryher Pool. It may still be in Great Pool, Tresco but needs confirmation. Appears to tolerate salinities that are almost as high as seawater. Fluctuates considerably in abundance and can often be found with *Potamogeton pectinatus*.

LILIACEAE – Lily family
Tulipa saxatilis Sieber ex Spreng.
Cretan Tulip
Tresco

Garden escape found outside Abbey Gardens and at Carn Near in 1996. Identified by Eric Clement as var. *townsendii*. Not seen recently (the photo was taken abroad).

ORCHIDACEAE – Orchid family
Spiranthes spiralis (L.) Chevall.
Autumn Lady's-tresses
Inhabited islands

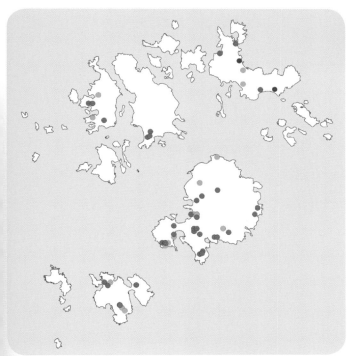

Native. Locally occasional to frequent. Numbers of flowering spikes can fluctuate markedly from year to year. In 'good' years there will be hundreds of flowering spikes and in other years few or none at the usual sites. The colony that grows along the turfy wall tops near the Woolpack on the Garrison can number hundreds some years if the wall tops are not strimmed! Also on the grass at Harry's Walls there can also be many hundreds of flowering spikes. In 'good' years the orchids can appear on roadside verges, garden lawns and almost any area of short grass. The basal rosettes can sometimes be found on short turf in winter, but also sometimes in spring. For example at Rushy Bay, Bryher in May 2012 a number of rosettes were found when searching for *Viola kitaibeliana* (but beware a superficial similarity to basal rosettes of *Centaurium erythraea*). Short grassland on heathlands, dunes, cricket fields, wall-tops, mown grassland around ancient monuments, and lawns.

Dactylorhiza praetermissa (Druce) Soó
Southern Marsh-orchid
St Mary's

Native. Currently found on Lower and Higher Moors. The numbers of flowering spikes vary from year to year, often responding positively to active management such as clearing away taller vegetation. Formerly recorded at Bant's Carn in 1974 (error?), around 1,000 plants were recorded in two areas on Lower Moors and in a damp field opposite Rose Hill in 1975. A report of the species on Samson in 1986 is now believed to have been in error for *Anacamptis pyramidalis*. Marshy areas.

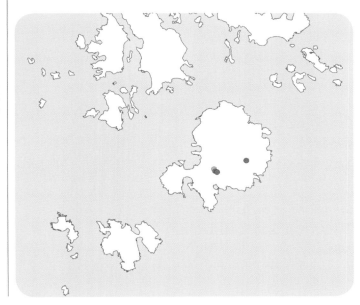

MONOCOTS - Monocotyledons

ORCHIDACEAE – Orchid family
Anacamptis pyramidalis (L.) Rich.
Pyramidal Orchid
Samson

Native. Rare. In 1997 the first plants were recorded in the dunes at the north end of Samson. It is likely they had been there in 1986 but were misidentified. It is presumed the plants originated from windblown seed. The colony has since spread over the area within the dune hollows and under the taller, bramble-dominated vegetation. The plants have also responded positively to management such as path cutting and are still spreading with around 100 flowering spikes counted in 2012. Dune grassland.

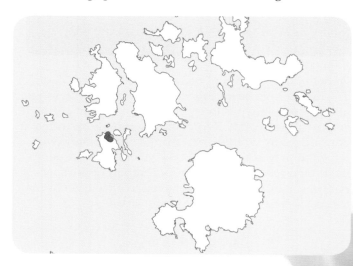

Gymnadenia borealis (Druce) R.M. Bateman, Pridgeon & M.W. Chase.
Fragrant Orchid
Extinct
Recorded at Great Bay, St Martin's in 1974 as *G. conopsea*, but more likely to have been this species, last seen 1977.

Dactylorhiza fuchsii (Druce) Soó
Common Spotted-orchid
Extinct
Recorded from Tresco in 1969, but apparently did not persist.

IRIDACEAE – Iris family
Homeria collina (Thunb.) Salisb.
Cape-tulip
St Mary's, St Martin's and Tresco

Neophyte. Garden escape with recent records from Thomas Porth, St Mary's; outside the Abbey Gardens and on a nearby rubbish dump, Tresco; and Middle Town, St Martin's in 2014. Dune and waste ground.

IRIDACEAE – Iris family
Aristea ecklonii Baker
Blue Corn-lily
Tresco and St Mary's

Neophyte. Escape from cultivation that has become established in places. On Tresco it occurs near the Abbey Gardens, Abbey Hill and several places on Appletree Banks. On St Mary's it has been recorded on the track to Salakee Farm, by Tremelethen House and Carreg Dhu. Field edges, dunes, woodland and roadsides.

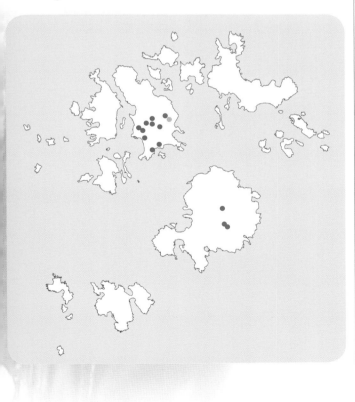

IRIDACEAE – Iris family
Libertia formosa Graham
Chilean-iris
St Mary's and Tresco

Neophyte. Garden escape naturalised in a few places. Most plants are not far from gardens, but on St Mary's it has recently been spreading around the coast on Salakee Down and around Old Town Bay. It is presumed the plant is being spread by birds, perhaps from garden waste? Coastal heath, roadsides and cliff slopes.

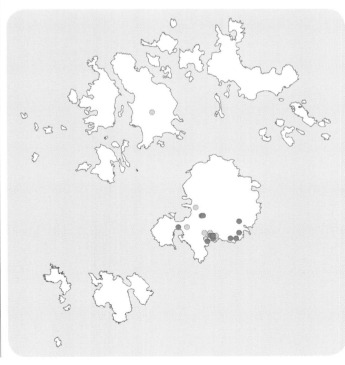

MONOCOTS - Monocotyledons

IRIDACEAE – Iris family
Sisyrinchium californicum (Ker Gawl.) W.T. Aiton
Yellow-eyed-grass
Tresco

Neophyte. Naturalised in Appletree Banks dunes, last seen 1992. May have been lost when the heliport was laid out.

IRIDACEAE – Iris family
Iris pseudacorus L.
Yellow Iris
Tresco, St Mary's, St Martin's, also Samson

Native. Locally abundant. The plant is most common on St Mary's with the main stands at Newford Duck Ponds and Watermill Lane; from Porthloo to Lower Moors and in fields between, and Holy Vale to Porthellick Pool. On Tresco it is found around Great Pool and on St Martin's it grows in a pool near Lawrence's Bay. On Samson there is a patch of irises near the ruined houses that may have originally been planted in a pond by the former inhabitants. Pools, wet fields (where it sometimes is locally dominant), ditches and marshes.

IRIDACEAE – Iris family
Iris foetidissima L.
Stinking Iris
Inhabited islands, also Samson and Teän

Native. Locally frequent. Lousley (1971) describes it as uncommon, and only recorded it from St Agnes, Tresco, St Mary's and Samson. So it is not clear whether there has been a recent considerable expansion in range or whether the plant had been previously overlooked in some places. There appears to be two types of the iris; those that can be seen to have followed tracks and roadways on the inhabited islands and those living in more 'natural' habitats such as on uninhabited islands and more remote places on the inhabited islands. It is well-established on Teän (Lousley may have just missed it here?) and Samson both in the dunes and slopes of the hills. Roadsides (where the seeds are easily spread), sand dunes, field edges and coastal slopes.

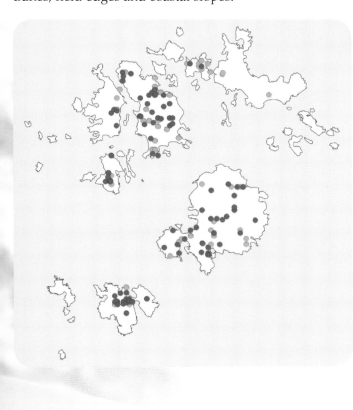

IRIDACEAE – Iris family
Iris latifolia (Mill.) Voss
English Iris
St Mary's

Neophyte. Garden escape in Old Town in 1996.

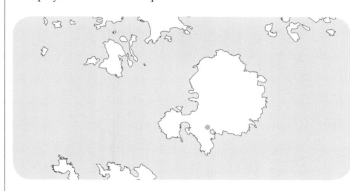

Iris xiphium L.
Spanish Iris
St Martins and St Mary's

Neophyte. Occasional relict of cultivation found in bulbfields or on disturbed ground.

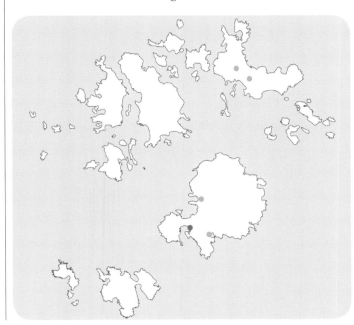

MONOCOTS - Monocotyledons

IRIDACEAE – Iris family
Iris x *hollandica* hort.
Dutch Iris
St Martin's

Neophyte. Relict of cultivation still found occasionally in bulbfields.

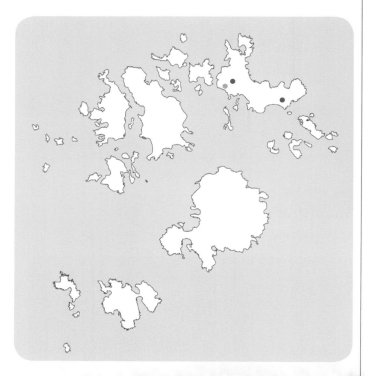

IRIDACEAE – Iris family
Watsonia borbonica (Pourr.) Goldblatt
Bugle-lily
Tresco and St Mary's

Neophyte. Occasional garden escape, established in a few places on Tresco especially near the Abbey Gardens and in the dunes (where it may have originally been planted). There is a 1995 record from Watermill Lane, St Mary's. Dunes and hedgebanks.

IRIDACEAE – Iris family
Gladiolus communis ssp. *byzantinus* (Mill.) R.C.V. Douin
Eastern Gladiolus (Whistling Jacks)
Inhabited islands and Gugh

IRIDACEAE – Iris family
Ixia campanulata Houtt.
Red Corn-lily
St Martin's and St Mary's

Neophyte. Frequent to locally abundant. A well-established relict of cultivation on the inhabited islands, less common on Tresco. One of the bulbous species originally introduced as a flower crop and later abandoned. The small corms are very persistent and grow wherever they fall or are thrown! Lately there has been a small resurgence of interest in the plant, probably due to visitors having seen the extraordinary display of the gladiolus in some fields. Corms are now sold commercially on a small scale. The local name derives from the squeaking noise that children made by blowing on a leaf held between their thumbs. Bulbfields, roadsides, bracken fields, dunes and disturbed ground.

Neophyte. Relict of cultivation. Has also has been found on Toll's Island where it may persist. Rejected corms are often thrown out onto hedgebanks. Roadsides and bulbfields.

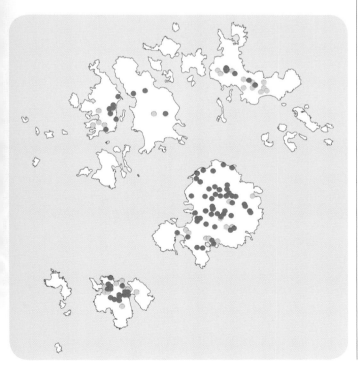

MONOCOTS - Monocotyledons

IRIDACEAE – Iris family
Ixia paniculata D. Delaroche
Tubular Corn-lily
St Martin's, Tresco and St Mary's

Neophyte. Relict of cultivation that persists in places. Roadsides, dunes and bulbfields.

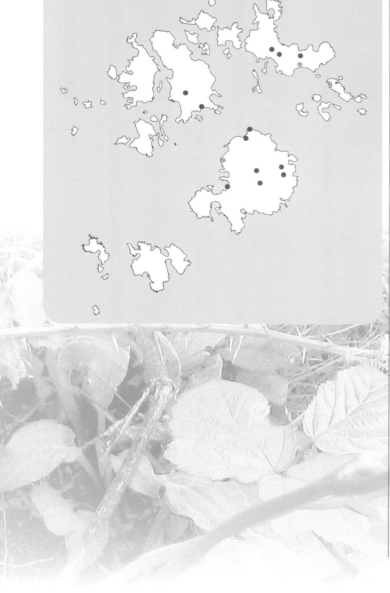

IRIDACEAE – Iris family
Sparaxis grandiflora (D. Delaroche) Ker Gawl.
Plain Harlequinflower
St Martin's, St Mary's and St Agnes

Neophyte. Relict of cultivation found on roadsides, fields and waste ground.

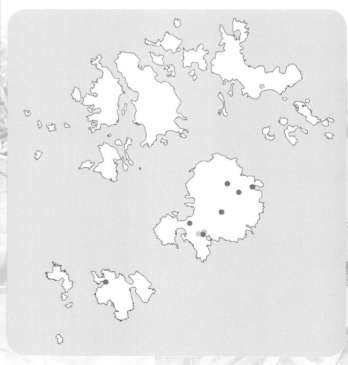

IRIDACEAE – Iris family
Freesia x *hybrida* L.H. Bailey
Freesia
Tresco and St Mary's

Neophyte. Relict of cultivation occasionally found in fields and roadsides.

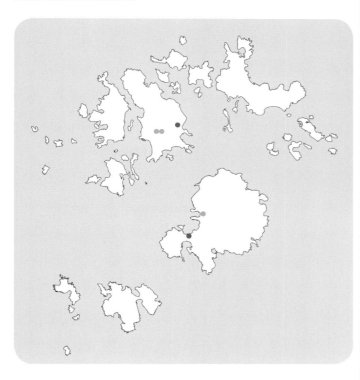

Crocosmia paniculata (Klatt) Goldblatt
Aunt-Eliza
Tresco

Neophyte. Garden escape recorded east of Great Pool in 2007.

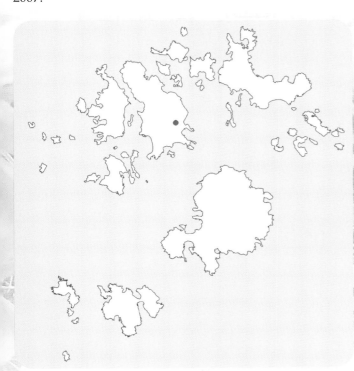

IRIDACEAE – Iris family
Crocosmia x *crocosmiiflora* (Lemoine) N.E. Br.
Montbretia
Inhabited islands

Neophyte. Naturalised and locally frequent, but appears less invasive than on the mainland. Apparently uncommon on Bryher. Hedgebanks, dunes and among bracken.

MONOCOTS - Monocotyledons

IRIDACEAE – Iris family
Chasmanthe bicolor (Gasp. Ex Ten.) N.E. Br.
Chasmanthe
Tresco, St Agnes, St Mary's and St Martin's

Neophyte. A garden escape that was first established on St Mary's and St Agnes and is now found on the other islands, although not apparently yet on Bryher. Roadsides and hedgebanks.

Chasmanthe aethiopica (L.) N.E. Br.
Extinct?

Included by Lousley (1971) as a garden escape on Tresco, it has not been found there since. There is some doubt over the identification.

XANTHORRHOEACEAE – Asphodel family
Kniphofia uvaria (L.) Oken
Red-hot-poker
Tresco, St Martin's and St Mary's

Neophyte. Naturalised in dunes and fields, sometimes from original plantings.

XANTHORRHOEACEAE – Asphodel family
Kniphofia x *praecox* Baker
Greater Red-hot-poker
Tresco, St Mary's and Bryher

Neophyte. Naturalised in a few places mostly in dunes or coastal.

XANTHORRHOEACEAE – Asphodel family
Phormium tenax J.R. & G. Forst.
New Zealand Flax
Inhabited islands

Neophyte. Very frequently planted on all the inhabited islands. As it self-seeds readily it is not always obvious which plants have been planted deliberately and which have established naturally. Dunes, tracks, heathland and coastal sites.

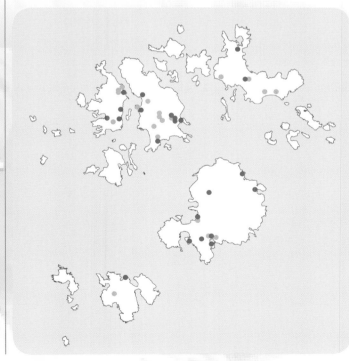

MONOCOTS - Monocotyledons

XANTHORRHOEACEAE – Asphodel family
Phormium cookianum Le Jol.
Lesser New Zealand Flax
St Martin's, Tresco, Bryher and St Mary's

Neophyte. Occasionally planted on the inhabited islands. Many hundreds of naturalised plants of this species are spreading very rapidly in the Great Bay area of St Martin's. The IoSWT are trying to control their spread, but so far no effective and practical method has been found due to the huge numbers of plants and their extremely fibrous leaves that resist cutting with chain saws. Dunes and heathland.

ALLIACEAE – Onion family
Allium roseum L.
Rosy Garlic
St Martin's and St Mary's

Neophyte. Only found on the above two islands, it was already naturalised there when Lousley saw it in 1939. A Mediterranean species, it could also have been introduced accidentally or as a garden plant and does not appear to have spread far from where originally found. It was being grown as a crop at Borough Farm, St Mary's in 2013. The variety growing on Scilly has red bulbils. Dunes, roadsides, grasslands, bracken fields and bulbfields.

ALLIACEAE – Onion family
Allium neapolitanum Cirillo
Neapolitan Garlic
St Martin's and St Mary's

Neophyte. An escape from cultivation, probably overlooked due to its similarity to other white-flowered *Allium* species and brief flowering season. Found in just a few fields, hedgebanks and tops of walls.

ALLIACEAE – Onion family
Allium triquetrum L.
Three-cornered Garlic
Inhabited islands, also Samson

Neophyte. Widespread and abundant on the inhabited islands. Perhaps the most ubiquitous weed in Scilly. In Spring the field edges and roadsides are lined by this and *Oxalis pes-caprae*. Recently found on Samson after vegetation was cleared from around the ruined buildings. A comment in Leland who wrote in about 1533 that 'diverse of islettes bereth wyld garlyk' (Leland [1535 - 1543] in Smith, 1906 - 1910) seemed to be dismissed by Borlase who said he had been informed that 'Wild Garlick' grew on some of the off-islands, but he met with none (Borlase, 1756). Later Woodley stated 'Garlick is much cultivated, although it also grows wild' (Woodley, 1822). This is puzzling as Wild Garlic, that is Ramsons *Allium ursinum,* is not known from Scilly and none of the other *Allium* species seem a likely candidate. But is it possible this species had been introduced to Scilly before the 1890s when it is supposed have been grown as a garden plant? Finding the plant on Samson, growing by the ruined houses is also interesting as it suggests it may have been cultivated there before the inhabitants were removed in 1855.

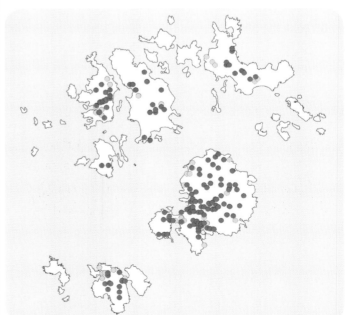

MONOCOTS - Monocotyledons

As with many other plants on Scilly Three-cornered Garlic originated from the Mediterranean. The pretty white flowers are sometimes mistaken for white bluebell and would have seemed an attractive plant for the garden. It is also sometimes used in cooking for its flavour. Locally cattle will graze the plant which can taint their milk. Bulb and other arable fields, hedgebanks, roadsides, walls and even in the wetlands and beside pools.

Based on Lousley's account (1971) the leek has expanded its range since 1939. It is easily spread by bulbils. On St Martin's where the plant seems particularly common, trailer-loads have sometimes been dumped on the shore. Bulbfields, field edges, dunes, bracken-covered cliff slopes and unmanaged ground.

Allium ampeloprasum var. *babingtonii* (Borrer)
Babington's Leek
Inhabited islands, also Teän

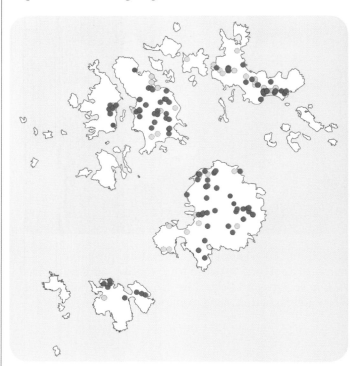

Archaeophyte. Locally abundant. The abundance of this species varies from island to island, being less common on St Agnes and Bryher than elsewhere. On some farms the plant is tolerated, at least until the stage when bulbfields need to be covered in plastic sheeting when it is necessary to cut down the tall fruiting stems. On other farms plants are pulled up and removed completely.

ALLIACEAE – Onion family
Allium ampeloprasum var. *ampeloprasum* L.
Wild Leek
St Mary's

Archaeophyte. Known from just three places on The Garrison, St Mary's. Had previously been overlooked or confused with var. *babingtonii* and not recorded until about 1991. The plants that grew on the campsite may have gone recently. The large stand near the Woolpack building (SV89651030) appears to be increasing. There is another group of plants among tall vegetation in a former cultivated 'square' (SV900099) reported by L. Askins. Could this have been the 'wild garlic' referred to by Woodley (1822) and others? Grassy edges.

ALLIACEAE – Onion family
Allium vineale L.
Wild Onion
St Martin's and St Mary's

Native. Only found in a few places including near Lunnon Farm, St Mary's and grass verges on St Martin's. Plants seen in flower had heads consisting entirely of bulbils. Fields and grassy verges.

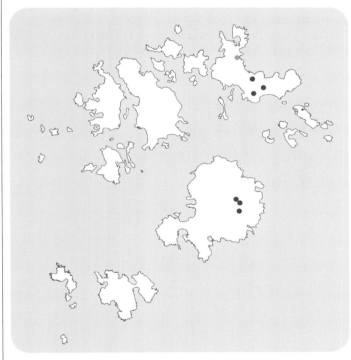

MONOCOTS - Monocotyledons

ALLIACEAE – Onion family
Nothoscordum borbonicum Kunth
Honeybells
Tresco, Bryher, St Martin's and St Mary's

Neophyte. An occasional garden escape, with scattered records around the islands. On Tresco it is found near the Abbey Gardens from where it probably originated, as well as on Appletree Banks and at New Grimsby. On Bryher it has been found by the quay and on St Martin's by the cricket field pool. Sites on St Mary's include Salakee Lane, Porthcressa, the road to Carn Friars and Porth Minick. Roadsides, dunes and grassland.

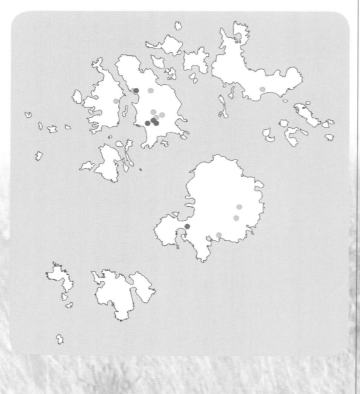

ALLIACEAE – Onion family
Agapanthus praecox ssp. *orientalis* (F.M. Leight) F.M. Leight
African Lily
Inhabited islands

Neophyte. Locally frequent to abundant. An escape from cultivation now well established mostly on St Agnes, St Mary's and especially Tresco. The plants seed freely and rhizomes are often thrown out from gardens where the clumps have become too dense. The plants on Appletree Banks, Tresco have been present since about 1939 and have spread throughout to such an extent they have changed the character of the dunes. Most plants have blue flowers, but white-flowered plants also occur. Grows in dunes, hedgebanks, and rough ground near the coast.

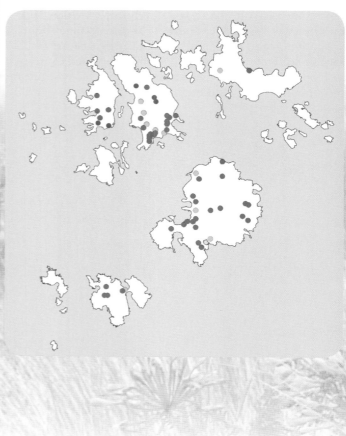

ALLIACEAE – Onion family
Tristagma uniflorum (Lindl.) Traub
Spring Starflower
St Mary's & St Agnes

ALLIACEAE – Onion family
Amaryllis belladonna L.
Jersey Lily
Tresco, Bryher, St Mary's and St Martin's

Neophyte. Sometimes called 'naked ladies' they are an occasional escape from cultivation or throw-outs from bulbfields where they are still cultivated.

Neophyte. Garden escape and bulbfield weed. On St Mary's it had been established in fields near Thomas Porth (possibly now lost?) and a verge at Porthloo for many years. Other records, from Tolman Point and Rocky Hill on St Mary's and Perglis bank, St Agnes are probably more recent escapes from cultivation. Verges, former bulbfields and dumps of garden material near the shore.

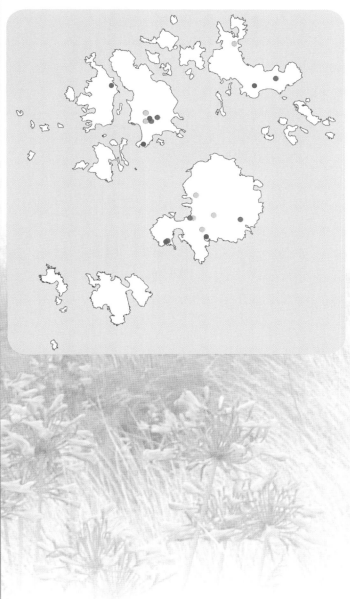

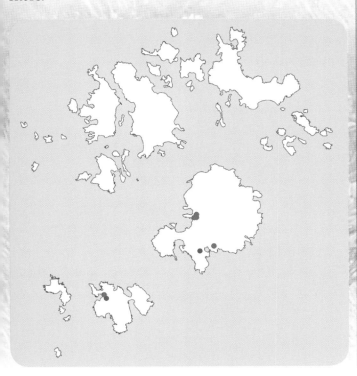

MONOCOTS - Monocotyledons

ALLIACEAE – Onion family
Nerine sarniensis (L.) Herb.
Guernsey Lily
St Mary's

Neophyte. Occasional garden escape or from thrown-out bulbs from where they are still cultivated.

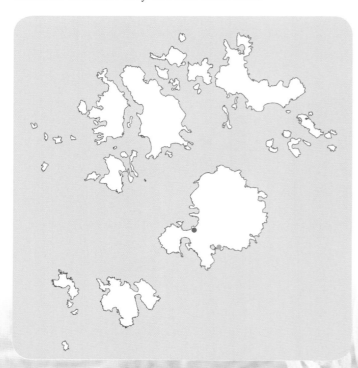

Nerine bowdenii Watson
Bowden Cornish Lily
St Mary's

Neophyte. Garden escape recorded by Colin French in 2009 Hugh Town.

ALLIACEAE – Onion family
Leucojum aestivum ssp. *pulchellum* (Salisb.) Briq.
Summer Snowflake
Tresco, St Mary's and St Agnes

Neophyte. Escape from cultivation. Lousley recorded the species from a number of localities but recent records include the following; beside a ditch north of Great Pool, Tresco where they have been for more than thirty years, Lower Town, St Agnes and in Watermill Lane, St Mary's. Ditches and hedgebanks.

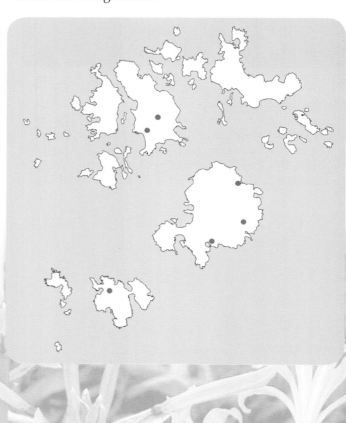

ALLIACEAE – Onion family
Narcissus L.
Daffodils
Inhabited islands

Neophyte. No attempt has been made to map or record *Narcissus* in Scilly. Many different varieties of cultivated *Narcissus* have become naturalised on the inhabited islands. Unwanted or 'rogue' bulbs are thrown out onto coasts and waste ground and may persist for some time. Donald Pett has published a useful book (Pett, 2009) that may in the future help sort out the hundreds of established 'wild' plants. Coasts, roadsides, hedgebanks and dunes.

ASPARAGACEAE – Asparagus family
Ornithogalum umbellatum L.
Star-of-Bethlehem
St Agnes, St Mary's and St Martin's

Neophyte. Includes ssp. *campestre* that was recorded as *O. angustifolium* from Porthlow Farm. Occasionally found in bulbfields where it has probably been accidentally introduced with other bulbs. May persist for a few years. Cultivated fields and waste ground.

MONOCOTS - Monocotyledons

ASPARAGACEAE – Asparagus family
Scilla verna Huds.
Spring Squill
Bryher, St Mary's and St Martin's

Native. Frequent, even abundant in places, along the west and southern coast of Bryher with smaller colonies on north-east coast of St Mary's and Peninnis Head. Records from Merchant's Point, Tresco and from Jacky's Point, White Island (St Martin's) need confirmation. The recent management work has been beneficial to the plant with colonies on Gweal Hill, Bryher and around Trenears's Rock, St Mary's increasing noticeably. Grassland and heathland mainly on coast.

ASPARAGACEAE – Asparagus family
Scilla peruviana L.
Portuguese Squill
St Mary's

Neophyte. Garden escape near Carn Thomas 2003, also cultivated as a crop, for example on Peninnis Farm.

ASPARAGACEAE – Asparagus family
Hyacinthoides non-scripta (L.) Chouard ex Rothm.
Bluebell
Inhabited islands, also St Helen's, Northwethel, Teän, White Island (off St Martin's), Samson, Annet, Nornour, Great Ganilly, Great Innisvouls and Little Arthur

Native. Widespread and locally abundant. Grows in more open areas in Scilly than usually seen on the mainland. Found on heathy areas, under bracken and in open grassland on cliff slopes. Away from the coast it is found on hedgebanks and in pastures.

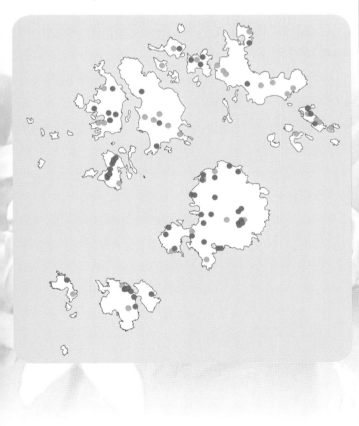

ASPARAGACEAE – Asparagus family
Hyacinthoides x massartiana Geerinck
(*H. non-scripta* x *H. hispanica*)
Hybrid Bluebell
Inhabited islands and Samson

Neophyte. Mainly found in bulbfields and cultivated areas. Believed to have been cultivated in the past. There has been confusion between *Hyacinthoides hispanica* Spanish Bluebell and the hybrid, and according to Dr Mark Spencer all the 'Spanish' bluebells on Scilly are in fact the hybrid. A small group of plants on Samson may also be this species that had escaped from a garden. It also seems that the hybrid may back-cross with the native species where they meet. Cultivated ground, plantations and verges. The map is an amalgamation of 'Spanish' bluebells and the hybrid.

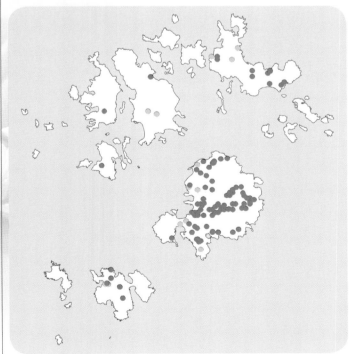

MONOCOTS - Monocotyledons

ASPARAGACEAE – Asparagus family
Muscari comosum (L.) Mill.
Tassel Hyacinth
St Martin's

Neophyte. Garden escape. Seen in 1967, 1990 and 2004 by Will Wagstaff and again in 2014 when he found two flowering plants in the dunes at (SV925157) near Lawrences's Bay.

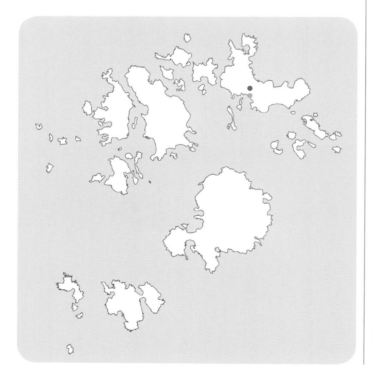

ASPARAGACEAE – Asparagus family
Ruscus aculeatus L.
Butcher's-broom
St Martin's, St Mary's, also Great Ganilly, Nornour and Great Ganinick

Native. A rare plant only found in a few places. It has been recorded from the area just north of English Island Point, St Martin's; from Nornour, Great Ganilly and Great Ganinick on the Eastern Isles; on the north coast of St Mary's from near Pendrathen quay to Innisidgen. Plants in the lane near Porthloo and in Hugh Town, St Mary's appear to be of recent garden origin. Lousley published a map of the distribution of *Ruscus* in his Flora (Lousley, 1971) which shows a similar distribution as today. Heathy places, rocky slopes and among coastal vegetation.

ASPARAGACEAE – Asparagus family
Yucca gloriosa L.
Spanish-dagger
Tresco

Neophyte. Garden escape (or originally planted) in dunes at Pentle Bay and Appletree Banks, Tresco.

ASPARAGACEAE – Asparagus family
Cordyline australis (G. Forst.) Endl.
Cabbage-palm
Inhabited islands

Neophyte. Planted on all the inhabited islands. Seedlings are uncommon and difficult to separate from planted specimens.

MONOCOTS - Monocotyledons

ASPARAGACEAE – Asparagus family
Agave americana L.
Centuryplant
St Agnes, St Mary's and Tresco

Neophyte. Sometimes found on walls and in gardens. There are no records of plants being naturalised, they all seem to have originally been planted.

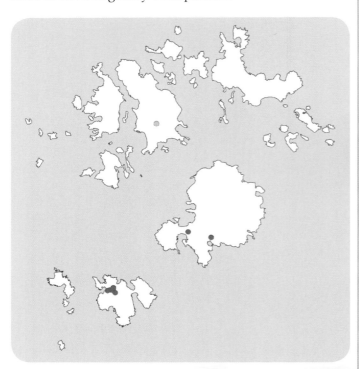

ASPARAGACEAE – Asparagus family
Furcraea longaeva Karw. and Zucc.
Furcrea
Tresco

Neophyte. Originally planted in the dunes near Pentle Bay but now spreading. Possibly this species, but several *Furcrea* species have been introduced to the Gardens.

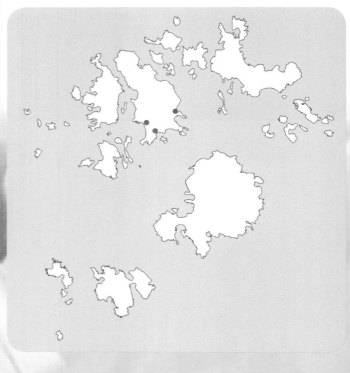

PONTEDERIACEAE – Pickerelweed family
Pontederia cordata L.
Pickerelweed
Tresco

Neophyte. Planted In Abbey Pool.

TYPHACEAE – Bulrush family
Sparganium erectum L.
Branched Bur-reed
St Mary's

Native. Not seen since 1988; formerly known from Lower Moors and the pond near Porthloo, but had always been very rare.

MONOCOTS - Monocotyledons

TYPHACEAE – Bulrush family
Typha latifolia L.
Bulrush
Tresco, St Mary's and Bryher

Native. Was only known from Great and Abbey Pools, Tresco until 1967 when 12 'maces' were on Porthellick pool (Lousley, 1971), since then it has spread into the Higher Moors. A few plants that appeared in Little Pool, Bryher before 2000 were possibly a deliberate introduction, but the plant quickly invaded both parts of the pool and despite attempts to remove it they have persisted and may have overwhelmed the former flora. Pools.

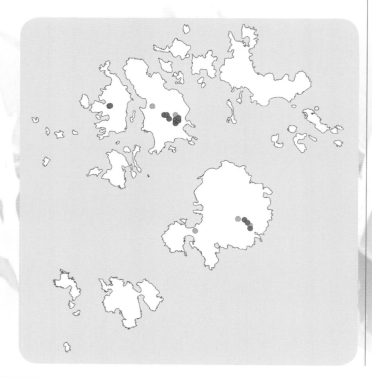

BROMELIACEAE – Rhodostachys family
Fascicularia bicolor (Ruiz & Pav.) Mez
Rhodostachys
Tresco, St Agnes, Bryher and St Mary's

Neophyte. Frequently planted and also naturalised in a few places notably in the dunes on Tresco. Most plants on St Mary's and Bryher have been planted so it is not always obvious which may be spreading naturally. Jack Oliver saw a Herring Gull drop a viable fragment of this plant which suggests this may be another way it can spread.

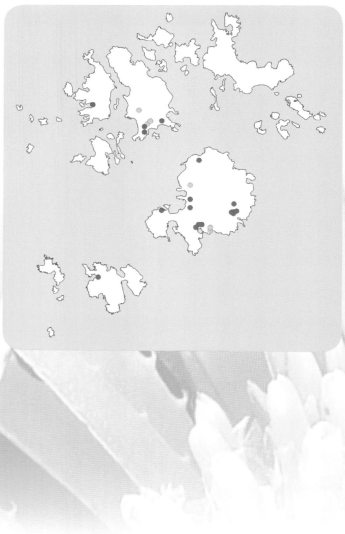

BROMELIACEAE – Rhodostachys family
Ochagavia carnea (Beer) L.B. Sm. & Looser
Tresco Rhodostachys
Tresco

Neophyte. Naturalised in dunes on Appletree and Rushy Banks. Originally planted but apparently now established.

JUNCACEAE – Rush family
Juncus articulatus L.
Jointed Rush
St Mary's, Bryher and St Martin's

Native. Most records are from St Mary's where plants are found in wet fields from the Lower Moor Extension through to Porthloo, also in Lower and Higher Moors, on Salakee Downs and in a wet flush near Bar Point (SV914127). On Bryher it occurs by the Little Pool, and on St Martin's in dunes above Great Bay (SV924163) and a small pool behind Lawrences Bay (SV925155). Had previously been recorded by Lousley from the SW corner of Great Pool, Tresco. Found in wet fields, around pools and damp flushes in dunes.

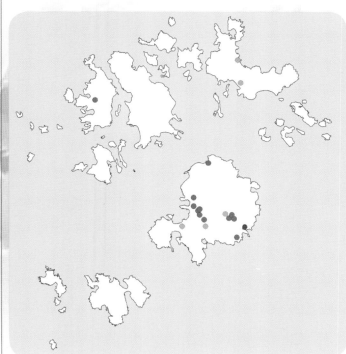

MONOCOTS - Monocotyledons

JUNCACEAE – Rush family
Juncus acutiflorus Ehrh. ex Hoffm.
Sharp-flowered Rush
St Mary's

Native. A rare plant that is only known from Higher and Lower Moors and the Carreg Dhu gardens. It has been known on the Higher and Lower Moors since at least Lousley's time. Wet grassland and marshy areas.

JUNCACEAE – Rush family
Juncus bulbosus L.
Bulbous Rush
St Mary's, St Martin's and St Agnes

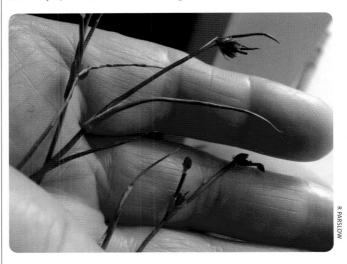

Native. A rare plant on Scilly, only now known from a very few localities. On St Mary's it is found in Higher Moors and also in some of the small seasonally-wet craters on Salakee Down (SV924102). On St Martin's it has been found in Coldwind Pit and also in the dunes (SV917160). It was also found in a wet flush on Wingletang, St Agnes (SV884074) in 1986. Grows in and beside pools, sometimes forming a mat over the surface, and on wet ground.

JUNCACEAE – Rush family
Juncus maritimus Lam.
Sea Rush
Bryher, St Mary's, Tresco

Native. Some plants on Scilly, notably on St Mary's at Lower and Higher Moors, can be ascribed to the variety *atlanticus* J.W. White which is confined to Scilly. It has lax panicles and a shorter lower bract (Sell and Murrell, 1996). The species is co-dominant with reed in parts of the 'reed' beds on Lower and Higher Moors. There are smaller stands of the rush elsewhere on the islands, including Abbey Pool, Tresco and around Bryher Pool. It is also found as small clumps in damp places, pools and wet flushes on coastal heath, for example on Castle Down, Tresco and Peninnis Head, St Mary's. Wetlands and pools.

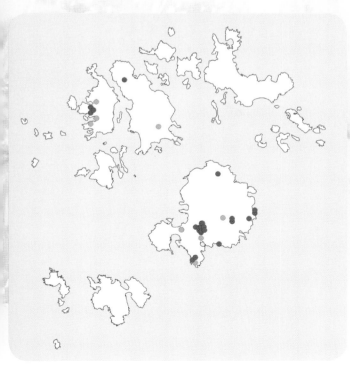

JUNCACEAE – Rush family
Juncus tenuis Willd.
Slender Rush
Tresco

Neophyte. Just one record from a wet field near Great Pool (SV89551462) in May 2008.

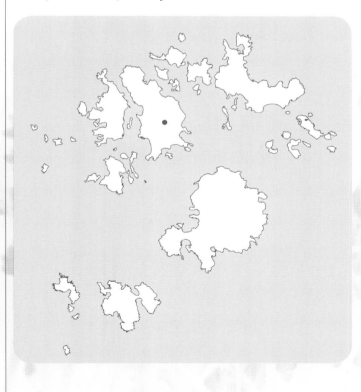

MONOCOTS - Monocotyledons

JUNCACEAE – Rush family
Juncus gerardii Loisel
Saltmarsh Rush
Bryher and St Agnes

Native. Rare, only recorded from around Big and Little Pools, St Agnes and on Bryher from beside the leat, the Pool and the Water Meadow Pool (behind Popplestone Bay). Lousley also recorded it from St Mary's and by Great Pool, Tresco but there are no recent records from there. Pool margins and marshy ground.

JUNCACEAE – Rush family
Juncus bufonius L.
Toad Rush
Inhabited islands, also Samson

Native. Widespread, and occasionally locally abundant on all the inhabited islands. Plants were found in the area cleared around the ruined houses on South Hill, Samson in 2007. The plants can be very variable, especially in size. They occur in two main types of habitat; cultivated fields and gardens, and in wetland sites including tracks on heathland.

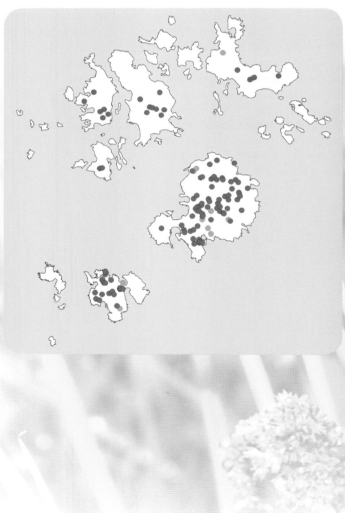

JUNCACEAE – Rush family
Juncus inflexus L.
Hard Rush
St Mary's and Tresco

Native. This is a rare plant in the Isles of Scilly. Only recorded from Tresco near Abbey Pool and on School Green in 2002, also on St Mary's in Higher Moors. Wet grassland and pool edges.

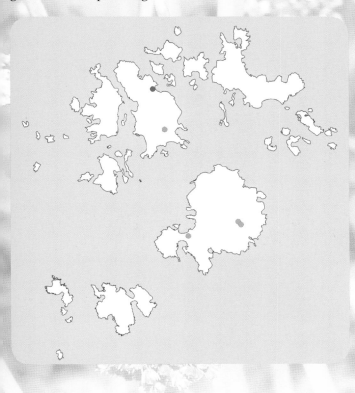

JUNCACEAE – Rush family
Juncus effusus L.
Soft-rush
St Mary's, Tresco, St Martin's and St Agnes

Native. Locally abundant on Tresco and St Mary's, elsewhere just scattered records. On St Martin's it has been recorded from Coldwind Pit and on St Agnes from the Meadow. The variety *subglomeratus* has also been recorded from a few places. Associated with wetland places including damp flushes and other places on moist soils.

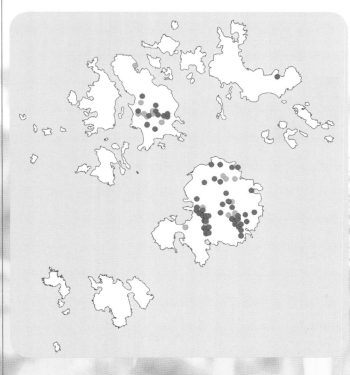

MONOCOTS - Monocotyledons

JUNCACEAE – Rush family
Juncus conglomeratus L.
Compact Rush
St Mary's

Native. Although Lousley recorded this species from Tresco there is only one recent record from near Porthellick, St Mary's (SV924104) in 1995.

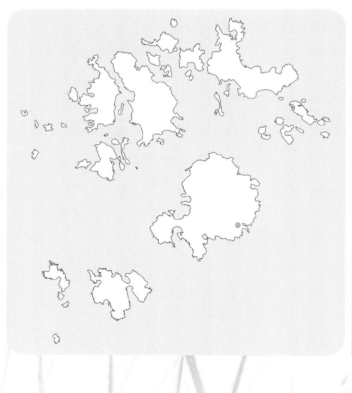

JUNCACEAE – Rush family
Luzula campestris (L.) DC.
Field Wood-rush
Inhabited islands, also Samson, St Helen's, Teän and Great Ganilly

Native. Frequent on most islands on heathy and grassy places, often near the coast, also on wall tops.

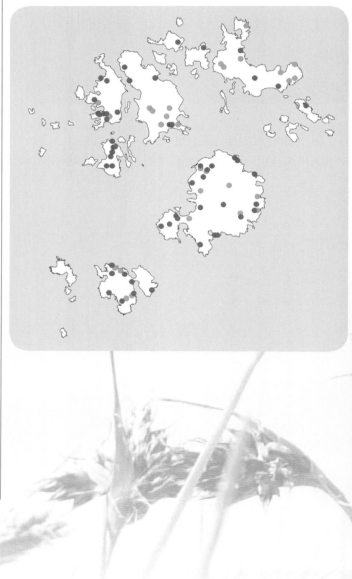

JUNCACEAE – Rush family
Luzula multiflora (Ehrh.)
Heath Wood-rush
Tresco and St Mary's

Native. Rare in Scilly. The only records from Tresco are from the Abbey Gardens and Middle Down (SV895158). There are a few more records from St Mary's; Higher Moors, the beach at Porth Hellick and Old Town churchyard. May be overlooked due to confusion with *L. campestris*. Heaths and wet meadows.

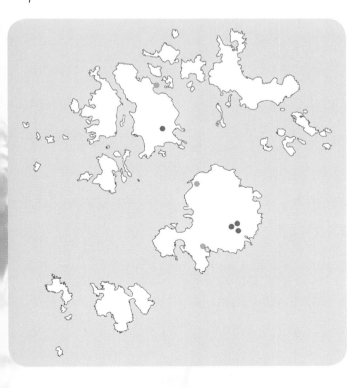

CYPERACEAE – Sedge family
Bolboschoenus maritimus (L.) Palla
Sea Club-rush
Bryher, Tresco, St Mary's and St Agnes, also Annet

Native. Only found in a few scattered localities including Lower and Higher Moors, St Mary's; Abbey and Great Pools, Tresco; Bryher Pool and leat; Big Pool on St Agnes and in seepages along the shore on Annet. Marshy places, around pools and in wet seepages.

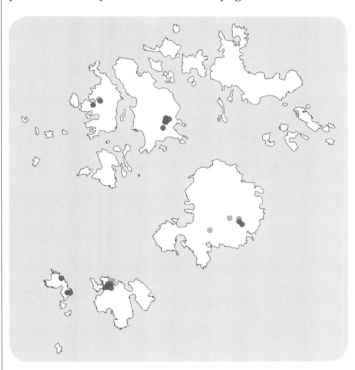

Eriophorum angustifolium Honck.
Common Cottongrass
Extinct?

Native. Last seen on Higher Moors, St Mary's but no records since 1998. The acid bog at the start of the trail to Holy Vale has long gone, and the place in Higher Moors (SV922109) where the plant last grew has since dried out. Attempts more recently by the IoSWT to restore the area and a higher water table in the hope the plant may reappear from buried seed have so far been unsuccessful.

MONOCOTS - Monocotyledons

CYPERACEAE – Sedge family
Eleocharis palustris (L.)
Common Spike-rush
Inhabited islands, also Samson

Native. Scattered sites around the islands. It is recorded on Tresco from beside Great Pool (SV898143-4) and the dump at (SV892141). On Bryher the species is frequent around Little Pool, the leat and Pool. On St Martin's it is found at Coldwind Pit and was recorded from a place in dunes (SV917160) in 1988. On St Agnes it is only known from around Big Pool and nearby areas. Lousley considered all the Scilly plants were ssp. *vulgaris*. On Samson it grows at Southward Well. The species is sometimes found growing with other *Eleocharis* species. Beside pools, flushes and in marshes.

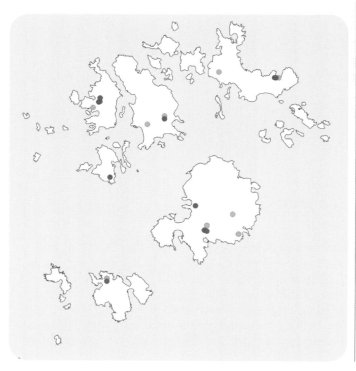

CYPERACEAE – Sedge family
Eleocharis uniglumis (Link) Schult
Slender Spike-rush
Tresco, St Mary's and St Agnes

Native. Only recorded from a few sites; around Big Pool, St Agnes, near Abbey Pool, Tresco; a wet field near Porthloo and in a water-filled pit on Salakee Down, St Mary's. Lousley commented that the plants on Scilly were not typical, they may be worth further investigation. Wet places around pools and in wet grassland.

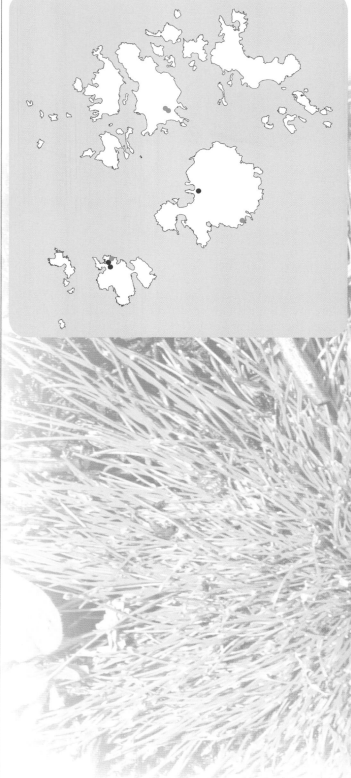

CYPERACEAE – Sedge family
Eleocharis multicaulis (Sm.) Desv.
Many-stalked Spike-rush
Tresco, St Mary's and St Agnes

Native. Also an uncommon species. Found in some of the same places as some other *Eleocharis* species. Locations include wet grassland and ruts near Big Pool, St Agnes; by the lakes on Tresco, and on St Mary's in a wet pit on Salakee Down and on Lower Moors, including the field at Old Town (SV91251032). This latter site has now been lost, when the field was used as dump for materials from the new school build in 2010. Plants have not been seen on Higher Moors since 1955. Lakesides, seasonal pools and wet grassland.

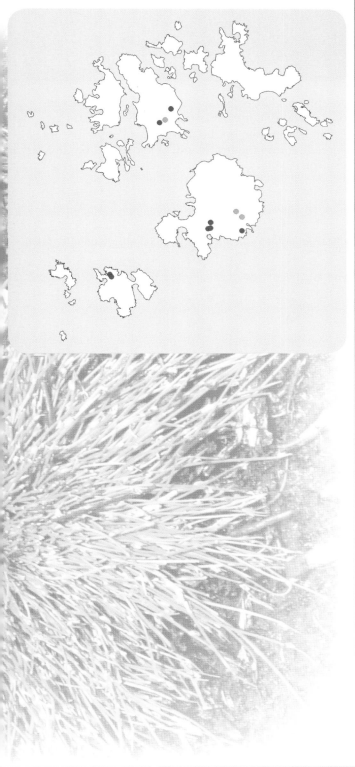

CYPERACEAE – Sedge family
Isolepis setacea (L.) R. Br.
Bristle Club-rush
Tresco, St Martin's, St Mary's and St Agnes

Native. Found in a few scattered sites, but often difficult to find as the places where it grows often dry up in summer. Recorded on Tresco by Abbey Pool (SV897142), and could still be there. On St Martin's it has been found at Coldwind Pit and on Tinkler's Hill (SV917166). On St Agnes the species has been recorded frequently on The Meadow and around Big Pool, also in a wet flush on Wingletang (SV884074). There are more localities on St Mary's including Lower Moors and the wet fields between there and Porthloo. Other sites are in the water-filled craters on Salakee Down and on Water Rocks Down. The plant appeared in great profusion on the draw-down zone of Porth Hellick Pool in 2009 with a number of other wetland species. In the past there has been confusion between this and *I. cernua*, due to the apparent variability in the length of the main bract. So it is important to check the nuts to confirm identification. Damp places such as wet cart ruts and flushes on heaths.

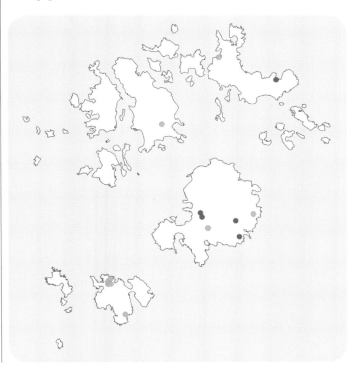

MONOCOTS - Monocotyledons

CYPERACEAE – Sedge family
Isolepis cernua (Vahl) Roem. & Schult.

Slender Club-rush

Tresco, St Martin's, St Mary's and St Agnes

Native. Apparently rarer than *I. setacea* but found in similar habitats and even the same places. Lousley considered *I. setacea* to be very rare and *I. cernua* to be frequent. But this does not seem the case now. It has not been seen near Great Pool on Tresco recently, but could still be there. On St Agnes it also occurs on The Meadow and around Big Pool, the damp flush on Wingletang near Beady Pool and a similar area on Horse Point. On St Mary's, it is found at Lower and Higher Moors and the wet fields between Lower Moor and Porthloo. It was also found on the drawdown around Porth Hellick Pool in 2009. The only record for St Martin's was from Chapel Down in 1988. In dry years the plants disappear early. Wet places usually on mud, ditches and seepages.

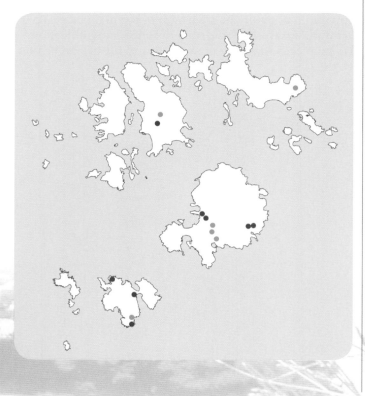

CYPERACEAE – Sedge family
Cyperus eragrostis Lam.

Pale Galingale

Tresco and St Mary's

Neophyte. Garden escape. Found in several places on Tresco outside the Gardens around SV892141. Plants established on the dump near Pentle Bay are spreading into the heathland. One record from St Mary's (SV913103) still there 2015.

Cyperus longus L.

Galingale

Tresco and St Mary's

Native in the British Isles but neophyte on Scilly. Galingale has been recorded as a casual or garden escape on Tresco (SV895149) and at Newford Duck Ponds, St Mary's.

CYPERACEAE – Sedge family
Carex paniculata L.
Greater Tussock-sedge
St Mary's

Native. Only known from Higher and Lower Moors and adjacent fields. The tussocks at Higher Moors are impressive plants, many are more than two metres tall. There are also many tussocks in the field adjoining Higher Moors, part of Lunnon Farm. Work to open up the area around the tussocks has been very beneficial and the plants are thriving. It is hard to believe that there were attempts to kill the plants with chemicals and by setting them on fire in the past (Lousley, 1971). Fortunately this was ultimately unsuccessful and now young plants are appearing. The matrix of the sedges include reeds and bracken to form the 'trunk'. Ferns and other plants grow epiphytically on them. Marshy areas where the water table is high.

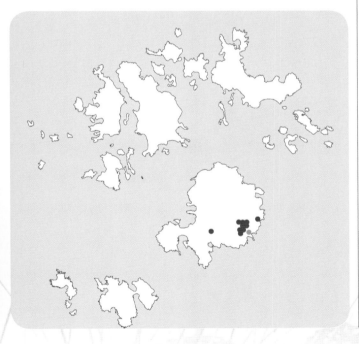

CYPERACEAE – Sedge family
Carex otrubae Podp.
False Fox-sedge
St Mary's

Native. A rare sedge in Scilly. Just a few individual plants have been found on Higher and Lower Moors where it grows around pools, grassy edges of paths and in ditches in the marsh.

MONOCOTS - Monocotyledons

CYPERACEAE – Sedge family
Carex divulsa ssp. *divulsa* Stokes
Grey Sedge
Tresco and St Mary's

Native. Locally frequent on Tresco but only known from Bar Point on St Mary's. On Tresco the species is found in grassland near the sluice at the end of Abbey Pool, through the Abbey Wood, Abbey Gardens, and from Abbey farm and the west end of Great Pool to Dolphin Town and New Grimsby. Under trees, roadsides and in open grassland.

Carex muricata ssp. *pairae* (F.W. Schultz) Celak.
Prickly Sedge
Extinct

Native. Only one record. Found by D.E. Allen (as *C. muricata* ssp. *lamprocarpa*) in 1975 in the former Tresco hotel grounds, on the verge of lane to Gimble Porth.

CYPERACEAE – Sedge family
Carex arenaria L.
Sand Sedge
Inhabited islands, also most uninhabited islands except the smallest rocky ones

Native. A considerable feature in the Isles of Scilly; widespread and locally abundant. In Scilly the sedge is not restricted to coastal habitats, but also occurs well away from the coast in heathland and on walls. On the island of Annet a very tall form of the sedge forms a windblown 'grass-like' sward that extends over large areas. The variability of this species in Scilly leads to occasional misidentifications for other species. The plant can range from very tiny dwarf plants to those in dunes with obvious far-creeping rhizomes in bare sand, and to tall grassy plants, often with reduced spikelets, away from the coast. In dunes and dune grassland it is often the first pioneer in unstable sand. Coastal grassland, dunes, heaths, inland verges and on walls.

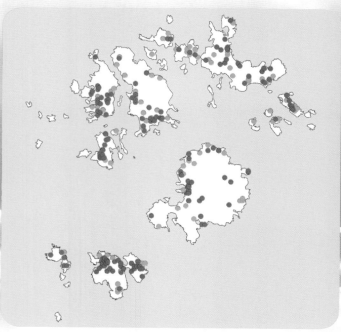

CYPERACEAE – Sedge family
Carex remota L.
Remote Sedge
Tresco

Native. Known from Abbey Hill (SV89151440) and near New Grimsby (SV889154). Path edge and under trees.

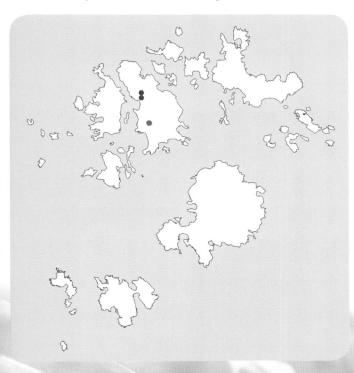

Carex echinata Murray
Star Sedge
Extinct

Native. Recorded from a bog on Higher Moors by Lousley. This was probably the acid bog that has since been lost.

CYPERACEAE – Sedge family
Carex riparia Curtis
Greater Pond-sedge
Tresco and St Mary's

Native. A rare sedge, only known from Tresco and St Mary's. On Tresco it was found near Abbey Drive (SV894145-SV895145) in 1997 and also on a garden rubbish dump (SV892141) in 1999. On St Mary's it still occurs on Higher Moors (SV922109). Marshes and waterlogged ground.

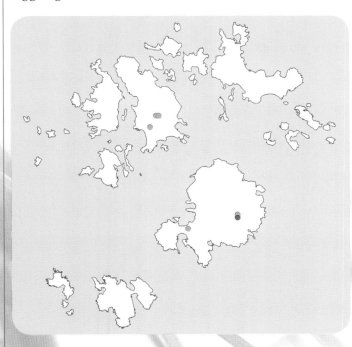

Carex pseudocyperus L.
Cyperus Sedge
Tresco

Neophyte. Plants found growing near Abbey Arch on Abbey Drive in 2011 had been planted.

MONOCOTS - Monocotyledons

CYPERACEAE – Sedge family
Carex pendula Huds.
Pendulous Sedge
Tresco, St Mary's and St Martin's

Native. This species has apparently extended its range since Lousley's Flora. Lousley considered it may be an established alien. But although some plants on St Martin's may have arisen from garden escapes, others are growing in more 'native' situations in Holy Vale, Lower and Higher Moors as well as in ditches, along pathsides on St Mary's and Tresco. Woodland, verges and ditches.

CYPERACEAE – Sedge family
Carex sylvatica Huds.
Wood-sedge
Tresco and St Mary's

Native. Found in the area known as Tresco Wood and beside the path to the Abbey, most recently in 1999, but probably still there. This is where several other woodland species occur. Also recorded in the Higher Moors, St Mary's by Michael Davies in 2007. Damp ground under trees.

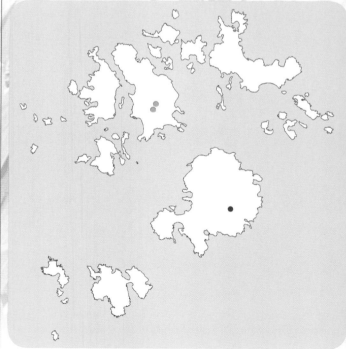

CYPERACEAE – Sedge family
Carex laevigata Sm.
Smooth-stalked Sedge
St Mary's

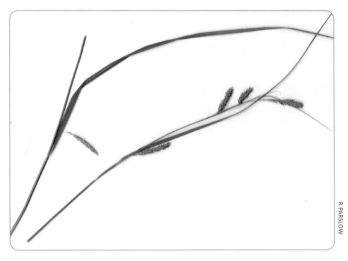

Native. Very rare. Only known from Higher Moors, St Mary's (SV923109 and SV92141083) in the open marshes.

Carex flacca Schreb.
Glaucous Sedge
Extinct

Native. The sole record is a herbarium specimen from maritime turf, Tresco, May 1929, collected by Col. R. Meinertzhagen.

Carex panicea L.
Carnation Sedge
Extinct

Native. Formerly St Mary's? Lousley records this species, but probably based the record on a herbarium specimen from 1923. There are no other records.

CYPERACEAE – Sedge family
Carex binervis Sm.
Green-ribbed Sedge
St Mary's

Native. Scattered mainly on sites near the coast, from the golf course and Halangy Down in the northwest of the island, round to Innisidgen, down the east of the island to Salakee. Heathland, often in grassy patches and along paths.

MONOCOTS - Monocotyledons

CYPERACEAE – Sedge family
Carex demissa Hornem.
Common Yellow-sedge
Tresco, St Mary's and St Agnes including Gugh

Native. A rare sedge that is only known from a few scattered sites, St Agnes on Wingletang (SV883074-SV884074) and Gugh (SV888080) in 2007; St Mary's in Lower Moors in 2000 (SV913105) and most recently Shooter's Pool in 2011 (SV91301078). It has also been found at Tresco Abbey Pool (SV897142) in 1995. Wetlands and flushes, but is nowhere common.

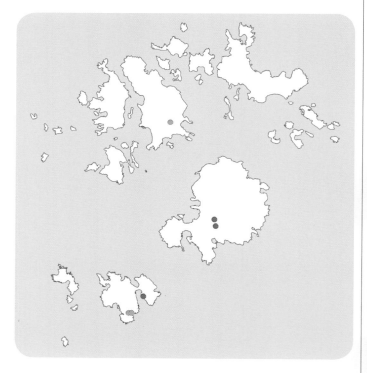

Carex nigra (L.) Reichard
Common Sedge
Extinct?

Native. There are records from St Martin's and Tresco in 1864. Lousley described it as 'locally plentiful' in Lower Moors. Puzzlingly there are no recent records.

CYPERACEAE – Sedge family
Carex pilulifera L.
Pill Sedge
Tresco, St Mary's, St Martin's, St Agnes and Gugh

Native. Lousley did not personally find this species in Scilly but he quotes an 1879 record from 'Higher Marsh', but considered it 'unsatisfactory' (Lousley, 1971). The first recent record was a specimen collected from St Mary's campsite by Mrs L. Knight in 1988. Since then the species has been found on all the inhabited islands except Bryher. On St Agnes it has been recorded on Wingletang Down and on Gugh. On St Martin's there are records from Great Bay and in the dunes at Lower Town. The records from Tresco are all on Castle Down; from St Mary's plants have been recorded several times from around the Innisidgen Burial Chambers, also from Mount Todden Down and from Salakee Down. Plants are found scattered in small numbers and easily overlooked, nowhere very common. Found on heathland and sandy tracks.

POACEAE – Grass family
Sasa veitchii (Carrière) Rehder
Veitch's Bamboo
St Mary's

Neophyte. In a hedge at Bar Point (SV915126) in 1996.

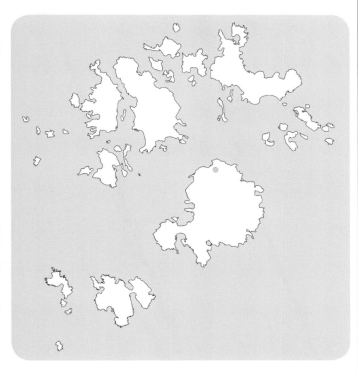

Pseudosasa japonica (Siebold & Zucc. Ex Steud.) Makino ex Nakai
Arrow Bamboo
St Mary's and St Agnes

Neophyte. An occasional garden escape. May still be at Lower Town, St Agnes and Rocky Hill, St Mary's although not recorded at either site since 1987. Still found on Toll's Island, St Mary's among bracken. Found at Higher Trenoweth in 2006 and Telegraph Road 2012.

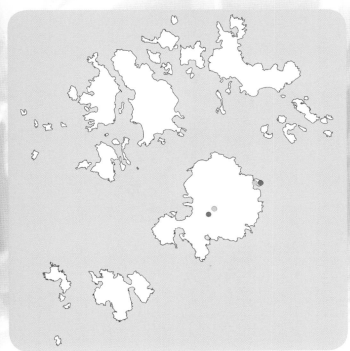

POACEAE – Grass family
Nassella trichotoma (Nees) Hack. Ex Arechav.
Serrated Tussock
St Agnes

Neophyte. Casual. Found in 2012 with *Vulpia bromoides* on the edge of a bulbfield (SV885082). Determined by T.B. Ryves.

MONOCOTS - Monocotyledons

POACEAE – Grass family
Schedonorus arundinaceus (Schreb.) Dumort.
Tall Fescue
Bryher

Native on the mainland but probably a neophyte on Scilly. Only known from a grass field just behind Rushy Bay, Bryher (SV87751415) in 2004. Possibly introduced with grass seed.

POACEAE – Grass family
Lolium perenne L.
Perennial Rye-grass
Inhabited islands, also Samson

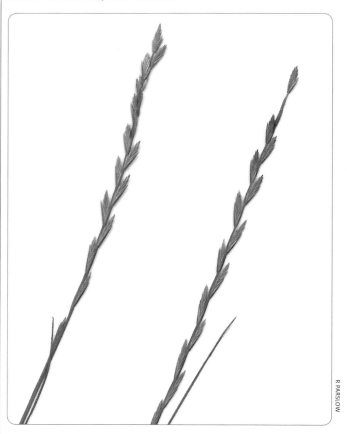

Native. Widespread and locally frequent to abundant in places. Lousley recorded this species from Teän, but not from Samson. On Samson it has been found both by Southward Well and the well on South Hill, perhaps accidentally introduced there, or it may be a relict from former cultivated fields. Grasslands, roadsides and along pathways.

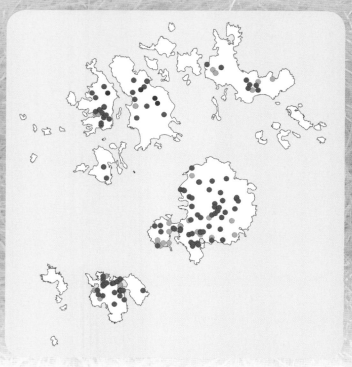

POACEAE – Grass family
Lolium multiflorum Lam.
Italian Rye-grass
Bryher, St Martin's, St Agnes and St Mary's

Neophyte. Not known from Tresco, occasionally recorded elsewhere. It was sown in bulbfields as a break crop, but less commonly now than formerly. Has been associated with agriculture in Scilly since at least 1870 when it was sown on barley fields with clover as a break crop (Lousley, 1971).

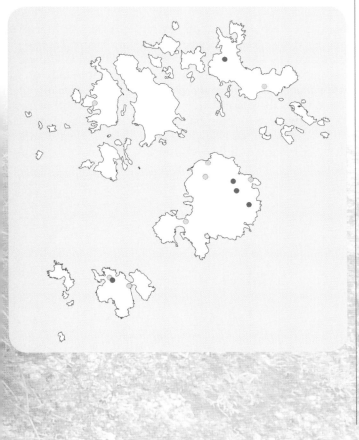

POACEAE – Grass family
Festuca rubra L.
Red Fescue
Inhabited islands, also the uninhabited islands except the Western and Norrard Rocks

Native. The most widespread and abundant grass in Scilly. In many of the plant communities where *F. ovina* might be expected it is replaced by *F. rubra*. Some of the coastal forms of the grass are very distinctive, including the dense 'mattresses' around the coast and on uninhabited islands. Subspecies *juncea* (Hack.) K. Richt. has been recorded from around the coast. Although the common grass is *F. rubra* ssp. *rubra*, inland some varieties may have arisen from introductions in seed mixtures. All grasslands.

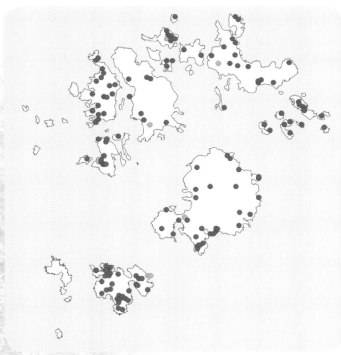

MONOCOTS - Monocotyledons

POACEAE – Grass family
Festuca ovina L.
Sheep's-fescue
Inhabited islands and Samson

Native. Mainly recorded on cliffs and dry grasslands. Although generally replaced by *F. rubra* in many habitats the very few records suggest it is under-recorded. Grasslands, often coastal.

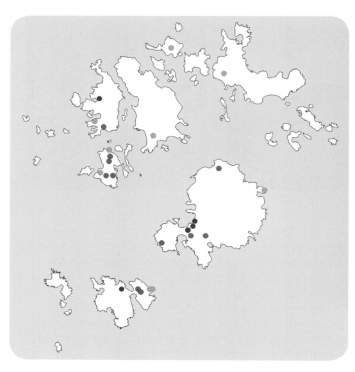

Vulpia fasciculata (Forssk.) Fritsch
Dune Fescue
Samson

Native. The only record of this species from the dunes on Samson (SV879131) in 2009 during CeC surveys needs confirmation.

POACEAE – Grass family
Vulpia bromoides (L.) Gray
Squirreltail Fescue
Inhabited islands, also Samson

Native. Locally frequent on Bryher, St Agnes, St Martin's and St Mary's, but only scattered records from Tresco. Plants can be very variable in size; in bulbfields, dunes and grasslands plants are often 50 cm or more tall, in drier places such as sandy ground or on the tops of stone walls the plants can be tiny, just a few centimetres tall. Found in two main habitats; in the bulbfields and cultivated places or elsewhere in grasslands, heathland, dune grassland and on wall-tops.

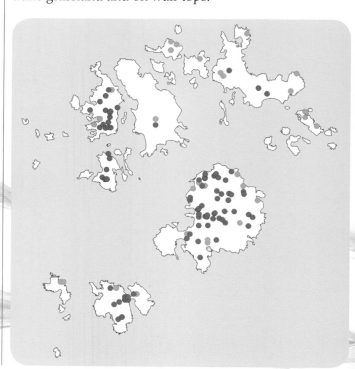

POACEAE – Grass family
Vulpia myuros (L.) C.C. Gmel.
Rat's-tail Fescue
Bryher, St Martin's, St Agnes and St Mary's

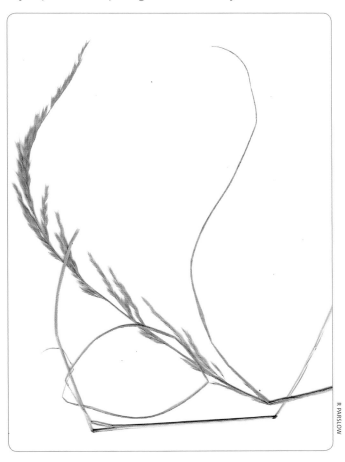

Archaeophyte. Occasional. Much less common than the previous species but often in the same areas in cultivated habitats and field margins. Cultivated fields and disturbed ground.

POACEAE – Grass family
Cynosurus cristatus L.
Crested Dog's-tail
Inhabited islands, also St Helen's

Native. Locally frequent except St Martin's. On St Martin's it is only known from a few places on Chapel Down, a suite of fields between Chapel Down and Higher Town and on Tinkler's Hill. The grass has an interesting history having been used in the 1800s to make plaited straw hats. It also seems to have been introduced in seed mixes so it is not possible to distinguish native plants from those derived from introduced seed. There is a record from St Helen's in 1983, a possible accidental introduction. Grasslands, roadsides and paths on heathland.

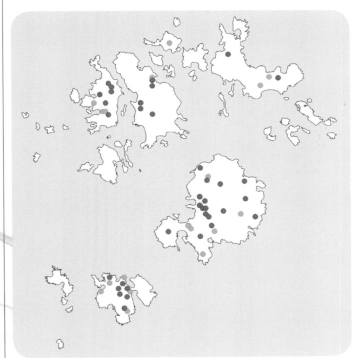

MONOCOTS - Monocotyledons

POACEAE – Grass family
Cynosurus echinatus L.
Rough Dog's-tail
St Mary's

Neophyte. This distinctive grass is only known from Buzza Hill and the immediate area. It is usually common beside the path up the hill from Porthcressa and has survived the recent resurfacing of the path. Records from Holy Vale in 2000 and Telegraph Lane in 2008 were probably accidentally transferred there. There are no recent records from Tresco where Lousley recorded the grass near Back Lane in 1940. It is surprising it has not travelled further. Verges of paths.

POACEAE – Grass family
Puccinellia maritima (Huds.) Parl.
Common Saltmarsh-grass
Bryher

Native. The only known site is from beside Bryher Pool. Lousley refers to a record from St Mary's in 1879, but the area near Porthellick is no longer brackish marsh. On Bryher the grass grows in the salt marsh strip between the Pool and the leat. Saltmarsh.

POACEAE – Grass family
Briza minor L.
Lesser Quaking-grass
Inhabited islands

Archaeophyte. Locally abundant. It is a pretty little grass that is common throughout the inhabited islands. Lousley called it 'dainty' which it is. Bulbfields and other cultivated and disturbed ground.

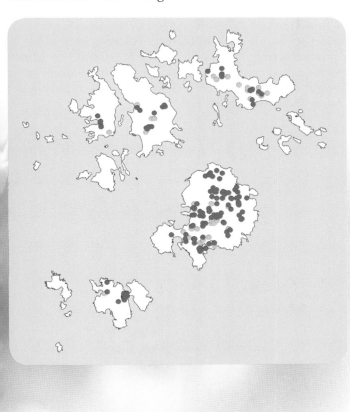

POACEAE – Grass family
Briza maxima L.
Greater Quaking-grass
St Agnes, St Martin's, St Mary's and Tresco

Neophyte. Possibly originally introduced, although there is no record of it being grown commercially. There are scattered records, mostly from St Mary's, where the main concentration is around Hugh Town, both in the town and especially on Buzza Hill, and out as far as Rocky Hill. On St Agnes it is found mainly on the bulb farm above the Bar and at Lower Town. The only recent record from Tresco is from Dolphin Town. On St Martin's it used to be common on the tops of stone hedges. Bulbfields, wall tops, roadsides and waste land.

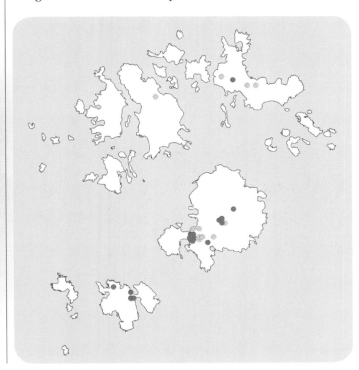

MONOCOTS - Monocotyledons

POACEAE – Grass family
Poa infirma Kunth
Early Meadow-grass
Inhabited islands

Native. Widespread and locally frequent on all the inhabited islands. The species had not been recognised in Scilly until discovered by John Raven in 1950 (Raven, 1950), but it had been known from the Channel Islands much earlier. In Scilly it appears very early in the year (a few plants just starting to flower were found on St Agnes in late January 2015), and although plants may linger in some places until May, it is usually at its best in March to early April. Very rarely, a few plants may be found in the autumn, presumably from seed that has germinated in late summer. It can be separated from *P. annua* by its short anthers and spikelets, the plants also are yellowish and tend to spread out, star-like, from the centre (as *P. annua* also can). From the 1970s the plant had been spreading, especially in places where vehicles and people pass. Where visitors congregated in places such as outdoor cafés, the grass could often be found growing underneath the tables! Since then there has been something of a reversal as it has disappeared from former footpath sites when they became too overgrown. Sandy and bare paths, roadside verges and shallow turf.

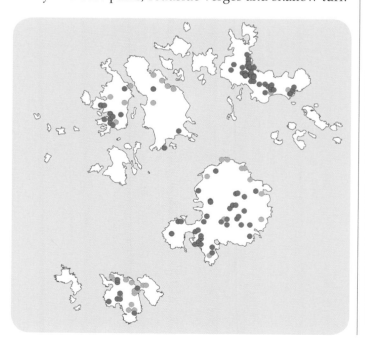

POACEAE – Grass family
Poa annua L.
Annual Meadow-grass
Inhabited islands, also Samson and Northwethel

Native. Abundant. A very common, widespread and variable grass. It may occur on some of the other uninhabited islands as it was recorded on Little Ganilly in 1987 and there are records from Northwethel in 1992. There are more recent records from Samson in 2008. On tracks and sandy ground the plants are small and yellowish, they often grow alongside (and can easily be mistaken for) *P. infirma*. They can flower all year round, including early in the year before *P. infirma*. On cultivated ground, especially in the bulbfields, the plant can be tall and spreading and appear very different in appearance. Tracks, gateways, grassland, pavement cracks and cultivated ground.

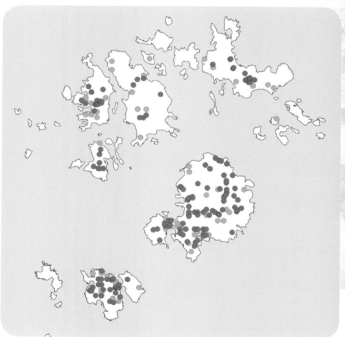

POACEAE – Grass family
Poa trivialis L.
Rough Meadow-grass
Inhabited islands, also Samson (including Puffin Island), Northwethel, St Helen's and Teän

Native. Frequent. Last recorded on Northwethel in 1992 and Teän in 1997, but likely to be still there. Grasslands, roadsides, cultivated fields and wetlands.

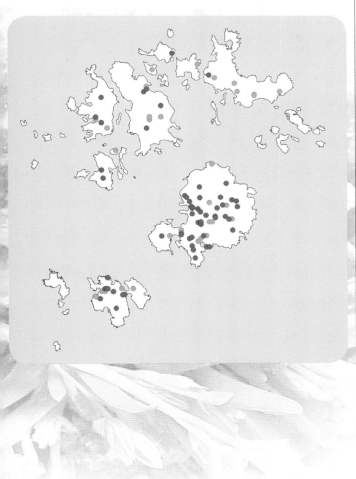

POACEAE – Grass family
Poa humilis Ehrh. ex Hoffm.
Spreading Meadow-grass
Inhabited islands, also Samson, St Helen's and Teän

Native. Occasional to locally frequent. Scattered records from the inhabited islands, also found on Samson near the ruins in 2002 and Teän in 2007. It had been recorded on St Helen's in 1983. On Tresco it has been found in a field by Great Pool (SV894146); on St Martin's it has been found on the cricket field. Mostly in damp places such as tracks and beside pools.

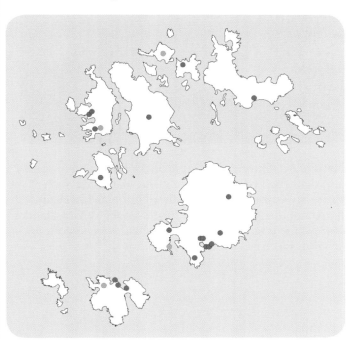

Poa pratensis L.
Smooth Meadow-grass
Inhabited islands, also Northwethel, Teän and St Helen's

Native. Widespread but scattered throughout the islands. Last recorded on Northwethel in 1992 and Teän in 1996. Grasslands, grassy places in heathland, on tops of walls, roadsides and dunes.

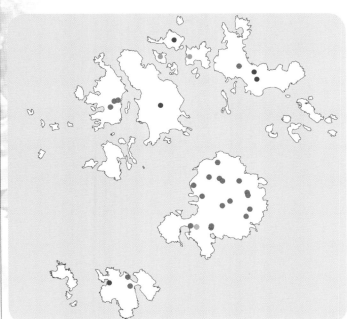

MONOCOTS - Monocotyledons

POACEAE – Grass family
Dactylis glomerata L.
Cock's-foot
Inhabited islands, also most uninhabited islands except Western and Norrard Rocks

Native. An abundant species found in a range of grassy and disturbed habitats, some quite surprising such as the rocky Guthers Island. Although it was last recorded on Annet in 1983 it is likely to still be there. Grasslands, around the edges of cultivated fields, on roadsides and around the coast.

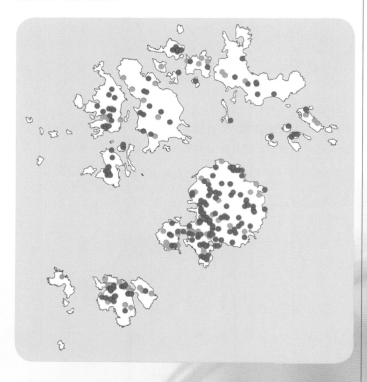

POACEAE – Grass family
Catapodium rigidum (L.) C.E. Hubb.
Fern-grass
Inhabited islands, also Samson

Native. Occasional. There is a small concentration of records around Hugh Town, St Mary's and near the Great Pool and Abbey Gardens, Tresco. Elsewhere there are individual records from Higher Town, St Martin's (SV929155); Rushy Bay, Bryher; Higher Town, St Agnes (SV884083) and by the ruins on Samson (SV87821246). The var. *majus* has been recorded once from the Abbey Gardens, Tresco (in 1999) and has been found several times on walls and roadsides in the area in Hugh Town, St Mary's between Church Road, Church Street and the alleyway by the incinerator. Roadsides, on walls and rocky places on heaths.

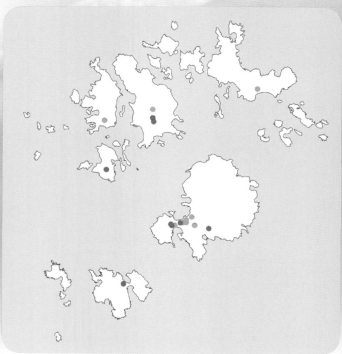

POACEAE – Grass family
Catapodium marinum (L.) C.E. Hubb.
Sea Fern-grass
Inhabited islands, also Samson, St Helen's, Northwethel, Round Island and Teän

Native. Locally frequent, but a patchy and gappy distribution. This species is more coastal than *C. rigidum*. Sandy and rocky places and often just above HWM on the coast.

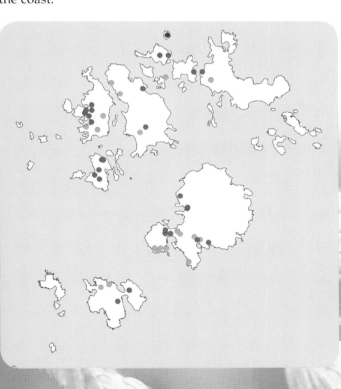

POACEAE – Grass family
Avenula pubescens (Huds.) Dumort.
Downy Oat-grass
Tresco

Native in the British Isles but a neophyte on Scilly. A flowering plant found near the sluice from Great Pool, Tresco (SV889148) in 2007 would have been an accidental introduction.

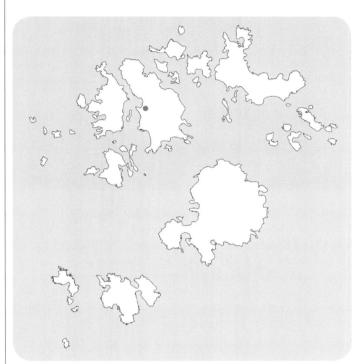

MONOCOTS - Monocotyledons

POACEAE – Grass family
Arrhenatherum elatius (L.) P. Beauv. Ex J. & C. Presl.

False Oat-grass

Inhabited islands, also Samson, St Helen's, Northwethel, Teän, Little Arthur and Great Ganilly

Native. Occasional to locally abundant. The var. *bulbosum* 'Onion Couch' has been recorded from many islands and is probably the common variety on Scilly, especially on the uninhabited islands. Lousley (1971) describes the species as 'rather rare', but it is now common and widespread, especially so on St Mary's. On Samson the grass appears to grow in former cultivated fields. Roadsides, grassland and shores.

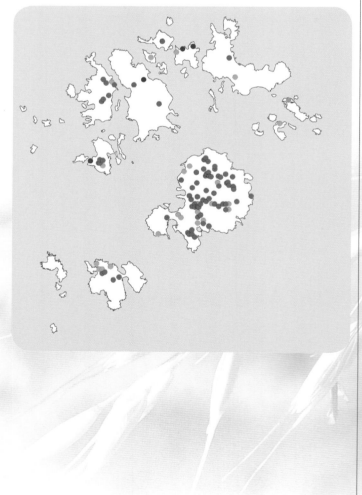

POACEAE – Grass family
Avena fatua L.

Wild-oat

St Martin's, St Agnes and St Mary's

Archaeophyte. Rare. An uncommon grass with a few records from St Mary's; Salakee 2005 and Lunnon 2004 and 2013; St Agnes (SV884084) in 2004; and from St Martin's (SV925155 and SV928158). Cultivated fields and on roadsides.

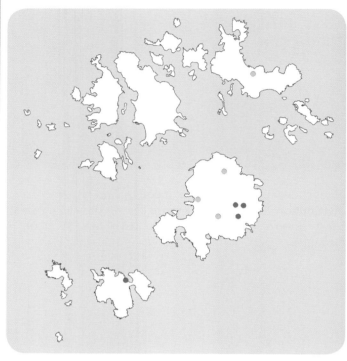

Avena sterilis L.

Winter Wild-oat

St Mary's

Neophyte. A casual record from a bulbfield near Porthellick House by J. Dony in 1976.

POACEAE – Grass family
Avena sativa L.
Oat
Tresco, St Mary's and St Martin's

Neophyte. A rare casual recorded at New Grimsby, Tresco 1998; Carron Farm 1999 and the rubbish pit, St Martin's 1997; and on St Mary's near Old Town 2009. Probably accidental introductions or from animal feed. Cultivated fields.

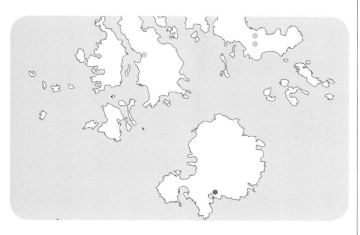

Trisetum flavescens (L.) P. Beauv.
Yellow Oat-grass
Extinct

Native. Formerly only known from Teän where it was seen by Lousley in 1967. The first recorded sighting had been in 1952 (Grose, in Lousley, 1971) of a dozen plants in a grassy clearing about 30 yards north of the pump, since that time the area is now dense bracken. A recently discovered record of several plants on a grass verge just off Pelistry Lane by P.J.O. Trist in 1975 may give credence to the story that *Trisetum* was said to be one of the grasses used locally in the straw plait industry which flourished for a short time in Scilly from 1840. Perhaps relics of earlier cultivation?

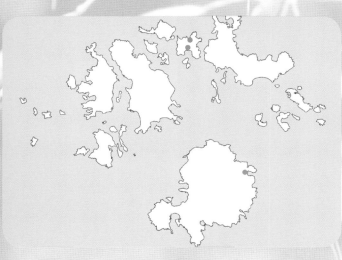

Rostraria cristata (L.) Tzvelev
Mediterranean Hair-grass
Extinct

Recorded by Lousley from bulbfield at Northward, Tresco in 1957.

POACEAE – Grass family
Koeleria macrantha (Ledeb.) Schult.
Crested Hair-grass
Bryher, Tresco, Teän and St Agnes

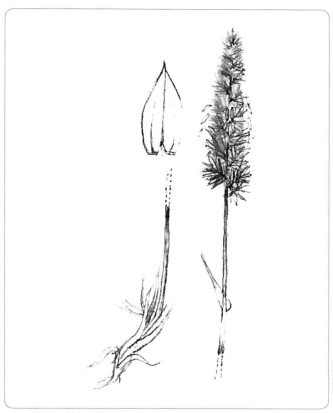

Native. Rare. Almost certainly under-recorded, several new records were made during surveys by CeC botanists in 2007. On Bryher the grass has been found near the Pool, Great Porth and on Shipman Head. It has also been recorded on Tresco just outside the Gardens (SV894143) and on St Agnes there are records from the Meadow and from Wingletang. CeC also found it on Teän. Sandy grassland by the coast and on heaths.

MONOCOTS - Monocotyledons

POACEAE – Grass family
Holcus lanatus L.
Yorkshire-fog
Inhabited islands, also all uninhabited islands except the Norrards and Western Rocks

Native. Locally dominant. A very widespread grass found almost everywhere, in a range of habitats. In the bulbfields the plants are large, often with the panicles purple in colour. On some uninhabited islands, for example Annet, there are large stands of the grass, possibly where there had been a fire in past. In bulbfields it often is very common after the crop has died down. Cultivated ground, roadsides, waste land, grasslands and on heathy cliff slopes.

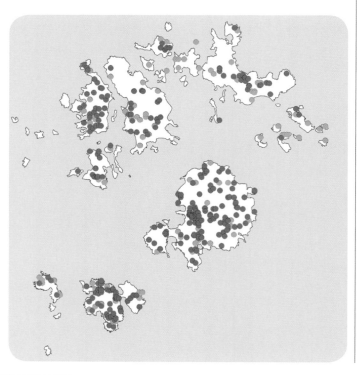

POACEAE – Grass family
Holcus mollis L.
Creeping Soft-grass
Bryher, St Martin's, Tresco and St Mary's

Native. An uncommon grass which does not appear to have spread very much since first recorded by D.E. Allen from the Abbey Gardens, Tresco in 1975. At the time Lousley was dubious about the identification. Since then there has been just a handful of records since 2000 from St Martin's; St Mary's in 2003; and Bryher in 2011. There are no subsequent records from Tresco. Bulbfields and cultivated ground also on some wall tops on St Mary's.

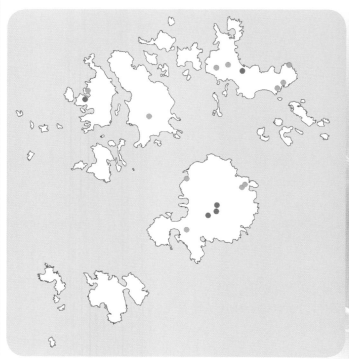

POACEAE – Grass family
Aira caryophyllea L.
Silver Hair-grass

Inhabited islands, also Samson, St Helen's, Teän, Nornour, Little Ganilly and Great Ganilly

Native. Widespread and locally frequent. There are records from both Annet and Little Ganilly in 1983, so it may still occur there. The plants on heathland, coastal habitats and walls tend to be very tiny compared with those growing in arable habitats. Although Stace (2010), and Cope and Gray (2009) are minded to reject the subspecies, plants attributable to ssp. *multiculmis* occur in bulbfields and disturbed ground on St Agnes, St Mary's and St Martin's. Dry grassland, sandy fields, heathland, dunes, wall-tops and rocky ground.

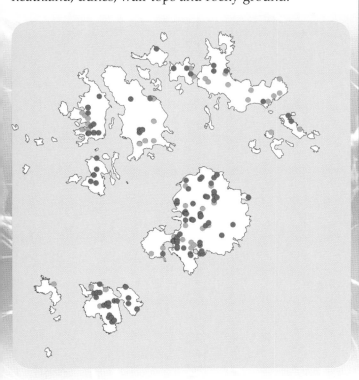

POACEAE – Grass family
Aira praecox L.
Early Hair-grass

Inhabited islands, also Samson, St Helen's, Teän and Great Ganilly

Native. Widespead and locally frequent. May occur in the same places as *A. caryophyllea*. Usually the plants on bare ground and wall tops are very tiny, as little as a centimetre, so they hardly show among mosses and lichens. In arable habitats the plants can be up to 30 cm tall and look very different. The very small size and early flowering may cause it to be often overlooked, although the dead heads can persist until late summer. Heathland, dunes, bare ground, coastal grassland, pavement cracks, wall tops and sandy arable fields.

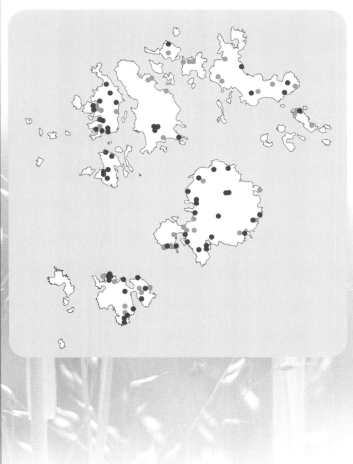

MONOCOTS - Monocotyledons

POACEAE – Grass family
Anthoxanthum odoratum L.
Sweet Vernal-grass

Inhabited islands, also Samson, St Helen's, Northwethel, Teän, Little Arthur, Little Ganilly and Great Ganilly

Native. Widespread and locally abundant on most islands. It was last recorded on Little Ganilly in 1983, but could still be present. Coastal heathland, grasslands, dunes, roadsides and bulbfields.

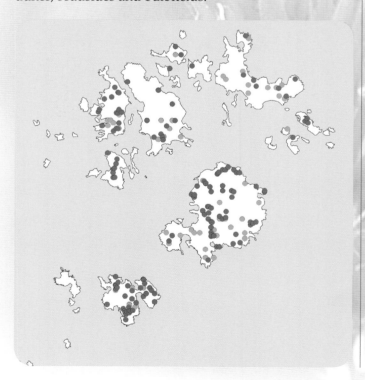

POACEAE – Grass family
Phalaris arundinacea L.
Reed Canary-grass

St Mary's

Native. A very rare grass in Scilly that is only known from St Mary's. There is only one record with a precise location, Lower Moors, SV913105 in 2002. It was not recorded by Lousley.

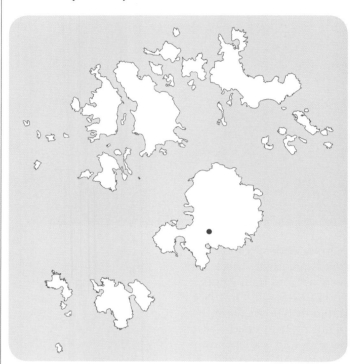

POACEAE – Grass family
Phalaris canariensis L.
Canary-grass
St Mary's and St Agnes

Neophyte. Casual. Recorded from Church Road, Hugh Town in 2000, Holy Vale (SV920115) in 2006 and near the quay, St Agnes in 2008. Roadsides, cultivated fields and waste ground.

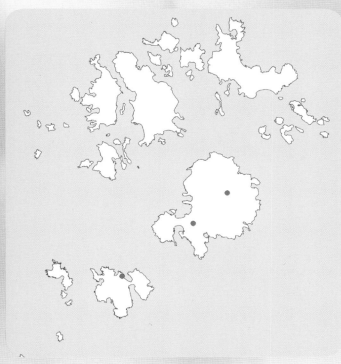

POACEAE – Grass family
Phalaris minor Retz.
Lesser Canary-grass
St Mary's

Neophyte. Occasional. Lousley (1971) describes this as 'locally abundant' and as a weed in bulbfields on St Mary's and St Martin's. It is also listed as one of the bulbfield species by Silverside (1977) so was common then. Although Cope and Gray (2009) describe it as 'in the Isles of Scilly - well-established in sandy, cultivated ground', it now only appears to have been recorded from a few places around Old Town, the road to the Airport and Hugh Town, most recently in Church Road in 1999, so seems to have undergone a substantial decline. Fields, pathways and roadside.

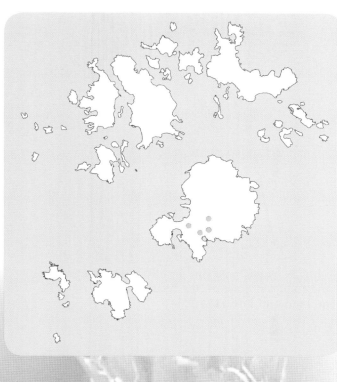

MONOCOTS - Monocotyledons

POACEAE – Grass family
Agrostis capillaris L.
Common Bent
Inhabited islands, also Teän, St Helen's, Samson and Little Arthur

Native. A widespread and locally abundant grass. It is the common grass on the heathlands. Grassland, heathland and farmland.

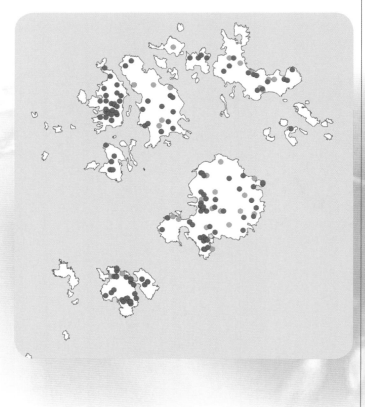

POACEAE – Grass family
Agrostis gigantea Roth
Black Bent
Inhabited islands

Archaeophyte. There are only a few scattered records of this grass. On Tresco it was last recorded at Borough Farm in 1986; Bryher, Hillside Farm in 2004; St Martin's it was found at Middle Town (SV919162) in 2004; on St Agnes it was found in a field off Barnaby Lane (SV88000818) in 2005. There are more records from St Mary's including Higher Moors in 2002, Parting Carn in 2003 and Newford Duck Ponds in 2008. Arable fields, roadsides and footpaths.

POACEAE – Grass family
Agrostis stolonifera L.
Creeping Bent
Inhabited islands, also most uninhabited islands except Western and Norrard Rocks

Native. Widespread and locally abundant. Found frequently on the uninhabited islands. Grows in many habitats, including wet fields, marshes, heath, coastal sites, roadsides and uncultivated ground.

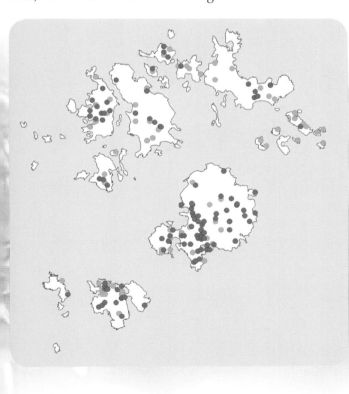

POACEAE – Grass family
Agrostis canina L.
Velvet Bent
Inhabited islands, also Samson and Great Ganilly

Native. Lousley described this species as 'locally plentiful' in marshy meadows on St Mary's. It is still mainly associated with wet grassland and damp habitats on St Mary's, but has subsequently been recorded on other islands. There had been some confusion in the past with *A. vinealis* Brown Bent which is more often found in bulbfields and dry habitats. Possibly occurs in drier habitats than on the mainland. Wetlands, damp grassland and roadsides.

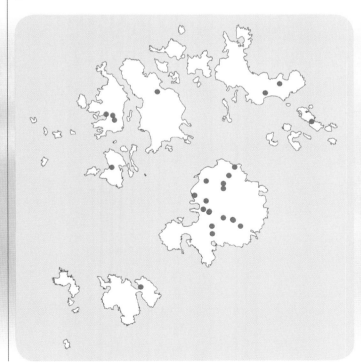

MONOCOTS - Monocotyledons

POACEAE – Grass family
Agrostis vinealis Schreb.
Brown Bent
St Agnes and Gugh, Bryher, Tresco and St Mary's, also St Helen's

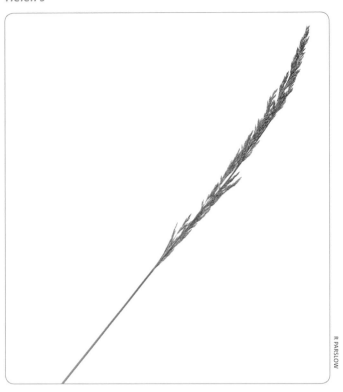

Native. Scattered records, although probably under-recorded due to confusion with *A. canina*. Apparently absent from St Martin's. Usually found in drier habitats than *A. canina*. Heathland and arable fields.

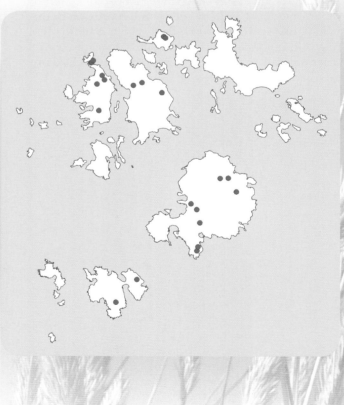

POACEAE – Grass family
Calamagrostis epigejos (L.) Roth
Wood Small-reed
St Martin's and St Mary's, also Samson, St Helen's, Teän, Great and Little Ganilly, Great Ganinick and Little Arthur

Native. Locally frequent in a few discrete locations in the north and east of the archipelago. It is most common on the Eastern Isles and Teän. A few more sites have been found since Lousley (1971) mapped the species (for example on St Mary's), but they are still within the same approximate area except for a 2015 record of a large stand of the grass on Samson where it may have been brought to the area by gulls. Among bracken and rough vegetation.

POACEAE – Grass family
Ammophila arenaria (L.) Link
Marram
Inhabited islands, also most larger uninhabited islands

Native. Locally abundant along sandy beaches. Although native to the islands, it is known that Augustus Smith planted additional Marram in the dunes in the 1830s to stabilise the sand. At that time sand blows caused great distress to the islanders, blowing into their houses and smothering their fields. It is now a characteristic species of sand dunes and sandy coast.

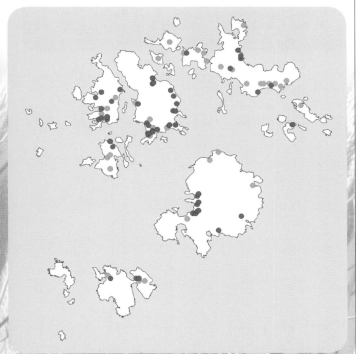

POACEAE – Grass family
Lagurus ovatus L.
Hare's-tail
St Mary's and Bryher

Neophyte. Casual records from Church Street and Church Road, Hugh Town in 1997. The finding of a group of at least five flowering plants apparently well established on open heathland on the west side of Shipman Head (SV874162), Bryher in September 2014 is unexplained. In 2016 a few plants were found in a border outside the Post Office where it was probably planted. Heathland or near habitations.

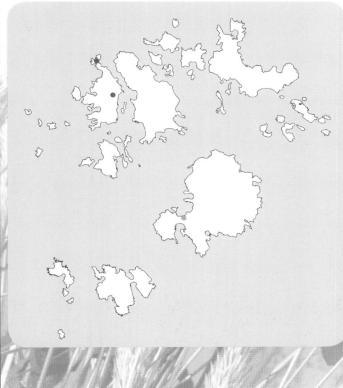

Gastridium ventricosum (Gouan) Schinz & Thell.
Nit-grass
Extinct

A record from a roadside in Old Town, St Mary's in 1922 was accepted by Lousley, but there have been no records since.

MONOCOTS - Monocotyledons

POACEAE – Grass family
Polypogon viridis (Gouan) Breistr.
Water Bent
Bryher

Neophyte. Casual. Usually a species of built-up and man-made habitats on the mainland, the Isles of Scilly records seem atypical. There are several records from the Badplace Hill area on Shipman Head growing on the rocky headland in 1991 and 1992, specimens from the area were confirmed by Eric Clement. It was still there in 2009. Open, rocky habitats.

POACEAE – Grass family
Alopecurus pratensis L
Meadow Foxtail
St Mary's

Native. Rare. Only known recently from near Hugh Town (SV905105) in 1988; from Higher Moors (SV923109) in 1997; and from fields at Parting Carn (SV916108) in 2003.

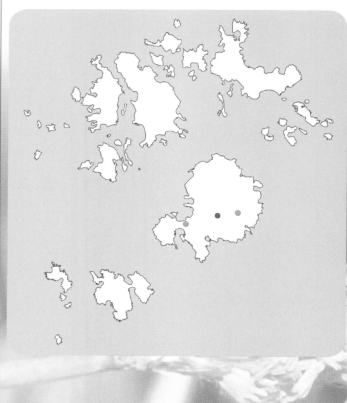

POACEAE – Grass family
Alopecurus geniculatus L.
Marsh Foxtail
Tresco, St Martin's and St Mary's

Native. Occasional and local. Found in a few wet grasslands where it often grows with *Agrostis stolonifera*. On Tresco it has been found on School Green in 1975 and the wet edge of a field north of Great Pool (SV89551462) in 2008. On St Martin's the grass can usually be found in the seasonally wet area beside the cricket field pool and has also been recorded in a wet area in the dunes at Lower Town. Populations on St Mary's occur in wet fields from near Porthloo to Lower Moors, and from Holy Vale, and from Lunnon Farm to Salakee. The grass frequently disappears when the sites dry out. Wet grassland, seasonal pools and ditches.

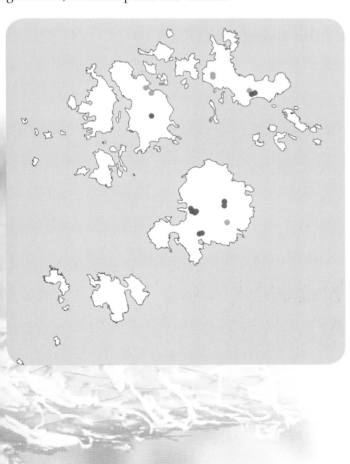

POACEAE – Grass family
Alopecurus myosuroides Huds.
Black-grass
Tresco

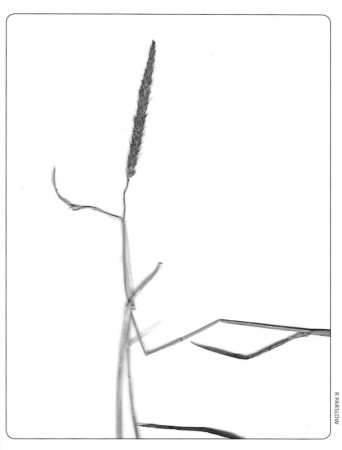

Archaeophyte. Rare. There is only one record of this species; it was found in 2008 growing in a wet corner of an arable field north of Great Pool (SV89551462). Whether it has been previously overlooked or was a recent introduction is unknown.

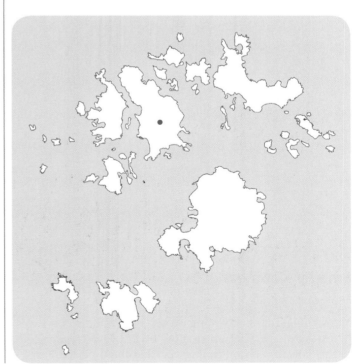

MONOCOTS - Monocotyledons

POACEAE – Grass family
Phleum pratense L.
Timothy
St Mary's and St Martin's

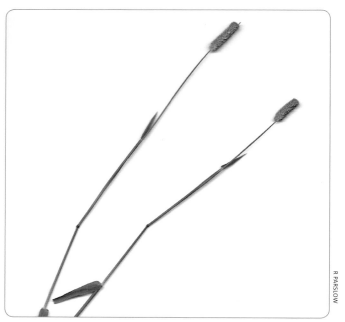

Native. Rare. Lousley (1971) comments that it was introduced as seed in temporary grass leys. This presumably no longer happens as there are just two recent records, one from near the Pottery on The Garrison in 1975 and from a field on St Martin's (SV926159) in 2000.

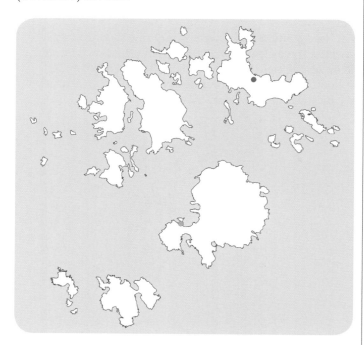

Phleum bertolonii DC.
Smaller Cat's-ear
St Mary's

Native. The plants growing on top of The Garrison walls (SV901106-SV901103) in shallow turf have been ascribed to this species, although it is possible they were dwarf specimens of *P. pratense*.

POACEAE – Grass family
Glyceria fluitans (L.) R. Br.
Floating Sweet-grass
St Mary's

Native. Rare. Only now known from Lower Moors. Sites in a nearby wet field and a ditch at Salakee (SV922107) are believed to have been lost; the first under stored building materials and Salakee ditch dried out when the stream was diverted. Wet fields and ditches.

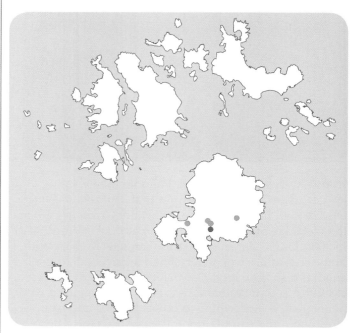

Glyceria x *pedicellata* F. Towns.
(*G. fluitans* x *G. notata*)
Hybrid Sweet-grass
St Mary's

Native. Rare. There is just one record of this grass, from a wet field (SV913103), near Lower Moors in 2003.

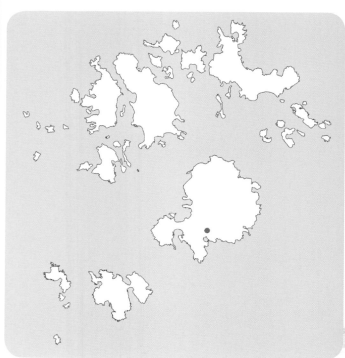

POACEAE – Grass family
Glyceria declinata Bréb.
Small Sweet-grass
St Mary's

Native. Rare. Found in similar places as other *Glyceria* species, sometimes even the same ones. In a few pools and wet areas in Lower Moors, in a ditch nearby (SV91251029), the wet fields between Lower Moors and Porthloo and formerly at Newford duck ponds (in 1988). The site in the field at SV91251032 near Lower Moors used for storing building materials may have been lost. Wet fields and ditches.

POACEAE – Grass family
Glyceria notata Chevall.
Plicate Sweet-grass
Tresco

Native. Recorded from north of Great Pool (SV895514) in May 2008 by David Bevan and R. Parslow. This is the first record for Scilly.

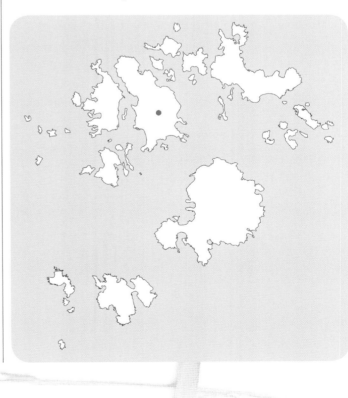

MONOCOTS - Monocotyledons

POACEAE – Grass family
Bromus racemosus L.
Smooth Brome
St Mary's and Bryher?

Native. There is a record from a bulbfield between Hugh Town and Old Town, St Mary's by R.C.L. Howitt in 1954 (determined by C.E. Hubbard). A plant identified as *Bromus commutatus* x *B. racemosus* was recorded from a bulbfield on St Mary's in 1996 by D.T. Holyoak. However Cope and Gray (2009) find separation of *B. racemosus* and *B. commutatus* unreliable, and Stace (2010) demotes *B. commutatus* to a subspecies of *B. racemosus*. A record from Bryher in 2012 has not been confirmed.

POACEAE – Grass family
Bromus hordeaceus L.
Soft-brome
Inhabited islands, also St Helen's

Native. Locally abundant. This very variable grass is common in Scilly. Some subspecies have been identified by Spalton. But as plants range in size from over a metre down to tiny plants with only one spikelet that grow on wall tops and dry banks, separating subspecies has not been very successful. The ssp. *hordeaceus* has been recorded from St Martin's, St Mary's and St Agnes. *B.* x *pseudothominei* Lesser Soft-brome found at Buzza hill by R.J. Murphy in 1995 is now included in this species (Cope and Gray, 2009). Bulbfields, grasslands, roadsides and disturbed ground.

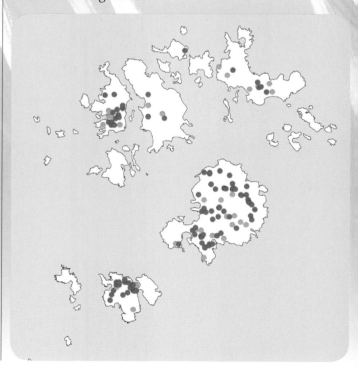

POACEAE – Grass family
Bromus secalinus L.
Rye Brome
St Mary's

Archaeophyte. A few clumps of the grass found in a field at Porthlow Farm in 2003, had possibly arrived as a contaminant in seed mix.

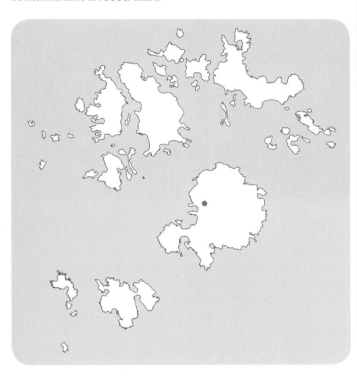

POACEAE – Grass family
Anisantha diandra (Roth) Tutin ex Tzvelev
Great Brome
Inhabited islands

Neophyte. This large, handsome grass is locally abundant, both along verges and in bulbfields on St Mary's, St Agnes and areas of St Martin's and Bryher, but uncommon on Tresco. On Tresco the only records are from a field at Town's Hill and New Grimsby. Bulb and other arable fields, roadsides and disturbed ground.

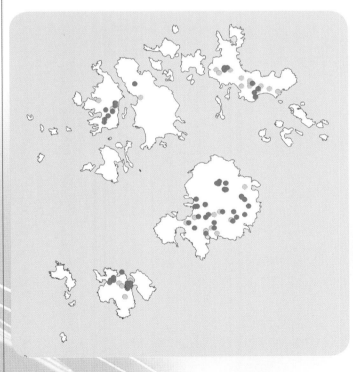

Bromus lepidus Holmb.
Slender Soft-Brome
Extinct

A specimen found by R.C.L. Howitt at Rose Hill in May 1954 was confirmed by C.E. Hubbard. There have been no subsequent records.

MONOCOTS - Monocotyledons

POACEAE – Grass family
Anisantha rigida (Roth) Hyl.
Rip-gut Brome
Inhabited islands

Neophyte. Due to its similarity to *A. diandra* this grass is almost certainly under-recorded. It may be better treated as a subspecies of *Anisantha diandra* as it appears to be closely related and intermediates occur. There is a 1973 record from Bryher (P.J.O. Trist) but nothing more since. It was not recorded by Lousley. Found in similar places to *A. diandra*, bulbfields, roadsides and on walls.

POACEAE – Grass family
Anisantha sterilis (L.) Nevski
Barren Brome
Inhabited islands, also Samson

Archaeophyte. This grass has a very patchy distribution on St Mary's, Bryher and St Agnes, occasional elsewhere. There can be confusion with depauperate forms of *A. diandra*, especially as they grow in similar places. The records from Samson in 2000 and 2009 could be an accidental transfer of seed by birds or by people visiting the island. Cultivated fields, roadsides and disturbed ground.

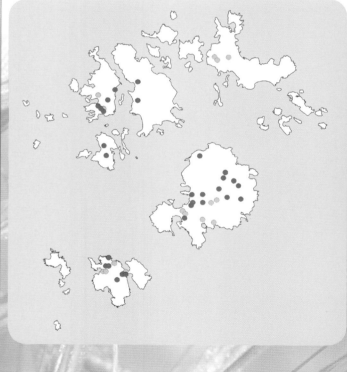

POACEAE – Grass family
Anisantha madritensis (L.) Nevski
Compact Brome
Bryher, St Martin's, St Mary's and St Agnes

Neophyte. Rare and local with few recent records. First recorded by Lousley on Bryher in 1939 when he considered it a recent introduction. Fortunately unlike the previous three species it is quite distinctive and easily identified. The years between 1995 and 1997 seem to have been when it was most often recorded. On Bryher it is known from two places on the west of the island (SV874150) near the Hotel in 2001 and near Rushy Bay (SV87645) in 2012. Former sites included on St Martin's in fields near Higher Town; St Mary's near Telegraph, Porth Mellon and the lane to the Airfield; and on St Agnes in fields above the Bar and Lower Town. Bulbfields, other cultivated fields and roadsides.

POACEAE – Grass family
Ceratochloa cathartica (Vahl) Herter
Rescue Brome
St Martin's and St Mary's

Neophyte. Locally frequent in a few discrete localities. According to Lousley (1971) the grass was first found on a roadside at Old Town before 1922 and had gradually spread on St Mary's and then St Martin's. It is now found from localities on both islands. Bulb and other cultivated fields.

MONOCOTS - Monocotyledons

POACEAE – Grass family
Brachypodium sylvaticum (Huds.) P. Beauv.
False Brome
Inhabited islands, also St Helen's, Northwethel, Teän, Samson, and in the Eastern Isles, Great Ganilly, Nornour and Ragged Island

Native. Common and locally abundant. It is a feature in many places such a roadsides and some heathlands. Heathland, cliff slopes, bulbfield margins, coasts and roadsides.

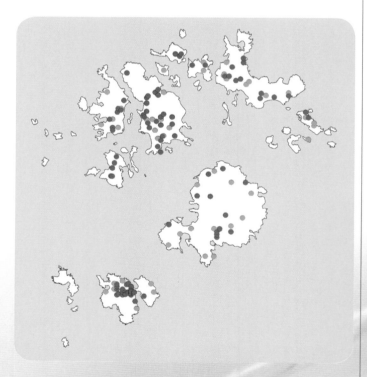

Elymus caninus (L.) L.
Bearded Couch
St Martin's

Native in the British Isles but if confirmed, a neophyte on Scilly. Frequently misidentified for the awned spikelet form of *Elytrigia repens*. One unconfirmed record from Lower Town area in 1990.

POACEAE – Grass family
Elytrigia repens (L.) Desv. Ex Nevski
Common Couch
Inhabited islands, Teän and the Arthurs

Native. Scattered records. Plants of ssp. *arenosus* have been recorded on Teän and the form *aristata* has been recorded from near the coast on St Agnes. The hybrid between this species and *Hordeum secalinum*, x *Elytrordeum langei*, was recorded by P.J.O. Trist from SV8815 near the former Post Office, Bryher, identified by C.E. Hubbard some time between 1960-1980. Found in fields, roadsides and shores.

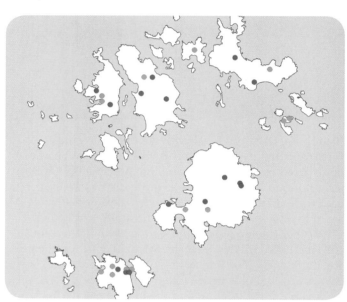

Elytrigia atherica (Link) Kerguélen
Sea Couch
St Agnes, Tresco, Bryher, St Mary's, Samson, Teän, White island (St Martin's), Little Arthur, Round Island and Great Ganilly.

Native. Scattered records from around the islands. Sites on Bryher include Stony Porth, Great Porth and Rushy Bay; also Porth Warna, St Agnes, White Island off St Martin's, both east and west coasts of Tresco, and Porth Minick and Old Town Bay, St Mary's. Dune edges.

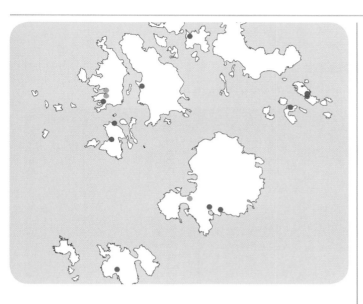

Elytrigia juncea ssp. *boreoatlantica* (Simonet & Guin.) Hyl.
Sand Couch
Inhabited islands, also St Helen's, Round Island, Teän, Samson (and Puffin Island), Annet and Great Ganilly

Native. Locally frequent. Probably under-recorded. Sandy shores above HWM.

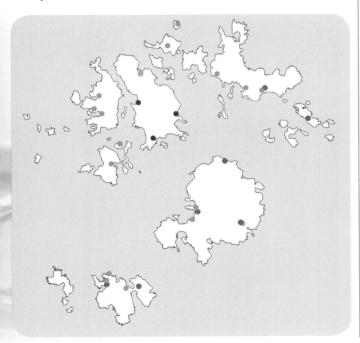

POACEAE – Grass family
Leymus arenarius (L.) Hochst.
Lyme-grass
St Mary's and Samson

Native. Not recorded until 2007, first from dunes at Pelistry Bay, St Mary's and then in 2008 from the dune at the north of Samson. It may have been previously overlooked?

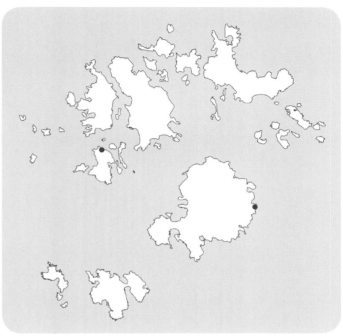

Hordeum murinum L.
Wall Barley
St Mary's, St Martin's and Bryher

Archaeophyte. Only two recent records, one from Buzza Hill, St Mary's in 2003 and Timmy's Hill, Bryher in 2004. The only other record was from Little Arthur Farm, in St Martin's in 1995.

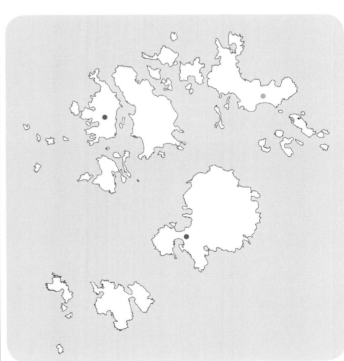

MONOCOTS - Monocotyledons

POACEAE – Grass family
Hordeum secalinum Schreb.
Meadow Barley
St Martin's

Native in the British Isles but possibly native on Scilly. Only one record from the corner of Pool meadow in 1995. Lousley describes the species as locally plentiful but it appears to have now disappeared.

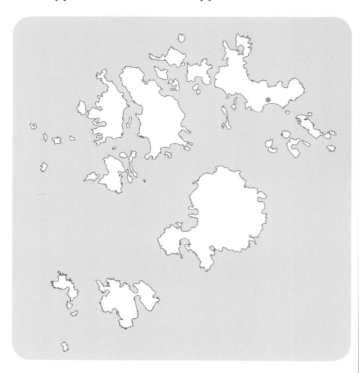

Triticum aestivum L.
Bread Wheat
St Mary's

Neophyte. Casual in fields at Lunnon in 2004 and Salakee in 2005. In Hugh Town in 2012.

POACEAE – Grass family
Danthonia decumbens (L.) DC.
Heath-grass
Inhabited islands, also Samson

Native. Locally abundant on all the inhabited islands. There is a 1983 record from St Helen's. A 1948 record from Annet (Lousley, 1971) is almost certainly an error. The grass is described as growing between *Calluna* bushes, but there are no records of *Calluna* there at the time or since. Perhaps a slip of the pen for Samson? *Danthonia* forms very distinctive patches of flattened tufts of straw-coloured plants especially on paths and between heather clumps on heathlands and around pools. Heathland, coastal grassland, paths, by pools and in damp places.

POACEAE – Grass family
Cortaderia selloana
(Schult. & Schult.) Asch. & Graebn.
Pampas-grass
St Mary's, St Martin's and Tresco

Neophyte. The plants growing on the rubbish dump on St Martin's and in Holy Vale, St Mary's in 1987 are originally of garden origin. The line of plants established from near Abbey Pool across Appletree Banks and dunes on Tresco may have originally been planted, but are now spreading naturally in the dunes. In 2016 a large, flowering plant was found in a field near Lower Moors (SV91251032) that may have spread from the dump? Heathland, dunes and rubbish dumps.

POACEAE – Grass family
Molinia caerulea (L.) Moench
Purple Moor-grass
Tresco, St Mary's and St Martin's

Native. On St Mary's, St Martin's and Tresco it is found in a few areas where it may be locally frequent. The plants on Scilly are not often as tall as on the mainland and where trampled down on pathways are often hardly recognisable. Water-logged ground on heathland and near pools.

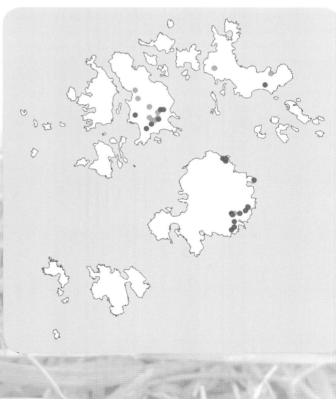

MONOCOTS - Monocotyledons

POACEAE – Grass family
Phragmites australis (Cav.) Trin. Ex Steud.
Common Reed
St Mary's and Tresco (St Agnes)

Native. On Tresco there are extensive reedbeds around Great Pool and a reed fringe around Abbey Pool. On St Mary's there are large areas of reed (and mixed reed/sea rush) swamp on Lower and Higher Moors, also in wet fields at Salakee, Lower Moors extension, Shooter's Pool, the wet fields and the pond near Porthloo and along some ditches. On St Agnes, Adrian Pearce showed me two prostrate reeds about 2.5 metres long that he found growing in a flower field lying along a strip of Mypex (the impermeable membrane used to exclude weeds). Presumably, when they had started to grow they were unable to continue to support themselves vertically, but continued growing horizontally! Besides reed swamp, reeds grow in wet fields, ditches and around pools.

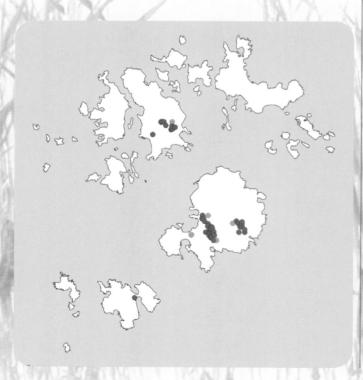

POACEAE – Grass family
Eragrostis cilianensis (All.) Vignolo ex Janch.
Stink-grass
St Mary's

Neophyte. Birdseed alien Hugh Town, 2000.

POACEAE – Grass family
Eleusine indica (L.) Gaertn.
Yard-grass
St Mary's

Neophyte. Casual. A plant found at Trenoweth, by A. Tompsett in 1998.

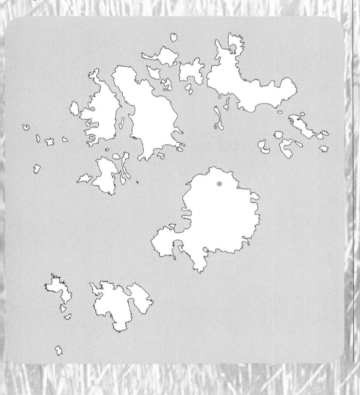

POACEAE – Grass family
Cynodon dactylon (L.) Pers.
Bermuda-grass
Tresco and St Mary's

Neophyte. There are old records in 1956 and 1967 from the Green at Old Grimsby and more recent ones from the Industrial Site in Hugh Town (SV909107) in 2002.

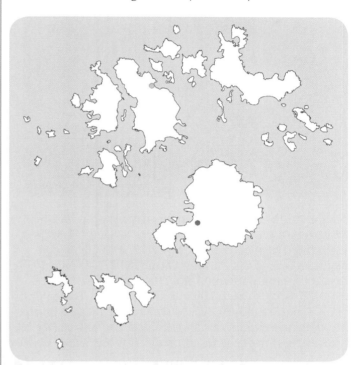

MONOCOTS - Monocotyledons

POACEAE – Grass family
Echinochloa crus-galli (L.) P. Beauv.
Cockspur
St Mary's, Tresco, St Martin's and Bryher

Neophyte. A rare casual that has turned up occasionally. Most recent records are from several sites on St Mary's, including Porthlow Farm (SV910115) and Old Town (SV912103) in 2005, Lunnon Farm (SV92531113) in 1998 and 2006, Hugh Town (SV90111049) and Old Town Bay (SV91361020) in 2009. The only record from Tresco was from a maize field (SV89561497) in 2011. In 2016 groups of plants were found in fields behind Lawrences Bay, St Martin's and near the former shop at Northward, Bryher. Found in arable fields or on disturbed ground.

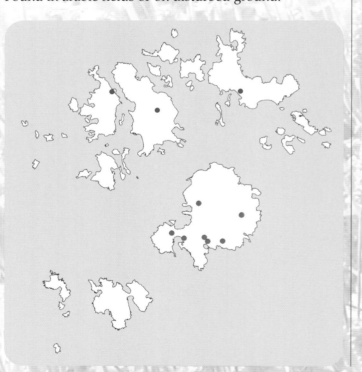

POACEAE – Grass family
Panicum miliaceum L.
Common Millet
Tresco and St Mary's

Neophyte. Casual. St Mary's, near Old Town 2014 and from near Pentle House, Tresco in 2015. Probably crop relict or birdseed alien.

POACEAE – Grass family
Setaria pumila (Poir.) Roem. & Schult.
Yellow Bristle-grass
St Mary's and St Martin's

Neophyte. Casual. This striking grass was presumed a birdseed alien when it was first found in a field at Old Town, St Mary's in 2005. Since then it has been found to be locally frequent in two areas on St Martin's; Middle Town (SV919161 - SV92101615) and Higher Town (SV93171562 - SV93181562). Arable fields.

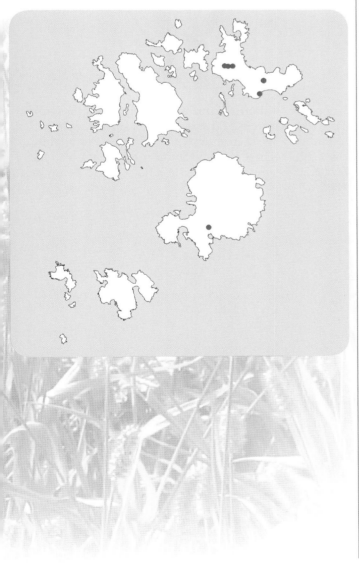

POACEAE – Grass family
Setaria verticillata (L.) P. Beauv.
Rough Bristle-grass
St Mary's

Neophyte. Casual. Presumed birdseed alien found in a field at Old Town in 2005. This is the same area where the previous species was found with a number of other unusual plants. The grasses were identified by Eric Clement, and are believed to be from a 'wild bird' food mix sold in St Mary's, the origin of which could not be traced. It is likely this may have been the source for several birdseed aliens found in St Mary's on 2000.

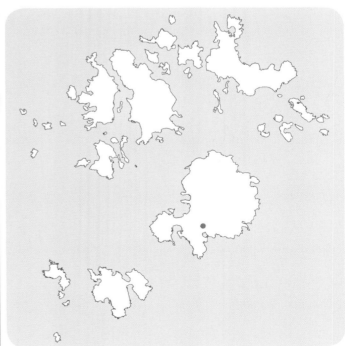

MONOCOTS - Monocotyledons

POACEAE – Grass family
Setaria viridis (L.) P. Beauv.
Green Bristle-grass
St Mary's

Neophyte. Birdseed alien, Ram's Valley, Hugh Town, St Mary's 2000, with other birdseed aliens.

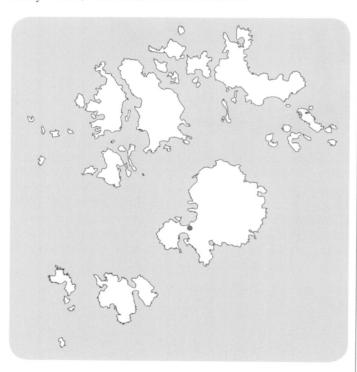

POACEAE – Grass family
Digitaria sanguinalis (L.) Scop.
Hairy Finger-grass
St Mary's

Neophyte. Although Lousley recorded this grass from Tresco as naturalised it has not been seen recently. Occurs as a weed in a garden by Thomas Porth (still there 2011) where it has been for many years, also has occurred as a birdseed alien in a field at Old Town in 2005.

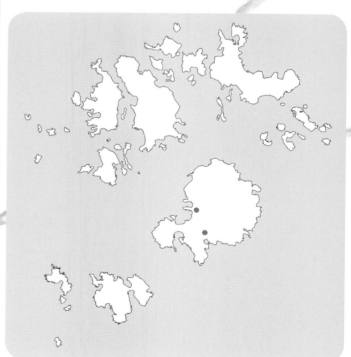

POACEAE – Grass family
Digitaria ciliaris (Retz.) Koeler
Tropical Finger-grass
St Mary's

Neophyte. Birdseed alien, Hugh Town, 2000.

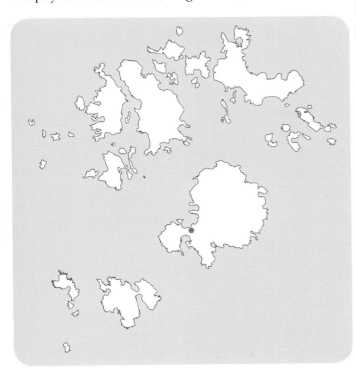

POACEAE – Grass family
Sorghum bicolor (L.) Moench
Great Millet
St Mary's

Neophyte. Birdseed alien, Hugh Town, 2000.

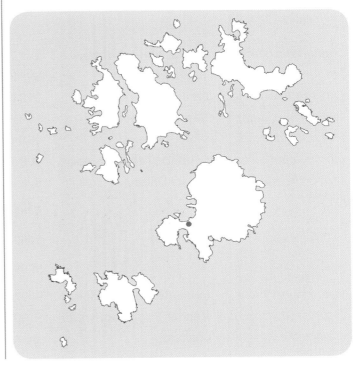

Garden escapes and species needing confirmation

Latin name	Common name	GR/Island	Date	Comment
Aconitum napellus	**Monk's-hood**	SV912102 Old Town, St Mary's	2008	Garden escape
Alchemilla spp.	**Lady's Mantle**	SV891143 Tresco	2009	Garden escape
Alstroemeria aurea	**Peruvian Lily**	SV894143 Tresco	1998	Garden escape
Amsinckia micrantha	**Common Fiddleneck**	SV914102 Old Town, St Mary's	2012	Needs confirmation
Arctotheca calendula	**Plain Treasureflower**	SV898107 St Mary's	1996	Garden escape
		SV889151 Tresco	1996	
		SV877147 Bryher	2006	
Argemone mexicana	**Mexican Poppy**	SV91701195 St Mary's	2012	Garden escape
Atriplex littoralis	**Grass-leaved Orache**	SV879149 Bryher	2014	Needs confirmation
Bupleurum subovatum	**False Thorow-wax**	SV893156 Tresco	1968	Garden escape
Cerinthe major	**Greater Honeywort**	SV888154 Tresco	2007	Garden escape
		SV879147 Bryher	2014	
Chaenomeles speciosa	**Japanese Quince**	SV891152 Bryher	2003	Garden origin
Clematis montana	**Himalayan Clematis**	Sv901145 Tresco	2011	Garden escape
Convolvulus sabatius	**Blue Ground Convolvulus**	SV915107 St Mary's	2008	Garden escape
Coriandrum sativum	**Coriander**	SV924121 St Mary's	2006	Garden escape
Cotoneaster x *watereri*	**Waterer's Cotoneaster**	SV928153	2011	Garden escape
Daphne laureola	**Spurge-laurel**	SV931154 St Martin's	2004	Garden origin
Eschscholzia californica	**Californian Poppy**	SV898108 St Mary's	1996	Garden escape
			1997	
Euphorbia characias	**Mediterranean Spurge**	SV920116 St Mary's	2014	Garden escape
Fragaria vesca	**Wild Strawberry**	SV91611118 St Mary's	2013	Origin unknown
Iberis sempervirens	**Perennial Candytuft**	SV915115 St Mary's	2009	Garden escape
Iberis umbellatum	**Garden Candytuft**	SV915102 St Mary's	2004	Garden escape
Ionopsidium acaule	**Violet Cress**	Weed in Abbey Gardens	1971	In Lousley
Ipomoea purpurea	**Common Morning-glory**	SV883084 St Agnes	2007	Garden origin
Leucanthemum x *superbum*	**Shasta Daisy**	SV881149 Bryher	2003	Garden origin or planted.
		SV880149 Bryher	2007	
		SV901104 St Mary's	2007	
		SV892154 Tresco	2007	
Lonicera japonica	**Japanese Honeysuckle**	SV927158 St Martin's	2011	Garden?
Lotus angustissimus	**Slender Bird's-foot-trefoil**	SV903104 St Mary's	2009	Needs confirming
Menyanthes trifoliata	**Bogbean**	Jac-A-Bar garden St Mary's	2006	Planted
Passiflora caerulea	**Blue Passionflower**	SV901105 St Mary's	2014	Garden origin

Persicaria capitata	**Pink-headed Persicaria**	SV916161 St Martin's	2015	Garden origin
Petunia x *hybrida*	**Petunia**	SV916161 St Martin's	2010/11	Garden escapes
		SV892141 Tresco	2015	
		SV904105 St Mary's	2015	
Pseudofumaria lutea	**Yellow Corydalis**	SV901100 St Mary's	2010	Garden escape
Stachys byzantina	**Lamb's-ear**	SV899141 Tresco	2011	Garden escape
Tradescantia zebrina	**Inch-plant**	SV919115 St Mary's	2014	Garden escape
Trifolium resupinatum	**Reversed Clover**	SV899147 Tresco	2014	Needs confirming
Urospermum dalechampii	**Smooth Golden Fleece**	A weed in Abbey Gardens		Not confirmed outside Gardens
Verbena bonariensis	**Argentinian Vervain**	SV904105 St Mary's	2015	Garden escape
Vicia parviflora	**Slender Tare**	SV881154 Bryher	2006	Possible error for *V. tetraspermum*
Vinca difformis	**Intermediate Periwinkle**	SV913102 St Mary's	2006	Garden origin
Viola x *contempta*		SV901107 St Mary's	2011	Needs confirmation

Alstroemeria aurea

Cerinthe major

Bibliography and useful references

Akeroyd, J. (1986). Oleg Vladimirovich Polunin (1914-1985). *Watsonia* 16: 105-7.

Akeroyd, J. (1992). Higher Plants, in Wildlife Reports. *British Wildlife* 3: 182.

Akeroyd, J. (2003). The Abbey Gardens of Tresco, Isles of Scilly. *Hortus* 65: 50-64.

Allen, D.E. (1997). *Rubus* L. (Rosaceae) in the Isles of Scilly: a revised list. *Watsonia* 21: 355-358.

Allen, D.E. (1998). Five new species of *Rubus* L. (Rosaceae) with transmarine ranges. *Watsonia* 22: 83-96.

Allen, D.E. (2000). *Rubus angusticuspis* Sudre (Rosaceae) in Scilly. *Watsonia* 23: 347-348.

Allen, D.E. (2001). Brambles (*Rubus* L. sect. *Rubus* and sect. *Corylifolii* Lindley, Rosaceae) of the Channel Islands. *Watsonia* 23: 421-435.

Anonymous. (n.d.). *A guide to the Geology of the Isles.* Isles of Scilly Museum Publication No. 11.

Anonymous. (n.d.). *Guide to the Natural History of Scilly – Nature trails and their habitat.* St. Ives.

Ashbee, P. (1974). *Ancient Scilly: From the First Farmers to the Early Christians*. David and Charles, Newton Abbot.

Baker, H. G. (1955). *Geranium purpureum* Vill. and *G. robertianum* L. in the British Flora: 1. *Geranium purpureum*. *Watsonia* 3: 160-167.

Bamber, R.N., Evans, N.J., Chimonides, P.J. & Williams, B.A.P. (2001). *Investigations into the hydrology, flora & fauna of the Pool of Bryher, Isles of Scilly.* Unpublished report to English Nature by NHM Consulting (Report no. ECM 759/01).

Barne, J.H., Robson, C.F., Kaznowska, S.S., Doody, J.P., Davidson, N.C., & Buck, A.L., eds. (1996). *Coasts and seas of the United Kingdom. Region 11 The Western Approaches: Falmouth Bay to Kenfig*. Peterborough, Joint Nature Conservation Committee.

Barrow, G. (1906). *The Geology of the Isles of Scilly*. HMSO, London. [With petrological contributions by J.S. Flett]

Bioret, F. & Daniels, R. (2006). Assessment of threats to populations of *Rumex rupestris* Le Gall (Shore Dock) in Britain and France, in Leach, S.J. & Page, C.N., Peytoureau, Y. & Sanford, M.N., eds. *Botanical Links in the Atlantic Arc*. Botanical Society of the British Isles, London.

Borlase, W. (1753). 'Of the great alterations which the Islands of Scilly have undergone since the time of the ancients'. *Philos. Trans.* 48: 55-67.

Borlase, W. (1756). *Observations on the Ancient and Present State of the Islands of Scilly*. Oxford.

Bowley, R.L. (1990). *The Fortunate Islands: The Story of the Isles of Scilly*, Rev. ed. 8. Bowley Publications Ltd., Isles of Scilly.

Boyden, H. (1890). The Flora of the Scilly Islands. *Trans. Penzance Nat. Hist. Antiq. Soc.* 3: 186.

Boyden, H. (1906). 'The Flora of the Isles of Scilly', in Tonkin, J.C. & Row, P. *Lyonesse: A Handbook for the Isles of Scilly*, ed. 4. Scilly and London.

Bramwell, D. & Bramwell, Z.I. (1974). *Wild Flowers of the Canary Islands*. Stanley Thornes, London.

British Geological Survey (1975). Isles of Scilly. *Geological Survey of England and Wales, New Series*, Sheet 357/360, 1:63,360/1:50,000. Ordnance Survey, Southampton.

Bull, J.C. & Kenyon, E.J. (2015). *Isles of Scilly eelgrass bed voluntary monitoring programme - 2014 Annual Survey*. Natural England Commissioned Report NECR178.

Burrows, R. (1971). *The Naturalist in Devon and Cornwall*. David & Charles, Newton Abbot.

Butcher, S.A., Biek, L., Charlesworth, D., Evans, J.G., Greig, J.R.A., Keeley, H., Miles, H., Thomas, A.C. & Turk, F.A. (1978). Excavations at Nornour, Isles of Scilly, 1969-73. *Cornish Archaeology* 17: 29-112.

Camidge, K., Charman, D., Johns, C., Meadows, J., Mills, S., Mulville, J., Roberts, H.M., and Stevens, T. (2010). *The Lyonesse Project: evolution of the coastal and marine environment of Scilly, Year 1 Report*. Truro (HE Projects, Cornwall Council).

Cornwall Biological Records Unit (1995). *The Isles of Scilly: NVC Mapping of Peninnis Head, The Plains & Great Bay and Gugh SSSIs*. A Preliminary report to EN Cornwall Office.

Chapman, D. (2008). *Wildflowers of Cornwall and the Isles of Scilly*. Alison Hodge, Penzance.

Charman, D., Bayliss, A., Camidge, K., Fyfe, R., Johns, C., Meadows, J., Mills, S., Mulville, J., Roberts, H.M., and Stevens, T. (2012). *The Lyonesse Project: evolution of the coastal and marine environment of Scilly, Draft Final Report*. Truro (HE Projects, Cornwall Council).

Cheffings, C.M. & Farrell, L., (eds.), **Dines, T.D., Jones, R.A., Leach, S.J., McKean, D.R., Pearman, D.A., Preston, C.D., Rumsey, F.J., Taylor, I.** (2005). The Vascular Plant Red Data List for Great Britain. *Species Status* 7: 1-116. Joint Nature Conservation Committee, Peterborough.

Christopher, K. (1961). Trends in the marketing of agricultural and horticultural produce in the Scilly Isles since 1945. *The Scillonian* 36 (147).

Cleator, B., Nunny, R., & Mackenzie, G. (1993). *Isles of Scilly 1993: expedition report. The status of the seagrass* Zostera marina *on the Isles of Scilly*. Coral Cay Conservation Sub-Aqua Club (CCCSA) & English Nature (EN). Report to the Isles of Scilly Marine Park Management Committee.

Clement, E.J. (1985). Hedge Veronica (*Hebe* x *franciscana*) and allies in Britain. *BSBI News* 41: 18.

Clement, E.J. (2001). *Aeonium* and *Aichryson* on the Isles of Scilly (v.c. 1b). *BSBI News* 88: 57-58.

Clement, E.J. (2003). Records of *Crassula multicava* from Scilly (v.c. 1b). *BSBI News* 92: 46.

Clement, E.J. (2004). *Senecio minimus* – a new weed on the Isles of Scilly. *BSBI News* 97: 48-49.

Clement, E.J. (2006). Notes on the Isles of Scilly Flora. *BSBI News* 103: 32-33.

Clement, E.J. & Foster, M.C. (1994). *Alien Plants of the British Isles*. Botanical Society of the British Isles, London.

Coleman, C. & O'Reilly, J. (2004). *Dwarf Pansy*. Unpublished report for the Isles of Scilly Wildlife Trust.

Coombe, D.E. (1961). *Trifolium occidentale*, a new species related to *T. repens* (L). *Watsonia* 5: 68-87.

Coombe, D.E. & Morisset, P. (1967). Further observations on *Trifolium occidentale*. *Watsonia* 6: 271-275.

Cooper, A. (2006). *The Secret Nature of the Isles of Scilly*. Green Books, Dartington.

Coulcher, P. (1999). *The Sun Islands - A Natural History of the Isles of Scilly*. The Book Guild Ltd., England.

Cope, T. & Gray, A. (2009). *Grasses of the British Isles*. Botanical Society of the British Isles, London.

Countryside Agency (2004). *The Isles of Scilly Area of Outstanding Natural Beauty Management Plan 2004–2009*.

Courtney, J.S. (1845). *A Guide to Penzance and its Neighbourhood, including the Isles of Scilly*. Rowe, Penzance.

Courtney, L.H. (1867). *A week in the Isles of Scilly*. Truro.

Crawford, O.G.S. (1927). Lyonesse. *Antiquity* 1: 5-14.

Curnow, W. (1876). A botanical trip to the Scilly Isles. *Hardwicke's Sci Gossip* 12: 162.

Dargie, T. (1990a). Isles of Scilly Dune vegetation Survey 1990. Vol 1 General Report. A report to the Nature Conservancy Council.

Dargie, T. (1990b). Isles of Scilly Dune vegetation Survey 1990. Vol 2 Individual Sites (except Tresco). A report to the Nature Conservancy Council.

Dargie, T. (1990c). Isles of Scilly Dune vegetation Survey 1990. Vol 3 Individual Sites, Tresco. A report to the Nature Conservancy Council.

Davey, F.H. (1902). *A tentative list of the flowering plants, ferns etc., known to occur in the county of Cornwall, including the Scilly Isles*. Penryn.

Davey, F.H. (1909). *Flora of Cornwall*. F. Chegwidden, Penryn.

Davis, R. (1999). *Species Action Plans for Plants: Shore Dock*. English Nature, Peterborough.

Dendy, W.C. (1857). *The beautiful islets of Britaine*. pp. 47, 51.

Dimbleby, G.W. (1977). A buried soil at Innisidgen, St Mary's, Isles of Scilly. *Cornish Studies* 415: 5-10.

Dorrien Smith, T.A. (1890). The Progress of the Narcissus Culture in the Isles of Scilly. *J. Roy. Hort. Soc* 12: 311-16.

Downes, A. (1957). Farming the Fortunate Islands. *Geography* 42: 105-112.

Edees, E.S. & Newton, A. (1988). *Brambles of the British Isles*. Ray Society, London.

Edmonds, E.A., McKeown, M.C. & Williams, M. (1975). *British Regional Geology: South-west England*. HMSO, London.

Environmental Records Centre for Cornwall and the Isles of Scilly (2013). *Isles of Scilly Land Cover 2005*. Environmental Records Centre for Cornwall and the Isles of Scilly, Truro.

Evans, J.G. (1984). Excavations at Bar Point, St Mary's, Isles of Scilly, 1979-80. *Cornish Studies*, 11: 28-30.

Ewing, P. (1897). Plants exhibited from Cornwall and the Scilly Isles. *Proc. & Trans. Nat. Hist. Soc. Glasgow* 4: 382.

Foster, I.D.L. (2006). Big Pool (SV878086), in Scourse, J.D., ed. *The Isles of Scilly: Field Guide*. Quaternary Research Association, London.

Fowler, J.A., ed. (1970). *Studies on the Flora and Fauna of St Agnes*. A report from City of Leicester Polytechnic.

Fowler, S.L. (1992). *Marine monitoring in the Isles of Scilly*. Report to Natural England.

French, C.N. (1985). The sub-fossil Flora of Cornwall and the Isles of Scilly. *Cornish Biological Records* No 8. Institute of Cornish Studies, Pool, Cornwall.

French, C.N., ed. (1994). *Check-list of the Flowering Plants and Ferns of Cornwall and the Isles of Scilly*. Cornish Biological Records Unit, Redruth, Cornwall.

French, C.N., Murphy, R.J. & Atkinson, M.G.C. (1991). *Flora of Cornwall*. Wheal Seton Press, Camborne.

Gibson, A.G. (1932). *The Isles of Scilly*, ed. 2. St Mary's, Isles of Scilly.

Gibson, F. (c. 1995) *Wild flowers of Scilly*. Beric Tempest, St. Ives.

Gillham, M.E. (1956). Ecology of the South Pembrokeshire Islands: V. Manuring by the colonial seabirds and mammals, with a note on seed distribution by gulls. *Journal of Ecology* 44: 429-454.

Gillham, M.E. (1970). Seed dispersal by birds, in Perring, F., ed. *The Flora of a Changing Britain*, pp. 90–98. Botanical Society of the British Isles, Conference Report 11. E. W. Classey, Hampton.

Graham Moss Associates (1984). *The Isles of Scilly, Comprehensive Land Use and Community Development Project*. London and Bristol.

Green, P.S. (1973). HEBE x FRANCISCANA (Eastwood) Souster, not *H*. x *lewisii* – Naturalized in Britain. *Watsonia* 9: 371-372.

Grigson, G. (1948). *The Scilly Isles*. Paul Elek, London.

Harvey, C. & Lousley, J.E. (1993). *Flowering Plants and Ferns in the Isles of Scilly*. Isles of Scilly Museum.

Heath, R. (1750). *A Natural History and Historical Account of the Islands of Scilly*. Manby & Cox, London.

Hencken, H. O'Neill (1932). *The Archaeology of Cornwall and Scilly*. Methuen, London.

Hiemstra, J., Evans, D.J.A., Scourse, J.D., McCarroll, D., Furze, M.F.A. & Rhodes. E. (2006). New evidence for a grounded Irish Sea glaciation on the Isles of Scilly, UK. *Quaternary Science Reviews* 25: 299-309.

Hipkin, C. (2003). Putting our alien flora into perspective. *British Wildlife* 14: 413-422.

Hunt, D. & Gibson, F. (1977a). *Wild Flowers of Scilly No 1, Coastal habitats*. St. Ives, Cornwall.

Hunt, D. & Gibson, F. (1977b). *Wild Flowers of Scilly No 2, Marshes, Stonewalls, Hedgerows & Windbreaks*. St. Ives, Cornwall.

Hunt, D. & Gibson, F. (1977c). *Wild Flowers of Scilly No 3, Wild flowers of the Bulb Fields*. St. Ives, Cornwall.

Jellicoe, G.A. (1965). *A landscape charter for the Isles of Scilly*. Report prepared for the Council of the Isles of Scilly.

Jermy, A.C., Arnold, H.R., Farrell, L. & Perring, F.H. (1978). *Atlas of the Ferns of the British Isles*. Botanical Society of the British Isles and British Pteridological Society, London.

Jury. S.L. (1977). The Herbarium of J.E. Lousley. *Watsonia,* 11: 312.

Kay, E. (1963). *Isles of Flowers*, ed. 2. Alvin Redman, London.

Kendall, M.A., Widdicombe, S., Davey, J.T., Somerfield, P.J., Austen, M.C.V. & Warwick, R.M. (1996). The biogeography of islands: preliminary results from a comparative study of the Isles of Scilly and Cornwall. *J. Mar. Biol. Assoc. U.K.* 76: 219-222.

King, R. (1985). *Tresco England's Island of Flowers*. Constable, London.

Kingsford-Curran, R. (1964). *Report of an archaeological & biological investigation of Great and Little Arthur*. 1-19.

Kirby, K. (2006). Table of Ancient Woodland Indicator plants (AWIs), in Rose, F. *The Wild Flower Key*, Rev. ed. 2. Warne.

Lancaster, R. (1984). The Garden. *J. Royal Hort. Soc.* 109(2): 69.

Land Use Consultants (1996). *Isles of Scilly Historic Landscape Assessment and Management Strategy*. Report prepared by Land Use Consultants in association with Cornwall Archaeological Unit for Duchy of Cornwall in partnership with Countryside Commission and MAFF.

Lawson, M.A. (1870). Additions to the flora of the Scilly Isles. *Journal of Botany* 8: 357.

L'Estrange, A.G. (1865). *Yachting Round the West of England*. Hurst and Blackett, London.

Le Sueur, F. (1976). *A Natural History of Jersey*. Phillimore.

Llewellyn, S. (2005). *Emperor Smith: the Man Who Built Scilly*. Dovecote Press, Dorset.

Locker, A. (1996). The Bird Bones, in Ashbee, P. *Cornish Archaeology* 35: 113-115. [A. Locker's contribution is part of the study of excavations on Halangy Down, St Mary's, Isles of Scilly 1964–1977)].

Lousley, J.E. (1939). Notes on the Flora of the Isles of Scilly 1. *J. Bot.* 77: 195-203.

Lousley, J.E. (1940). Notes on the Flora of the Isles of Scilly 2. *J. Bot.* 78: 153-160.

Lousley, J.E. (1960). Plant Notes (description of *Crassula decumbens*). *Proc. Bot. Soc. Brit. Isles* 4: 42-43.

Lousley, J.E. (1971). *The Flora of the Isles of Scilly*. David & Charles, Newton Abbot.

Lousley, J.E. (1973). 'Mesembryanthemums' established in the Isles of Scilly, in Green, P.S., ed. *Plants wild and cultivated*, pp. 83-85. London.

Lousley, J.E. & Kent, D.H. (1981). *Docks and Knotweeds of the British Isles*. Botanical Society of the British Isles, London.

Lousley's unpublished notebooks (in archives of Isles of Scilly Museum).

Magalotti, L. (1821). *Travels of Cosmo the Third, Grand Duke of Tuscany, through England, during the Reign of King Charles the Second, 1669*. J. Mawman, London.

Malloch, A.J.C. (1972). Salt-spray deposition on the maritime cliffs of the Lizard Peninsula. *J. Ecol.* 60: 103–112.

Margetts, L.J. & David, R.W. (1981). *A Review of the Cornish Flora 1980*. Institute of Cornish Studies, Redruth.

Margetts, L.J. & Spurgin, K.L. (1991). *The Cornish Flora Supplement 1981-1990*. Trendrine Press, Zennor.

Marquand, E.D. (1893). Further records for the Scilly Isles. *J.Bot.* 31: 265.

Mawer, D. (2002). A day in the Life of The Senior Conservation Warden. *Isles of Scilly Wildlife Trust Newsletter* No.1, Spring/Summer 2002.

Maybee, R. (1884). *Sixty-Eight Years' Experience on the Scilly Islands*. Penzance.

McClintock, D. (1975a). *The Wild Flowers of Guernsey*. Collins, London.

McClintock, D. (1975b). A note on *Fascicularia, Ochagavia* and *Rhodostachys*. Watsonia 10: 289-290.

McClintock, D. (1977) J.E. Lousley and plants alien in the British Isles. *Watsonia,* 11: 287-290.

McDonnell, E.J. & King, M.P. (2006). *Rumex rupestris* Le Gall (shore dock) in SW England: review of recent surveys and assessment of current status, in Leach, S.J., Page, C.N., Peytoureau, Y. & Sanford, M.N., eds. *Botanical Links in the Atlantic Arc*. Botanical Society of the British Isles, London.

Meredith, H.M. (1994). *Scrophularia scorodonia* L. (Balm-leaved Figwort) - an enigma, in Murphy, R.J., ed. *Botanical Cornwall* 6: 21-34. Cornish Biological Records Unit, Redruth.

Millett, L. & Millett, M. (1852). Wild Flowers and ferns of the Isles of Scilly observed in June and July. *Trans. Nat. Hist. Antiq. Soc. Penzance.* 2: 75-78.

Mothersole, J. (1910). *The Isles of Scilly: Their story, their folk and their flowers.* The Religious Tract Society, London.

Mulville, J. (2007). *Islands in a Common Sea Archaeological fieldwork in the Isles of Scilly 2006 - St Mary's & St Martin's.* Historic Environment Service, Truro.

Murphy, R.J., Page, C.N., Parslow, R.E. & Bennallick, I.J. (2012). *Ferns, Clubmosses, Quillworts and Horsetails of Cornwall and the Isles of Scilly.* Environmental Records Centre for Cornwall and the Isles of Scilly, Truro.

Murphy, R.J. (2009). *Fumitories of Britain and Ireland.* Botanical Society of the British Isles, London.

Murphy, R.J. (2016). *Evening-primroses (Oenothera) of Britain and Ireland.* Botanical Society of the British Isles, London.

National Meteorological Library - records for St Mary's Airport weather Station 1961-1990.

Neil, C.J., King, M.P., Evans, S.B., Parslow, R.E., Bennallick, I.J. & McDonnell, E.J. (2001). *Shore dock (Rumex rupestris): report on fieldwork undertaken in 2000.* Plantlife Report 175.

Nelhams, M. (2000). *Tresco Abbey Garden: A Personal and Pictorial History.* Dyllansow Truran, Truro.

Newton, A. & Randall, R.D. (2004). *Atlas of British and Irish Brambles.* Botanical Society of the British Isles, London.

North, I.W. (1850). *A Week in the Isles of Scilly.* Rowe, Penzance & Longman, London.

Ottery, J. (1996). *The Isles of Scilly Wild Flower Guide.* Julia Ottery & Amadeus Fine Arts (Scilly).

Page, C.N. (1988). *Ferns: Their habitats in the Landscape of Britain and Ireland.* Collins, London & Glasgow.

Page, C.N. (1997). *The Ferns of Britain and Ireland*, ed. 2. Cambridge University Press, Cambridge.

Parslow, R. (1985). A survey of the Shore dock *Rumex rupestris* in the Isles of Scilly, 1984. *Bull. Brit. Ecol. Soc.* 16: 92-95.

Parslow, R. & Colston, A. (1991). *Botanical Survey of English Heritage Sites in the Isles of Scilly.* Unpublished report to English Heritage.

Parslow, R. & Colston, A. (1994). *The current status of* Rumex rupestris *Le Gall in the Isles of Scilly.* Unpublished report as part of English Nature Species Recovery Programme.

Parslow, R. (1995). *Isles of Scilly: Proposed management strategy for Countryside Stewardship, with reference to important weed and wall species.* Unpublished report to Farming and Rural Conservation Agency.

Parslow, R. (1996). *Shore Dock* Rumex rupestris *Le Gall in the Isles of Scilly.* Unpublished report as part of English Nature Species Recovery Programme.

Parslow, R. (1999). *Species Action Plan for* Ophioglossum *species on Scilly.* Unpublished report to English Nature.

Parslow, R. (2001). Plants on Scilly. *Isles of Scilly Bird and Natural History Review 2000*, pp. 187-191.

Parslow, R. (2002a). *Management Plan for the Isles of Scilly.* Unpublished document for Isles of Scilly Wildlife Trust.

Parslow, R. (2002b). Plant Records for 2001. *Isles of Scilly Bird and Natural History Review 2001*, pp. 175-176.

Parslow, R. (2003). Plant Report for 2002. *Isles of Scilly Bird and Natural History Review 2002*, pp. 191-193.

Parslow, R. (2004). Some New and Interesting Plant Records for the Isles of Scilly in 2003. *Isles of Scilly Bird and Natural History Review 2003*, pp. 173-174.

Parslow, R. (2005a). Plant notes and records from the Isles of Scilly, in Murphy, R.J. & Bennallick. I.J., eds. *Botanical Cornwall* 13: 80-82.

Parslow, R. (2005b). Plant records for 2004. *Isles of Scilly Bird and Natural History Review 2004*, pp. 185-186.

Parslow, R. (2006a). An introduction to the flora of the Isles of Scilly, in Leach, S.J., Page, C.N., Peytoureau, Y. & Sanford, M.N., eds. *Botanical Links in the Atlantic Arc.* Botanical Society of the British Isles, London.

Parslow, R. (2006b). Maritime communities as habitats for *Ophioglossum* ferns in the Isles of Scilly in Leach, S.J., Page, C.N., Peytoureau, Y. & Sanford, M.N., eds. *Botanical Links in the Atlantic Arc.* Botanical Society of the British Isles, London.

Parslow, R. (2006c). Scilly Plant News 2005. *Isles of Scilly Bird and Natural History Review 2005*, pp. 173-175.

Parslow, R. (2007a). *The Isles of Scilly.* Collins, London. [New Naturalist Library No. 103]

Parslow, R. (2007b). Beneath your feet! *Isles of Scilly Bird and Natural History Review 2006*, pp. 152-153.

Parslow, R. (2008). Plant News from Scilly in 2007. *Isles of Scilly Bird and Natural History Review 2007*, p. 146.

Parslow, R. (2009a). *Checklist of Plants and Ferns in the Isles of Scilly.* Isles of Scilly Museum.

Parslow, R. (2009b). Plant Records from Scilly in 2008. *Isles of Scilly Bird and Natural History Review 2008*, pp.147-148.

Parslow, R. (2010a). *Arable plants of bulbfields and other arable fields in the Isles of Scilly.* Isles of Scilly Area of Outstanding Natural Beauty.

Parslow, R. (2010b). Plant Records from Scilly in 2009. *Isles of Scilly Bird and Natural History Review 2009*, pp. 159-161.

Parslow, R. (2011) Plant Records for 2010. *Isles of Scilly Bird and Natural History Review 2011*, pp. 144-145.

Parslow, R. (2012). Plants of Scilly in 2011. *Isles of Scilly Bird and Natural History Review 2012*, pp. 192-193.

Parslow, R. (2013). Scilly Ferns. *Isles of Scilly Bird and Natural History Review 2012*, pp. 205-206.

Parslow, R. (2014). Writing a new Flora for Scilly. *Isles of Scilly Bird and Natural History Review 2013*, pp. 207-209.

Parslow, R. (2015). A botanical tragedy – or is it? *Isles of Scilly Bird and Natural History Review 2014*, p. 196.

Paton, J.A. (1968). *Wild Flowers in Cornwall and the Isles of Scilly*. D. Bradford Barton, Truro.

Paul, A.M. (1987). The status of *Ophioglossum azoricum* (Ophioglossaceae: Pteridophyta) in the British Isles. *Fern Gaz.* 13(3): 173-187.

Pett, D.E. (2009). *Flower Growing on the Isles of Scilly*. Blue Printing Company. [Based on Tresco Abbey Archives with a checklist of daffodils]

Pett, D.E. (2012). *Horticulture in the Isles of Scilly: a documentary history and Source Book*. Mary Pett, Penryn.

Polunin, O. (1953). *Some Plant Communities of the Scilly Isles*. Duplicated typescript.

Preston, C.D. and Sell, P.D. (1988). The Aizoaceae naturalized in the British Isles. *Watsonia* 17: 217-245.

Preston, C.D., Pearman, D.A. & Dines, T.D., eds. (2002). *New Atlas of the British & Irish Flora*. Oxford University Press, Oxford.

Preston, C.D. & Arnold, H.R. (2006). The Mediterranean-Atlantic and Atlantic elements in the Cornish Flora, in Leach, S.J., Page, C.N., Peytoureau, Y. & Sanford, M.N., eds., *Botanical Links in the Atlantic Arc*. Botanical Society of the British Isles, London.

Preston, C.D., Pearman, D.A. & Hall, A.R. (2004). Archaeophytes in Britain. *Bot. J. Linn. Soc.* 145: 257-294.

Ralfs, J. (1876). *Materials towards a flora of the Scilly Isles*. Manuscript note book in Penzance Library.

Ralfs, J. (1879). *Manuscript flora of West Cornwall*. Penzance public library.

Randall, R.E. (2003). Biological Flora of the British Isles: *Smyrnium olustrum* L. *J. Ecol.* 91: 325-340.

Randall, R.E. (2004). Biological Flora of The British Isles: *Viola kitaibeliana* Schult(es). *J. Ecol.* 92: 361- 369.

Ratcliffe, J. & Straker, V. (1996). *The Early Environment of Scilly*. Cornwall Archaeological Unit, Cornwall County Council, Truro.

Ratcliffe, J. & Straker, V. (1997). The changing landscape and coastline of the Isles of Scilly: recent research. *Cornish Archaeology* 36: 64-76.

Raven, J.E. (1950). Notes on the flora of the Scilly Isles and the Lizard Head. *Watsonia* 1: 356-358.

Rodwell, J.S., ed. (1991a). *British Plant Communities, 1. Woodlands and Scrub*. Cambridge University Press, Cambridge.

Rodwell, J.S., ed. (1991b). *British Plant Communities, 2. Mires and heaths*. Cambridge University Press, Cambridge.

Rodwell, J.S., ed. (1992). *British Plant Communities, 3. Grasslands and montane communities*. Cambridge University Press, Cambridge.

Rodwell, J.S., ed. (1995). *British Plant Communities, 4. Aquatic communities, swamps and tall-herb fens*. Cambridge University Press, Cambridge.

Rodwell, J.S., ed. (2000). *British Plant Communities, 5. Maritime communities and vegetation of open habitats*. Cambridge University Press, Cambridge.

Rostański, K. (1982). The species of *Oenothera* L. in Britain. *Watsonia*: 14, 1-34.

Sargeant, P. (1993). *The great Scillionian storm versus the Dwarf Pansy*. The Proceedings of the National Forum for Biological Recording conference: Crises and Biological Recording.

Scaife, R.G. (1984). A History of Flandrian Vegetation in the Isles of Scilly: Palynological Investigations of Higher Moors and lower Moors peat mires, St Mary's. *Cornish Studies* 11: 33-47.

Scourse, J.D. (1991). Late Pleistocene stratigraphy and palaeobotany of the Isles of Scilly. *Philos. Trans., Ser. B* 334: 405- 448.

Scourse, J.D. (2005). Notes from Lecture in Isles of Scilly Museum 13 April 2005 (pers. comm. K. Sawyer).

Sell, P. & Murrell, G. (1996). *Flora of Great Britain and Ireland, Vol.5 Butomaceae – Orchidaceae*. Cambridge University Press, Cambridge.

Sell, P. & Murrell, G. (2006). *Flora of Great Britain and Ireland, Vol.4 Campanulaceae – Asteraceae*. Cambridge University Press, Cambridge.

Sell, P. & Murrell, G. (2009). *Flora of Great Britain and Ireland, Vol.3 Mimosaceae – Lentibulariaceae*. Cambridge University Press, Cambridge.

Sell, P. & Murrell, G. (2014). *Flora of Great Britain and Ireland, Vol.2 Capparaceae – Rosaceae*. Cambridge University Press, Cambridge.

Silverside, A. (1977). *A phytosociological survey of British arable weeds and related communities* (PhD thesis). Durham University.

Smith, L.T., ed. (1906-1910). *The Itinerary of John Leland in or about the years 1535-43*. Bell, London.

Smith, W.W. (1907). Note on a peculiar tussock-formation. *Trans. & Proc. Bot. Soc. Edinburgh*. 23: 234-235.

Smith, W.W. (1909). Notes of the flora of the Scilly Isles. *Trans. & Proc. Bot. Soc. Edinburgh*. 24: 36.

Sneddon, P. & Randall, R.E. (1994). *Coastal vegetated shingle structures of Great Britain: Appendix 3 Shingle sites in England*, pp.82-87. Joint Nature Conservation Committee, Peterborough.

Somerville, A. (1893). Additional Records for the Scilly Isles. *J. Bot.* 31: 118-120.

Spalding, A. & Sargeant, P. (2000). *Scilly's Wildlife Heritage*. Twelveheads Press, Cornwall.

Spalding, A., ed. (1997). *Red Data Book for Cornwall and the Isles of Scilly*. Croceago Press, Praze-an-Beeble.

Spencer, N. (1772). *The Complete English Traveller*. J Cooke, London.

Stace, C.A. (1961). Some studies in *Calystegia*: compatibility and hybridisation in *C. sepium* and *C. silvatica*. *Watsonia* 5: 88-105.

Stace, C.A. (2010). *New Flora of the British Isles*, ed. 3. Cambridge University Press, Cambridge.

Stace, C.A. & Crawley, M.J. (2015). *Alien Plants*. Collins, London. [New Naturalist Library No. 129]

Sterry, P. (2014). Know your ivy. *Isles of Scilly Bird and Natural History Review 2013*, p. 210.

Stewart, A., Pearman, D.A., & Preston, C.D., eds. (1994). *Scarce Plants in Britain*. Joint Nature Conservation Committee, Peterborough.

Stroh, P.A., Leach, S.J., August, T.A., Walker, K.J., Pearman, D.A., Rumsey, F.J., Harrower, C.A., Fay, M.F., Martin, J.P., Pankhurst, T., Preston, C.D. & Taylor, I. (2014). *A Vascular Plant Red List for England*. Botanical Society of Britain and Ireland, Bristol.

Teague, A.H. (1891). Plants growing in Tresco Abbey Gardens. *Trans. Penzance Nat. Hist. Antiq. Soc.* 3: 157-171.

Thomas, C. (1985). *Exploration of a drowned Landscape – Archaeology and History of the Isles of Scilly*. Batsford, London.

Thurston, E. & Vigurs, C.C. (1922). *A supplement to F. Hamilton Davey's Flora of Cornwall*. Royal Institution of Cornwall, Truro.

Thurston, E., & Vigurs, C.C. (1923). Note on the Cornish Flora. *J. Roy. Inst. Cornwall*. 70: 164-168.

Tonkin, J.C. & R.W. (1882). *Guide to the Isles of Scilly*, ed. 1. Penzance.

Tonkin, J.C. & R.W. (1887). *Guide to the Isles of Scilly*, ed. 2. Penzance.

Tonkin, J.C. & R.W. (1893). *Guide to the Isles of Scilly*, ed. 3. Penzance. [With a list of plants by W. Curnow and H. Boyden]

Townsend, F. (1864). Contributions to a Flora of the Scilly Isles. *J. Bot.* 2: 102-120. [A reprint with corrections and comments by Lousley is in the Isles of Scilly Museum archives]

Troutbeck, J. (1794). *A survey of the Ancient and Present State of the Scilly Islands*. Sherborne.

Turner, G. (1964). 'Some Memorialls towards a Natural History of the Sylly Islands (reproduction of a manuscript dated c.1695)'. *The Scillonian* 159: 154-157.

Vyvyan, C.C. (1953). *The Scilly Isles*. Robert Hale, London.

Wagstaff, W. (2005). Rare Plants on Scilly. *Isles of Scilly Bird and Natural History Review 2004*, pp. 187-190.

White, W. (1879). *A Londoner's Walk to the Land's End and a Trip to the Scilly Isles*, ed. 3. Chapman and Hall, London.

Wiggington, M.J., ed. (1999). *British Red Data Books 1. Vascular Plants,* ed. 3. Joint Nature Conservation Committee, Peterborough.

Wilson, P. & King, M. (2003). *Arable Plants – a field guide*. WILDguides Ltd., Hampshire.

Wilson, P.J. (1992). British Arable Weeds. *British Wildlife*, 3: 149-161.

Woodley, G. (1822). *A view of the present state of the Scilly Islands*. London.

Young, B. (1998). *Mapping Zostera beds in Special Areas of Conservation by Aerial Photography*. Report to English Nature by BKS Surveys Ltd.

Rosemary Parslow points out an interesting specimen

Glossary of terms

Alien	Plant not native to area, probably introduced by man
Altered rock	Rock that has been changed by heat or in some other way
Archaeophyte	Plant introduced before AD 1500
AWI	Ancient Woodland Indicator - classification based on plants present before 1600
BCG	Botanical Cornwall Group
Blown sand	Deposits of wind-blown sand
BPS	British Pteridological Society (Fern Society)
Braided paths	Where erosion causes a series of interlinking parallel paths
BSBI	Botanical Society of Britain and Ireland
Casual	Plant that is not native and not naturalised
CeC	Cornwall Wildlife Trust's consultancy
Cornubian Batholith	A series of granite intrusions that underlie the SW peninsula of Great Britain
Cupola/bosses	Granite domes
DEFRA	Department for Environment, Farming and Rural Affairs
Dykes	Mineral intrusions into the granite
EN	English Nature (successor to NCC)
Epicalyx	Calyx-like structure outside the calyx
ERCCIS	Environmental Records Centre for Cornwall and the Isles of Scilly
HWM	High Water Mark – highest point normally reached by tide
Introgression	Characters acquired by a species from another by hybridisation, and then back-crossing
IoSBG	Isles of Scilly Bird Group
IoSWT	Isles of Scilly Wildlife Trust
Kelp burning	Collecting and burning brown seaweeds to produce ash for glass and soap-making
Killas	Soft, shale-like rocks that have been altered by heat
Lenses	Small, sometimes lens-shaped deposits into another material
Loess	Fine-grained material, originally wind-blown
Lusitanian	Pertaining to the Iberian peninsula
Mesolithic	Period between Paleolithic and Neolithic
Midden	Rubbish pit. Contain useful domestic material for archaeological dating
Native	Plant that belongs here having arrived naturally, often long ago
NCC	Nature Conservancy Council (the previous government body dealing with nature)
NE	Natural England (the latest manifestation of NCC)
Neolithic	New Stone Age, beginning of farming and production of polished stone artifacts
Neophyte	Plant introduced after AD 1500
NHM	Natural History Museum
NVC	National Vegetation Classification. National classification of plant communities
OS	Ordnance Survey, UK Government official map makers

Penknife point	Convex-backed, small flint tool
Phase 1	Habitat survey method & classification
Ram or rabb	Concrete-like material formed from decomposed granite, used as mortar
Ram pits	Small quarries to extract ram for road surfacing and mortar, mostly now overgrown
RSPB	Royal Society for the Protection of Birds
Robbed out	Destruction of ancient buildings to re-use the stones
Seepages	Small trickles of freshwater
Species Recovery Programme	A project by EN and others to research decline of certain rare species and try to assist their recovery
Spring tides	Tides just after full or new moon - most extreme at Spring and Autumn equinox
SSSI	Protected site, designated for important flora, fauna, geology or physiographic features
Submarine	Under the sea
Tombolo	A sand bar connecting an island to the mainland – e.g. between St Agnes and Gugh
Tourmalised schist	Coarse-grained rock with tourmaline crystals
Tors	Rock outcrops known as carns on Scilly
Tundra	Open, tree-less cold landscape
VC recorder	Official recorder appointed to a vice-county
Vice-county	Geographical divisions used in biological recording
Votive	Sacred offerings often of high value
WFS	Wild Flower Society
Winter annuals	Plants that germinate in autumn and carry out their life cycle in spring

The tombolo between St Agnes and Gugh

List of recorders, photographers, artists and contributors

All the following have contributed to the recording, identification and understanding of the Scillonian Flora, past and present, and we are very grateful to them. We apologise if anyone has been omitted. Photographers and artists are acknowledged on the pages but indicated here by colour text.

Adams, S.
Akeroyd, J.R.
Allen, D.E.
Andrews, G.
Arneshaw, J.
Askins, E.
Askins, J.
Atkinson, M.G.C.
Atkinson, A.A.

Barkham, J.
Barrett, D.
Bathe, G.,
Beavis, I.C.
Bennallick, I.J.
Bevan, D.
Bevan, J.
Billinghurst, P.
Bishop, J.G. & W.
Blackwell, C.
Block, M.
Bloomfield, L.
Blunden, T.
Boothby, B.
Botanical Cornwall Group
Botanical Group 2011
Bowdrey, J.P.
Bowman, R.P.
Bowyer, J. & R.
Boyden, C.R.
Briggs, M.
Brightman, F.
Bristowe, W.S.
Bryson, K.
BSBI Field Meeting 2008
Buckingham, S.
Budd, P.
Bull, K.E.
Burges, R.
Burt, B.L.
Burton, R.

Butcher, A.
Cadbury, D.A.
Campbell, C
Cannon, J. F. M.
Carmichael, C.A.M.
Cavanagh, Mr & Mrs
Chapman, F.
Clarke, J.
Cleave, A.
Clement, E.J.
Clough, P.
Coleman, C.
Colls, H.
Colston, A.C.S.
Compton, S.
Cook, K.
Cooke, E.W.
Coombe, D.E.
Corbet, G.
Coulcher, P.
Cox, J. & P.
Crackles, E.
Craig, Miss A.
Crawley, M.J.
Cronk, Q.C.B.
Curnow, W.

Dallas, J.E.S.
Dandy, J.E.
Daniels, F.T.
Dargie, T.
David, J.C.
Davies, M.
Dawson, R.
Delve, J.
De Potier, A.
De Sausmarez, N.S.
Dicker, J.
Doody, P.
Dorrien-Smith, T.
Downes, H.

Dupree, D.

Ealey, P.
Edwards, B.
Elias, D.
Emmerich, S.

Farrant, R.
Farrell, L.
Fenton, G.
Fenwick, D. (APHOTO)
Finch, R.A.
Finucane, D.
Fitzgerald, Lady R.
Fowler, J.A.
French, C.N.

Gainey, P.
Gaze, M.
Gerrans, M.B.
Ghullam, M.
Gibbs, A.E.
Gibbs, D.G.
Gibson, A. & L.
Glasscock, J.
Gomm, F.R.
Goodey, M.
Gordon, V.
Goriup, P.
Graham, B.
Grant, G.
Green, P.
Griffiths, E.
Grigson, G.
Grimshaw, J.
Grose, J.D.
Grouse, D.J.
Groves, H. & J.
Groves, M.& S.
Gurr, A.
Gurr, M.

Hale, J.
Halliday, G.
Hanbury, F.J.
Hardaker, W.H.
Harley, J. & R.
Harley, R.M.
Harris, E.B.
Harrison, T.
Hart, D.
Harvey L.A.
Harvey, C.
Hathway, A.
Hathway, R.
Hawkins, W.

Hawksford, J.
Hayle, J.
Heaney, V.
Helm, S.
Herring, P.
Hick, W.
Hicks, F.
Hicks, J.
Hicks, M.
Hirons, G.
Holyoak, D.T.
Hornby, P.
Horton, A.
Howard, A.
Howcroft, M.
Howitt, R.C.L.
Hubbard, C.E.
Humphries, J.

Irving, R.J.

Jackson, V.
Jaques, J.M.
Johnson, H.
Jury, S.L.

Kay, E.
Kershaw, G.
Kitchener, G.D.
Knight, L.

Lamb, Miss A.V.
Lancaster, R.
Laney, B.G.
Laney, L.
Last, Mr
Latham, E.
Leach, S.
Lees, R.
Legg, E.
Leonard, A.
Little, D.A.J.
Little, J.E.
Longman, A.V.
Lording, T.
Lousley, D.W.
Lousley, J.E.
Love, J.
Love, M.
Lowe, K.
Lynes, M.

McCallum Webster, M.
MacKenzie, P.Z
Macneill, M.
Margetts, L.J.

Marks, K
Martin, J.P.
Mason, J.
Matheson, J.
Mawer, D.
Mawer, P.
Maycock, R.
Mayne, J.P.
McCrae, A.
Meinertzhagen, R.
Melling, T.
Meredith, H.M.
Middlehurst, J.
Miles, B.A.
Miller, G.B.
Millett, L.
Milne, B.
Monro & Nunney
Montegriffo, N.
Moore, A.
Morgan, B.M.C.
Mountford, O.
Mouritsen, L.
Murphy R.J.
Murrish, P.
Myles, S.

Nellist, D.
Nelson, D.
Newbould, C.
Nicholas, C.
Norman, E.
Norton, J.A.

O'Donnell, S.
Oliver, J.E.
O'Reilly, C & J.
Oswald, P.
Ottery, J. & S.

Page, C.N.
Page, R.
Palmer, J.R.
Parslow, H.
Parslow, J.L.F.
Parslow, J.M.L.
Parton, F.
Paton, J.A.
Paul, A.
Pavlou, S.
Payne, R.M.
Pearce, A. & M.
Pearman, D.A.
Pearson, J.
Perring, F.
Phillip, R.

Phillips, M.
Philp, E.G.
Pilkington, S.
Plumb, C.
Pogson, C.I.
Polunin, O.
Porley, R.
Pow, Miss
Precey, P.
Preston, C.D.
Price-Goodfellow, D.
Pugsley, H.

Quick, H.M.

Ralfs, J.
Ralphs, I.
Randall, R.E.
Raven, J.E.
Rayner, J.W.
Rees, E.A.
Ribbon, R.W.
Ribbons, R.J.
Rich, T.C.G.
Robbins, S.
Roberts, A.
Roberts, J.
Roper, C.
Roper, R.
Rumsey, F.
Russell, B.H.S.
Russell, J.

Salmon, M.
Sandwith, C.I.
Sargeant, P.
Sawyer, K.
Schofield, H & C.
Scott, D.
Scott, J.
Sears, E.
Selby, P.
Shaw, A.
Sherlock, R.&.N.E.
Silverside, A.
Simpson, F.W.
Sisam, C.
Smart, R.W.J.
Smith, W.W.
Somerville, A.
Spalding, A.
Spencer, J.
Spencer, M.
Spooner, B.M.
Sproull, J.
Spurgin, K.

Stace, C.
Stace, H.E.
Sterry, P.
Stewart, N.F.
Stokoe, R.
Stribley, M.
Stroh, P.
Sturdy, B.M.
Stuttard, R.

Tellam, R.V.
Tompsett, A.
Tompsett, P.E.
Tonkin, J.C.
Tout, M.
Townsend, F.
Trist, P.J.O.
Turberville, J.
Turk, S. M.
Turner, D.M.
Turner, K.A.

Underhill, A.

'Vagabondo'
Verrall, P.

Wagstaff, W.
Wake, A.J.
Walker, A.
Walker, V.
Walpole, M.
Walters, M.
Wanstall, P.J.
Ward, B.T.
Ward, S.
Waterhouse, G.
Watkins, F.
Watts, J.
Watts, M.
Webber, J.
Welch, R.C.
Wells, M.
West, C.
Westall, C.
Westwood, B.
White, A.
Wild, C.
Wilson, M.
Woodhall, A.
Woods, A.
Worth, J.
Worth, P.J.

Yeo, P.
Young, A.

A photographic record

Gazetteer

Abbey Hill, Tresco (SV8914)
Airfield, St Mary's (SV9110)
Anneka's Quay/Bar, Bryher (SV8815)
Annet (SV8608)
Annet Head (SV861089)
Appletree Banks, Tresco (SV8913)
Back Lane, Tresco (SV8915)
Badplace Hill, Bryher (SV875162)
Bakery, St Martin's (SV9315)
Bant's Carn, St Mary's (SV9012)
Bar Point, St Mary's (SV9112)
Barnaby Lane, St Agnes (SV8807)
Bathinghouse Pool, Tresco (SV8913)
Beacon Hill, Tresco (SV8815)
Benhams, Garrison, St Mary's (SV9010)
Big Pool, St Agnes (SV8708)
Blockhouse, Tresco (SV8915)
Bonfire Carn, Bryher (SV879142)
Borough Farm, St Mary's (SV8911)
Borough, Tresco (SV8914)
Bryher churchyard (SV879149)
Bryher Pool (SV873149)
Burnt Hill, St Martin's (SV936159)
Buzza Hill, St Mary's (SV9010)
Carn Adnis, St Agnes (SV882076)
Carn Irish, Annet (SV858087)
Carn Leh, St Mary's (SV9109)
Carn Morval Down, St Mary's (SV9011)
Carn Near, Tresco (SV8913)
Carn Thomas, St Mary's (SV9010)
Carreg Dhu, St Mary's (SV917110)
Carron Farm, St Martin's (SV9215)
Castle Bryher, Norrard Rocks (SV864139)
Castle Down, Tresco (SV8815)
Chad Girt, White Island, St Martin's (SV926174)
Chapel Down, St Martin's (SV9415)
Charles's Castle, Tresco (SV882162)
Churchtown Farm, St Martin's (SV9215)
Content Farm Lane, St Mary's (SV915119)

Covean, St Agnes (SV884083)
Crab's Ledge, Tresco (SV8913)
Cromwell's Castle, Tresco (SV8815)
Daymark, St Martin's (SV9416)
Deep Point, St Mary's (SV9311)
Dial Rocks, Tresco (SV8815)
Doiley Wood, St Mary's (SV911105)
Dropnose Porth, Gugh (SV8908)
East Porth, Samson (SV9016)
English Island Point, St Martin's (SV9315)
Ennor Castle, St Mary's (SV914103)
Fields above the Bar, St Agnes (SV805082)
Foreman's Island, near St Helen's (SV900160)
Garrison Arch, St Mary's (SV9010)
Garrison walls, St Mary's (SV9010)
Gimble Porth, Tresco (SV891160)
Golf Course, St Mary's (SV9111)
Gorregan, Western Rocks (SV848057)
Great Arthur, Eastern Isles (SV9413)
Great Bay, St Martin's (SV9216)
Great Ganilly, Eastern Isles (SV9414)
Great Ganinick, Eastern Isles (SV9313)
Great Hill, Teän (SV9116)
Great Innisvouls, Eastern Isles (SV9514)
Great Pool, Tresco (SV8815)
Green Bay, Bryher (SV880149)
Gugh Bar (SV8808)
Guthers, Eastern Isles (SV919144)
Gweal Hill, Bryher (SV8714)
Halangy Down, St Mary's (SV9192)
Hangman's Bay/Island, Bryher (SV8815)
Harry's Walls, St Mary's (SV9010)
Haycocks, Annet (SV9012)
Heathy Hill, Bryher (SV8714)
Hedge Rock, near Teän (SV907158)
Hell Bay Hotel, Bryher (SV8714)
Hell Bay, Bryher (SV8715)
Higher Moors, St Mary's (SV9210)
Higher Town, St Agnes (SV8808)
Higher Town, St Martin's (SV9315)
Higher Trenoweth, St Mary's (SV9112)
Hillside farm, Bryher (SV8714)
Holgate's Green, St Mary's (SV904106)

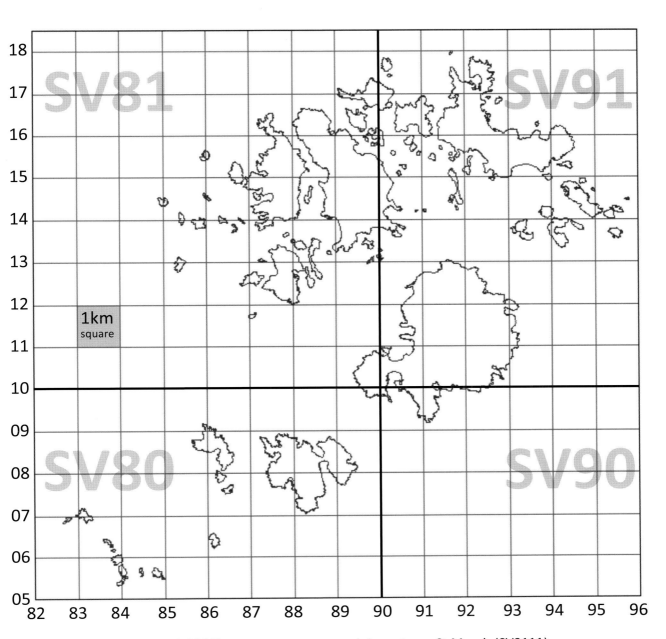

Holy Vale, St Mary's (SV9111-9211)
Hugh Town, St Mary's (SV9010)
Illiswilgig, Norrard Rocks (SV859139)
Innisidgen, St Mary's (SV9212)
Jac-a-Bar, St Mary's (SV9112)
Jacky's Point, St Mary's (SV931107)
King Edward's Road, St Mary's (SV9010)
Lawrence's Bay, St Martin's (SV9215)
Lifeboat Station, St Mary's (SV9010)
Little Arthur Farm, St Martin's (SV9315)
Little Arthur, Eastern Isles (SV9413)
Little Ganilly, Eastern Isles (SV9314)
Little Ganinick, Eastern Isles (SV9313)
Little Innisvouls, Eastern Isles (SV9514)
Little Pool, Bryher (SV876149)
Little Pool, St Agnes (SV8708)
Little Porth, St Mary's (SV902104)

Longstone, St Mary's (SV9111)
Low Pool, St Mary's (SV9110)
Lower Moors, St Mary's (SV9110)
Lower Town, St Agnes (SV8708)
Lower Town, St Martin's (SV9116)
Lunnon Farm, St Mary's (SV9211)
Maypole, St Mary's (SV9111)
McFarlands Down, St Mary's (SV913124)
Meadow/Cricket pitch, St Agnes (SV8708)
Melledgan, Western Rocks (SV862064)
Men-a-Vaur, near St Helen's (SV893163)
Menawethan, Eastern Isles (SV9513)
Merchant's Point, Tresco (SV8916)
Middle Arthur, Eastern Isles (SV9413)
Middle Down, Tresco (SV8914)
Middle Town, St Agnes (SV8808)
Middle Town, St Martin's (SV9216)

Mincarlo, Norrard Rocks (SV854129)
Monument, Tresco (SV892142)
Morning Point, St Mary's (SV9009)
Mount Todden, St Mary's (SV9213)
Nag's Head Down, St Agnes (SV8707)
New Grimsby, Tresco (SV8815)
New Inn, Tresco (SV8815)
Newford Duck Ponds, St Mary's (SV9112)
Newford Island (SV907112)
Newman's Battery, St Mary's (SV8910)
Normandy Down, St Mary's (SV931112)
Normandy, St Mary's (SV927111)
Nornour, Eastern Isles (SV944148)
North Hill, Samson (SV8712)
Northward, Bryher (SV8815
Northward, Tresco (SV8915)
Northwethel, near Tresco (SV8916)
Nowhere, St Mary's (SV9110)
Old Grimsby, Tresco (SV8915)
Old Man, Teän (SV904163)
Old Post Office, Bryher (SV882155)
Old Quay, St Martin's (SV9215)
Old Town Bay, St Mary's (SV9110)
Old Town, St Mary's (SV9110)
Oratory, St Helen's (SV902169)
Parting Carn, St Mary's (SV915108)
Pednbrose, Teän (SV911168)
Pelistry, St Mary's (SV928117)
Pendrathen, St Mary's (SV9112)
Peninnis, St Mary's (SV9109)
Pentle Bay, Tresco (SV9014)
Periglis, St Agnes (SV8708)
Plumb Island, Tresco (SV8814)
Pool Green/ Cricket field, St Martin's (SV9315)
Pool Road, Tresco (SV8914)
Pool, Bryher (SV8714)
Porth Hellick Down, St Mary's (SV9210)
Porth Hellick, St Mary's (SV9210)
Porth Mellon, St Mary's (SV9010)
Porth Minick, St Mary's (SV9110)
Porth Seal, St Martin's (SV9116)
Porth Wreck, St Mary's (SV9210)
Porthcressa, St Mary's (SV9010)

Porthellick Pool, St Mary's (SV9210)
Porthloo Duck Pond, St Mary's (SV9111)
Porthloo/Porthlow, St Mary's (SV9111)
Puffin Island, Samson (SV8813)
Pungies Lane, St Mary's (SV9112)
Ragged Island, Eastern Isles (SV946138)
Rocky Hill Lane, St Mary's (SV914110)
Rocky Hill, St Mary's (SV9111)
Rose Hill, St Mary's (SV9110)
Rosevear, Western Rocks (SV839059)
Round Island (SV9017)
Rushy Bank, Tresco (SV8913)
Rushy Bay, Bryher (SV8714)
Salakee Down, St Mary's (SV9210)
Salakee farm, St Mary's (SV9210)
Sally Port, St Mary's (SV901104)
Samson Hill, Bryher (SV8814)
Samson Well (SV878126)
Sandpit, Gugh (SV8808)
School Green, Tresco (SV8915)
Scilly Rock, Norrard Rocks (SV860155)
Seal Rock, Norrard Rocks (SV845140)
Seaways Farm, St Mary's (SV9112)
Shipman Head Down (SV8715)
Shipman Head, Bryher (SV8716)
Shooters' Pool, St Mary's (SV912107)
South Hill, Samson (SV8712)
St Helen's (SV8816)
St Martin's quarry/dump (SV928158)
Star Castle, St Mary's (SV9910)
Taylor's Island, St Mary's (SV906115)
Teän (SV9116)
Teän Plat Point, St Agnes (SV8807)
Telegraph Road, St Mary's (SV9111)
Telegraph, St Mary's (SV9111)
The Plains, St Martin's (SV9216)
The Town, Bryher (SV8815)
Timmy's Hill, Bryher (SV897152)
Tinkler's Hill, St Martin's (SV9116)
Tolls Island, St Mary's (SV9312)
Tolman Point, St Mary's (SV9110)
Tommy's Hill, Tresco (SV897152)
Towns Hill, Tresco (SV8815)

Tremelethen, St Mary's (SV9210)
Trenear Point, St Mary's (SV9212)
Trenoweth, St Mary's (SV919126)
Tresco churchyard (SV8915)
Trewince, St Mary's (SV9111)
Troy Town, St Agnes (SV8708)
Valhalla, Tresco (SV8914)
Veronica Farm, Bryher (SV8714)
Watch Hill, Bryher (SV8715)
Water Meadow Pool, Bryher (SV876151)
Water Rocks Down, St Mary's (SV9311)
Watermill Stream, St Mary's (SV9112)
Watermill, St Mary's (SV9212)
Well Cross Lane, St Mary's (SV9010)
West Porth, Samson (SV8712)
White Island, Samson (SV8712)
White Island, St Martin's (SV9217)
Wingletang Bay, St Agnes (SV8807)
Wingletang, St Agnes (SV8807)
Woolpack Battery, St Mary's (SV8910)
Woolpack walls, St Mary's (SV8909)

A view of Tresco from St Martin's

Index to species

Acacia falciformis 152
    melanoxylon 152
ACANTHACEAE 355
Acanthus mollis 355
Acer campestre 219
    pseudoplatanus 218
Achillea maritima 388
    millefolium 388
Aconitum napellus 514
Adder's-tongue 90
    Least 92
    Lesser 91
    Small 91
Adiantum capillus-veneris 96
Aegopodium podagraria 417
Aeonium spp. 127
    arboreum 127
    balsamiferum 127
    cuneatum 127
    haworthii 127
Aeonium 127
    Tree 127
Aesculus hippocastanum 218
Agapanthus praecox ssp. *orientalis* 448
Agave americana 456
Ageratum houstonianum 406
Agrostemma githago 264
Agrostis canina 493
    capillaris 492
    gigantea 492
    stolonifera 493
    vinealis 494
Aichryson laxum 130
    pachycaulon 130
Aira caryophyllea 489
    caryophyllea ssp. *multiculmis* 489
    praecox 489
AIZOACEAE 278
Ajuga reptans 345
Ake-ake 384
Akiraho 382
Albizia lophantha 152
Alchemilla spp. 514
Alder 175
    Grey 175
Alexanders 416
ALISMATACEAE 429
Alison, Sweet 232
Alkanet, Green 306
ALLIACEAE 444
Allium ampeloprasum var. *ampeloprasum* 447
    ampeloprasum var. *babingtonii* 446
    neapolitanum 445
    roseum 444
    triquetrum 445
    vineale 447
Allseed 194
    Four-leaved 261
Alnus glutinosa 175
    incana 175
Alopecurus geniculatus 497
    myosuroides 497
    pratensis 496
Alstroemeria aurea 514
Altar-lily 427
Amaranth, Common 277
    Green 277
AMARANTHACEAE 269
Amaranthus albus 277
    caudatus 277
    hybridus 277
    retroflexus 277
Amaryllis belladonna 449
Ambrosia artemisiifolia 402
Ammi majus 422
Ammophila arenaria 495
Amsinckia micrantha 514
Anacamptis pyramidalis 434
Anagallis arvensis ssp. *arvensis* 293
    arvensis ssp. *arvensis* var. *arvensis* 293

    arvensis ssp. *arvensis* var. *arvensis* f. *azurea* 293
    arvensis ssp. *arvensis* var. *arvensis* f. *carnea* 293
    tenella 292
Anchusa arvensis 305
    azurea 306
Anchusa, Garden 306
Angelica archangelica 423
    sylvestris 423
Angelica, Garden 423
    Wild 423
ANGIOSPERMS 111
Anisantha diandra 501
    madritensis 503
    rigida 502
    sterilis 502
Anthemis arvensis 390
    cotula 390
    punctata 389
    punctata ssp. *cupaniana* 389
Anthoxanthum odoratum 490
Anthriscus caucalis 415
    cerefolium 415
    sylvestris 415
Anthyllis vulneraria 131
Aphanes arvensis 163
    australis 163
APIACEAE 414
Apium graveolens 421
    graveolens var. *dulce* 421
    inundatum 422
    nodiflorum 421
APOCYNACEAE 302
Apple 156
    Crab 155
Apple-mint 350
    False 350
Apple-of-Peru 314
Aptenia cordifolia 279
AQUIFOLIACEAE 356
Arabidopsis thaliana 225
ARACEAE 427
ARALIACEAE 412
Arctium minus 359
    nemorosum 360
Arctotheca calendula 514
Arenaria serpyllifolia 254
    serpyllifolia ssp. *serpyllifolia* 254
Argemone mexicana 514
Aristea ecklonii 435
Armeria maritima 242
Armoracia rusticana 228
Arrhenatherum elatius 486
    elatius var. *bulbosum* 486
Artemisia absinthium 387
    maritima 387
    vulgaris 386
Arum italicum 427
    italicum ssp. *italicum* 427
    italicum ssp. *neglectum* 427
Ash 320
ASPARAGACEAE 452
ASPLENIACEAE 97
Asplenium adiantum-nigrum 97
    adiantum-nigrum × *A. obovatum* 98
    adiantum-nigrum × *A. obovatum* ssp. *lanceolatum* 98
    ceterach 103
    marinum 99
    obovatum ssp. *lanceolatum* 98
    ruta-muraria 100
    × *sarniense* 98
    scolopendrium 97
    trichomanes ssp. *quadrivalens* 99
Aster lanceolatus 378
    novi-belgii 377
    × *versicolor* 377
ASTERACEAE 359
Athyrium filix-femina 100
Atriplex glabriuscula 273
    glabriuscula × *A. prostrata* 273
    × *gustafssoniana* 273

    halimus 275
    hortensis 272
    laciniata 274
    littoralis 514
    longipes × *A. prostrata* 273
    patula 274
    prostrata 272
Aunt-Eliza 441
Avena fatua 486
    sativa 487
    sterilis 486
Avenula pubescens 485
Azolla filiculoides 94
Baldellia ranunculoides 429
Ballota nigra ssp. *meridionalis* 342
Balm 346
Balm of Gilead 187
Balsam-poplar, Eastern 188
Bamboo, Arrow 475
    Veitch's 475
Barbarea verna 226
    vulgaris 226
Barley, Meadow 506
    Wall 505
Bartsia, Yellow 354
Bean, Broad 137
Bear's-breech 355
Bedstraw, Heath 300
    Hedge 300
    Lady's 299
    Marsh 299
    Tree 298
Beech 171
Beet, Root 276
    Sea 275
Bellflower, Adria 357
    Ivy-leaved 357
    Rampion 357
    Trailing 357
Bellis perennis 384
    perennis flore pleno 384
Bent, Black 492
    Brown 494
    Common 492
    Creeping 493
    Velvet 493
    Water 496
Bergenia crassifolia 125
Bermuda-grass 509
Berula erecta 418
Beta vulgaris ssp. *maritima* 275
    vulgaris ssp. *vulgaris* 276
Betonica officinalis 341
Betony 341
Betula sp. 174
    pendula 174
BETULACEAE 174
Bidens tripartita 405
Bindweed, Field 311
    Hedge 312
    Large 313
    Sea 312
Birch 174
Bird's-foot 132
    Orange 133
Bird's-foot-trefoil, Common 131
    Greater 131
    Hairy 132
    Slender 514
Bistort, Amphibious 242
Bitter-cress, Hairy 230
    Wavy 229
Bittersweet 317
Black Parsley, Madeira Giant 424
Blackberry 159
Black-bindweed 248
Black-grass 497
Black-poplar, Hybrid 187
Blackthorn 155
Blackwood, Australian 152

BLECHNACEAE 101
Blechnum cordatum 101
    spicant 102
Blinks 287
Bluebell 453
    Hybrid 453
Blue-gum, Southern 215
Bogbean 359, 515
Bolboschoenus maritimus 465
Borage 307
    Slender 307
BORAGINACEAE 303
Borago officinalis 307
    pygmaea 307
Botrychium lunaria 93
Box 124
Boxwood, African 289
Brachyglottis × *jubar* 399
    monroi 399
    repanda 399
    'Sunshine' 399
Brachypodium sylvaticum 504
Bracken 95
Bramble, Elm-leaved 160
Brassica napus ssp. *oleifera* 233
    nigra 234
    rapa 233
    rapa ssp. *oleifera* 233
BRASSICACEAE 224
Bridewort 154
    Confused 154
Bristle-grass, Green 512
    Rough 511
    Yellow 511
Briza maxima 481
    minor 481
Broadleaf, New Zealand 410
Brome, Barren 502
    Compact 503
    False 504
    Great 501
    Rescue 503
    Rip-gut 502
    Rye 501
    Smooth 500
BROMELIACEAE 458
Bromus commutatus × *B. racemosus* 500
    hordeaceus 500
    hordeaceus ssp. *hordeaceus* 500
    lepidus 501
    × *pseudothominei* 500
    racemosus 500
    secalinus 501
Brooklime 323
Brookweed 294
Broom 150
    Spanish 151
Broomrape, Ivy 355
    Lesser 355
Buckler-fern, Broad 105
    Narrow 104
Buckwheat 244
Buddleja davidii 339
    globosa 339
Buddleja 339
Bugle 345
Bugle-lily 438
Bugloss 305
Bullwort 422
Bulrush 458
Bupleurum subovatum 420
    subovatum 514
Burdock, Lesser 359
    Wood 360
Bur-marigold, Trifid 405
Burnet, Fodder 162
Bur-reed, Branched 457
Butcher's-broom 454
Buttercup, Bermuda 181
    Bulbous 118

    Celery-leaved 121
    Corn 119
    Creeping 117
    Hairy 118
    Meadow 117
    Rough-fruited 120
    Scilly 120
    Small-flowered 120
    St Martin's 119
Buttonweed, Annual 393
BUXACEAE 124
Buxus sempervirens 124
Cabbage-palm 455
Cakile maritima 235
    maritima ssp. *integrifolia* 235
Calamagrostis epigejos 494
Calamint, Common 347
Calendula arvensis 401
    officinalis 400
CALLITRICHACEAE 333
Callitriche brutia 334
    hamulata 335
    obtusangula 334
    platycarpa 334
    stagnalis 333
    stagnalis ssp. *intermedia* 333
    stagnalis ssp. *major* 333
Calluna vulgaris 295
Caltha palustris 116
Calystegia × *lucana* 312
    sepium 312
    sepium ssp. *roseata* 312
    sepium ssp. *roseata* × *C. silvatica* 312
    sepium ssp. *sepium* 312
    silvatica 313
    silvatica ssp. *silvatica* 313
    soldanella 312
Campanula portenschlagiana 357
    poscharskyana 357
    rapunculus 357
CAMPANULACEAE 357
Campion, Bladder 264
    Red 266
    Sea 265
    White 265
Canary-grass 491
    Lesser 491
    Reed 490
Candytuft, Garden 241, 514
    Perennial 241, 514
CANNABACEAE 168
Cape-gooseberry 315
Cape-tulip 434
CAPRIFOLIACEAE 407
CAPRIFOLIACEAE 408
Capsella bursa-pastoris 225
Cardamine flexuosa 229
    hirsuta 230
    pratensis 229
Carduus tenuiflorus 360
Carex arenaria 470
    binervis 473
    demissa 474
    divulsa ssp. *divulsa* 470
    echinata 471
    flacca 473
    laevigata 473
    muricata ssp. *lamprocarpa* 470
    muricata ssp. *pairae* 470
    nigra 474
    otrubae 469
    panicea 473
    paniculata 469
    pendula 472
    pilulifera 474
    pseudocyperus 471
    remota 471
    riparia 471
    sylvatica 472
Carlina vulgaris 359

Carpobrotus acinaciformis 283
    edulis 284
    edulis var. *chrysophthalmus* 284
    edulis var. *edulis* 284
    edulis var. *rubescens* 284
    glaucescens 285
Carrot, Wild 426
CARYOPHYLLACEAE 254
Castanea sativa 172
Catapodium marinum 485
    rigidum 484
    rigidum var. *majus* 484
Catchfly, Night-flowering 266
    Small-flowered 266
Cat's-ear 364
    Smaller 498
Celandine, Lesser 123
CELASTRACEAE 176
Celery, Wild 421
Centaurea cyanus 363
    nigra 363
Centaurium erythraea 302
Centaury, Common 302
Centranthus ruber 410
Centunculus minimus 293
Centuryplant 456
Cerastium diffusum 259
    fontanum 258
    fontanum ssp. *holosteoides* 258
    glomeratum 258
    semidecandrum 259
    tomentosum 257
Ceratochloa cathartica 503
Cerinthe major 514
Chaenomeles speciosa 514
Chaffweed 293
Chamaemelum nobile 389
Chamerion angustifolium 211
Chamomile 389
    Corn 390
    Sicilian 389
    Stinking 390
Charlock 234
Chasmanthe aethiopica 442
    bicolor 442
Chasmanthe 442
Checkerberry 297
Chenopodium album 271
    bonus-henricus 269
    ficifolium 271
    hybridum 270
    murale 270
    polyspermum 269
    quinoa 271
    rubrum 269
Cherry, Wild 155
Chervil, Bur 415
    Garden 415
Chestnut, Sweet 172
Chickweed, Common 255
    Greater 256
    Lesser 256
Chicory 363
Chilean-iris 435
Chrysocoma coma-aurea 378
Cichorium intybus 363
Cineraria 398
Cinquefoil, Creeping 162
Circaea lutetiana 214
Cirsium arvense 362
    arvense var. *integrifolium* 362
    arvense var. *mite* 362
    palustre 361
    vulgare 361
Claytonia perfoliata 286
    sibirica 287
Cleavers 301
Clematis flammula 116
    montana 514
    vitalba 116

Index to species

Clematis, Himalayan 514
Clethra arborea 294
CLETHRACEAE 294
Clinopodium ascendens 347
Clover, Alsike 143
    Bird's-foot 141
    Clustered 144
    Crimson 147
    Hare's-foot 149
    Knotted 148
    Red 146
    Reversed 515
    Rough 148
    Strawberry 144
    Subterranean 149
    Suffocated 144
    Western 143
    White 142
    Zigzag 147
Clubmoss, Krauss's 90
    Mossy 90
Club-rush, Bristle 467
    Sea 465
    Slender 468
Cochlearia danica 240
    danica × *C. officinalis* 240
    officinalis 240
Cock's-foot 484
Cockspur 510
Codlins and Cream 207
Colt's-foot 399
Comfrey, Common 305
    Russian 305
Conium maculatum 420
Conopodium majus 417
CONVOLVULACEAE 311
Convolvulus arvensis 311
    sabatius 514
Convolvulus, Blue Ground 514
Conyza bonariensis 381
    canadensis 380
    floribunda 380
    sumatrensis 381
Coprosma repens 298
Cordyline australis 455
Coriander 514
Coriandrum sativum 514
Corncockle 264
Cornflower 363
Corn-lily, Blue 435
    Red 439
    Tubular 440
Cornsalad, Common 408
    Keel-fruited 408
    Tooth-fruited 408
Coronilla valentina 133
Correa backhouseana 219
Cortaderia selloana 507
Corydalis, Yellow 515
Corylus avellana 176
Cotoneaster cambricus 157
    horizontalis 157
    simonsii 157
    × *watereri* 514
Cotoneaster, Himalayan 157
    Wall 157
    Waterer's 514
    Wild 157
Cottongrass, Common 465
Cottonweed 388
Cotula australis 393
Couch, Bearded 504
    Common 504
    Onion 486
    Sand 505
    Sea 504
Crack-willow 188
    Hybrid 188
Crambe maritima 236
Crane's-bill, Cut-leaved 199

    Dove's-foot 200
    Druce's 198
    Pencilled 198
    Round-leaved 199
    Shining 201
    Small-flowered 200
Crassula decumbens 125
    multicava 126
Crassula, Fairy 126
CRASSULACEAE 125
Crataegus monogyna 158
Creeping-Jenny 291
Crepis biennis 369
    capillaris 370
    vesicaria ssp. *taraxacifolia* 370
Cress, Thale 225
    Violet 514
Crithmum maritimum 418
Crocosmia × *crocosmiiflora* 441
    paniculata 441
Crowfoot, Ivy-leaved 122
    Three-lobed 121
Cuckooflower 229
Cudweed, Common 372
    Heath 373
    Jersey 374
    Marsh 373
CUPRESSACEAE 109
Cupressus macrocarpa 109
Cuscuta epithymum 313
Cymbalaria muralis 329
    muralis ssp. *muralis* 329
    muralis ssp. *muralis* f. *alba* 329
Cynodon dactylon 509
Cynosurus cristatus 479
    echinatus 480
CYPERACEAE 465
Cyperus eragrostis 468
    longus 468
Cypress, Monterey 109
Cyrtomium falcatum 103
Cytisus scoparius ssp. *scoparius* 150
Dactylis glomerata 484
Dactylorhiza fuchsii 434
    praetermissa 433
Daffodils 451
Daisy 384
    a Cape 402
    Cape 401
    Oxeye 391
    Seaside 379
    Shasta 514
Daisy-bush 383
    Coastal 384
    Mangrove-leaved 382
Dandelions 369
Danthonia decumbens 506
Daphne laureola 514
Datura stramonium 315
Daucus carota 426
    carota ssp. *carota* 426
    carota ssp. *gummifer* 426
Dead-nettle, Cut-leaved 343
    Henbit 343
    Red 342
    White 342
Delairea odorata 398
DENNSTAEDTIACEAE 95
Dewberry 159
Dewplant, Deltoid-leaved 280
    Pale 279
    Purple 282
    Rosy 282
    Shrubby 280
    Sickle-leaved 281
Dianthus barbatus 268
    plumarius 268
Dicksonia antarctica 95
DICKSONIACEAE 95
Digitalis purpurea 322

Digitaria ciliaris 513
    sanguinalis 512
Diplotaxis muralis 233
    tenuifolia 232
DIPSACACEAE 410
Dipsacus fullonum
Disphyma crassifolium 282
Dock, Argentine 250
    Broad-leaved 253
    Clustered 251
    Curled 250
    Fiddle 253
    Shore 252
    Water 250
    Wood 251
Dodder 313
Dog's-tail, Crested 479
    Rough 480
Dog-violet, Common 190
    Early 191
    Heath 191
Drosanthemum floribundum 279
DRYOPTERIDACEAE 102
Dryopteris affinis 104
    affinis ssp. *affinis* 104
    affinis ssp. *borreri* 104
    carthusiana 104
    dilatata 105
    filix-mas 103
Duckweed, Common 428
    Least 429
Echinochloa crus-galli 510
Echium pininana 304
    plantagineum 304
    × *scilloniensis* 304
    vulgare 303
Eelgrass 430
ELAEAGNACEAE 165
Elaeagnus 'Ebbingei' 165
    pungens 165
ELATINACEAE 186
Elatine hexandra 186
Elder 407
Elecampane 375
Eleocharis multicaulis 467
    palustris 466
    palustris ssp. *vulgaris* 466
    uniglumis 466
Elephant-ears 125
Eleusine indica 509
Elm, Cornish 167
    Dutch 166
    English 167
    Wych 167
Elms 166
Elymus caninus 504
Elytrigia atherica 504
    juncea ssp. *boreoatlantica* 505
    repens 504
    repens f. *aristata* 504
    repens ssp. *arenosus* 504
    repens × *Hordeum secalinum* 504
×*Elytrordeum langei* 504
Enchanter's-nightshade 214
Epilobium ciliatum 210
    ciliatum × *E. obscurum* 210
    × *dacicum* 210
    hirsutum 207
    lanceolatum 209
    montanum 208
    obscurum 210
    obscurum × *E. parviflorum* 210
    obscurum × *E. tetragonum* 210
    × *palatinum* 208
    palustre 211
    parviflorum 208
    parviflorum × *E. tetragonum* 208
    × *semiobscurum*
    tetragonum 209
EQUISETACEAE 93

Equisetum arvense 93
  × *litorale* 93
Eragrostis cilianensis 508
Erepsia heteropetala 283
Erica arborea 296
  cinerea 296
  erigena 297
  vagans 297
ERICACEAE 295
Erigeron glaucus 379
  karvinskianus 379
Eriophorum angustifolium 465
Erodium cicutarium 204
  cicutarium ssp. *dunense* 204
  lebelii 204
  maritimum 203
  moschatum 203
Eryngium maritimum 414
Erysimum cheiranthoides 224
  cheiri 224
Escallonia macrantha 406
Escallonia 406
ESCALLONIACEAE 406
Eschscholzia californica 514
Eucalyptus spp. 216
  globulus 215
  pulchella 216
  urnigera 216
  viminalis 215, 216
Euonymus japonicus 176
Eupatorium cannabinum 405
Euphorbia amygdaloides ssp. *amygdaloides* 185
  characias 514
  cyparissias 185
  helioscopia 182
  lathyris 183
  mellifera 182
  paralias 184
  peplis 182
  peplus 183
  portlandica 184
EUPHORBIACEAE 181
Euphrasia 353
  anglica var. *gracilescens* 353
  brevipila 353
  confusa 353
  confusa × *E. nemorosa* 353
  curta 353
  micrantha 353
  tetraquetra 353
Evening primroses 212
Evening-primrose, Common 213
  Intermediate 212
  Large-flowered 212
  Small-flowered 212
Everlastingflower, Cape 375
  Silver-bush 374
Everlasting-pea, Broad-leaved 138
Eyebright 353
Eyebrights 353
FABACEAE 131
FAGACEAE 171
Fagopyrum esculentum 244
Fagus sylvatica 171
Fallopia baldschuanica 247
  convolvulus 248
  japonica 247
Fascicularia bicolor 458
Fat-hen 271
Fatsia japonica 413
Fatsia 413
Fennel 420
Fern, Chain 101
  Guernsey 98
  Kangaroo 107
  Maidenhair 96
  Ribbon 96
  Royal 94
  Water 94
Fern-grass 484

  Sea 485
Fescue, Dune 478
  Rat's-tail 479
  Red 477
  Squirreltail 478
  Tall 476
Festuca ovina 478
  rubra 477
  rubra ssp. *juncea* 477
  rubra ssp. *rubra* 477
Feverfew 385
Ficaria verna 123
  verna ssp. *fertilis* 123
  verna ssp. *ficariiformis* 123
Ficus carica 168
Fiddleneck, Common 514
Field-speedwell, Common 326
  Green 325
  Grey 325
Fig 168
Figwort, Balm-leaved 338
  Common 337
  Water 338
Filago vulgaris 372
Filipendula ulmaria 158
Finger-grass, Hairy 512
  Tropical 513
Fireweed, Toothed 397
Flax 193
  Fairy 194
  Pale 193
Fleabane, Argentine 381
  Bilbao 380
  Canadian 380
  Common 375
  Guernsey 381
  Mexican 379
Flossflower 406
Fluellen, Round-leaved 330
  Sharp-leaved 330
Foeniculum vulgare 420
Fool's-water-cress 421
Forget-me-not, Changing 310
  Creeping 308
  Early 310
  Field 309
  Tufted 309
  Water 308
Foxglove 322
Fox-sedge, False 469
Foxtail, Marsh 497
  Meadow 496
Fragaria vesca 514
Fraxinus excelsior 320
Freesia × *hybrida* 441
Freesia 441
Fuchsia 213
Fuchsia magellanica 213
  magellanica 'Corallina' 213
Fumaria 112
  bastardii 114
  bastardii var. *gussonei* 114
  bastardii var. *hibernica* 114
  capreolata ssp. *babingtonii* 113
  muralis ssp. *boroei* 114
  muralis var. *major* 114
  occidentalis 113
  officinalis 115
  officinalis ssp. *officinalis* 115
  officinalis ssp. *wirtgenii* 115
  purpurea 115
Fumitories 112
Fumitory, Common 115
Furcraea longaeva 456
Furcrea 456
Galingale 468
  Pale 468
Galinsoga parviflora 404
  quadriradiata 404
Galium album 300

  aparine 301
  palustre 299
  palustre ssp. *elongatum* 299
  palustre ssp. *palustre* 299
  saxatile 300
  verum 299
Gallant-soldier 404
Garlic, Neapolitan 445
  Rosy 444
  Three-cornered 445
Gastridium ventricosum 495
Gaultheria procumbens 297
  shallon 297
Gazania rigens 372
  rigens var. *uniflora* 372
GENTIANACEAE 302
GERANIACEAE 198
Geranium dissectum 199
  lucidum 201
  maderense 202
  molle 200
  × *oxonianum* 198
  purpureum 201
  pusillum 200
  robertianum 201
  rotundifolium 199
  versicolor 198
  yeoi 202 202
Geranium 205
  Peppermint-scented 205
German-ivy 398
Giant-rhubarb 124
  Brazilian 124
Gilia capitata 289, 514
Gladiolus communis ssp. *byzantinus* 439
Gladiolus, Eastern 439
Glaucium flavum 112
Glaux maritima 292
Glebionis segetum 390
Glechoma hederacea 345
Glyceria declinata 499
  fluitans × *G. notata* 498
  fluitans 498
  notata 499
  × *pedicellata* 498
Gnaphalium luteoalbum 374
  sylvaticum 373
  uliginosum 373
Golden Fleece, Smooth 515
Goldenrod 376
  Canadian 376
Goldilocks, Shrub 378
Good King Henry 269
Goosefoot, Fig-leaved 271
  Many-seeded 269
  Maple-leaved 270
  Nettle-leaved 270
  Red 269
Gorse 151
  Western 152
Grass-poly 206
Griselinia littoralis 410
GRISELINIACEAE 410
Ground-elder 417
Ground-ivy 345
Groundsel 396
  Heath 397
  Sticky 397
Guava, Chilean 216
Guelder-rose 407
Gum, Ribbon 215
  White Peppermint 216
Gunnera manicata 124
  tinctoria 124
GUNNERACEAE 124
Gymnadenia borealis 434
  conopsea 434
Gypsywort 348
Hair-grass, Crested 487
  Early 489

Index to species

Mediterranean 487
Silver 489
HALORAGACEAE 130
Hard-fern 102
Greater 101
Hare's-tail 495
Harlequinflower, Plain 440
Hart's-tongue 97
Hawkbit, Autumn 365
Lesser 366
Rough 365
Hawk's-beard, Beaked 370
Rough 369
Smooth 370
Hawkweed 371
Mouse-ear 371
Hawthorn 158
Hazel 176
Heath, Cornish 297
Irish 297
Heather 295
Bell 296
Tree 296
Heath-grass 506
Hebe 327
Hebe, Lewis's 327
Hedera helix 413
hibernica 413
hibernica cv Hibernica 413
Hedge-parsley, Knotted 426
Upright 425
Helianthus annuus 403
× *laetiflorus* 403
× *multiflorus* 403
petiolaris 403
Helichrysum bracteatum 374
petiolare 374
Heliotrope, Winter 400
Helminthotheca echioides 366
Hemlock 420
Hemp-agrimony 405
Henbane 314
Heracleum sphondylium 425
Herb-Robert 201
Giant 202
Greater 202
Hibiscus trionum 222
Hieracium umbellatum ssp. *bichlorophyllum* 371
Hippophae rhamnoides 165
Hogweed 425
Holcus lanatus 488
mollis 488
Holly 356
Holly-fern, House 103
Homeria collina 434
Honckenya peploides 255
Honesty 232
Honeybells 448
Honeysuckle 408
Japanese 514
Perfoliate 408
Honeywort, Greater 514
Hop 168
Hordeum murinum 505
secalinum 506
Horehound, Black 342
Horse-chestnut 218
Horse-radish 228
Horsetail, Field 93
Hottentot Fig 284
Humulus lupulus 168
Hyacinth, Tassel 454
Hyacinthoides hispanica × *H. non-scripta* 453
× *massartiana* 453
non-scripta 453
Hydrangea macrophylla 288
Hydrangea 288
HYDRANGEACEAE 288
HYDROCOTYLACEAE 414
Hydrocotyle ranunculoides 414

vulgaris 414
Hyoscyamus niger 314
HYPERICACEAE 195
Hypericum androsaemum 195
elodes 197
hircinum 196
humifusum 196
× *inodorum* 195
perforatum 196
pulchrum 197
Hypochaeris radicata 364
Iberis sempervirens 241, 514
Iberis umbellata 241, 514
Iceplant, Heart-leaf 279
Ilex aquifolium 356
Inch-plant 515
Inula helenium 375
Ionopsidium acaule 514
Ipomoea purpurea 514
IRIDACEAE 434
Iris foetidissima 437
× *hollandica* 438
latifolia 437
pseudacorus 436
xiphium 437
Iris, Dutch 438
English 437
Spanish 437
Stinking 437
Yellow 436
Isolepis cernua 468
setacea 467
Ivy, Atlantic 413
Common 413
Irish 413
Ixia campanulata 439
paniculata 440
Jasione montana 358
Jasmine, Summer 320
Jasminum officinale 320
JUNCACEAE 459
Juncus acutiflorus 460
articulatus 459
bufonius 462
bulbosus 460
conglomeratus 464
effusus 463
effusus var. *subglomeratus* 463
gerardii 462
inflexus 463
maritimus 461
maritimus var. *atlanticus* 461
tenuis 461
Kangaroo-apple 319
Karo 410
Kattegat Orache 273
Ketmia, Bladder 222
Kickxia elatine 330
spuria 330
Knapweed, Common 363
Kniphofia × *praecox* 443
uvaria 442
Knotgrass 246
Equal-leaved 246
Ray's 245
Sea 245
Knotweed, Japanese 247
Koeleria macrantha 487
Kohuhu 411
Lactuca serriola 368
Lady-fern 100
Lady's Mantle 514
Lady's-tresses, Autumn 433
Lagurus ovatus 495
Lamb's-ear 515
LAMIACEAE 340
Lamiastrum galeobdolon ssp. *argentatum* 342
Lamium album 342
amplexicaule 343
hybridum 343

purpureum 342
Lampranthus conspicuus 281
falciformis 281
roseus 282
Lapsana communis 364
Lathyrus annuus 138
japonicus ssp. *maritimus* 137
latifolius 138
pratensis 138
Laurustinus 407
Lavender-cotton 388
Leek, Babington's 446
Wild 447
Lemna minor 428
minuta 429
LEMNACEAE 428
Leontodon hispidus 365
saxatilis 366
Lepidium coronopus 231
didymum 231
heterophyllum 230
Leptospermum lanigerum 214
scoparium 214
Lettuce, Prickly 368
Leucanthemum × *superbum* 514
vulgare 391
Leucojum aestivum ssp. *pulchellum* 450
Leymus arenarius 505
Libertia formosa 435
Ligustrum ovalifolium 321
vulgare 321
vulgare var. *coombei* 321
LILIACEAE 432
Lily, African 448
Bowden Cornish 450
Guernsey 450
Jersey 449
Peruvian 514
Lily-of-the-valley-tree 294
Limosella aquatica 339
LINACEAE 193
Linaria vulgaris 330
Ling 295
Linum bienne 193
catharticum 194
usitatissimum 193
Little-Robin 201
Littorella uniflora 333
Lobelia erinus 359
Lobelia, Garden 359
Lobularia maritima 232
Lolium multiflorum 477
perenne 476
Lonicera caprifolium 408
japonica 514
periclymenum 408
Loosestrife, Yellow 291
Lords-and-Ladies, Italian 427
Lotus angustissimus 514
corniculatus 131
pedunculatus 131
subbiflorus 132
Lousewort 354
Love-lies-bleeding 277
Lucerne 140
Luma apiculata 217
Lunaria annua 232
Lupin, Tree 150
Lupinus arboreus 150
Luzula campestris 464
multiflora 465
Lycopus europaeus 348
Lyme-grass 505
Lysimachia nemorum 290
nummularia 291
punctata 291
vulgaris 291
LYTHRACEAE 206
Lythrum hyssopifolia 206
portula 207

salicaria 206
Madder, Field 298
    Wild 301
Male-fern 103
    Golden-scaled 104
Mallow, Common 220
    Dwarf 221
    Small 220
Malus pumila 156
    sylvestris 155
Malva arborea 221
    moschata 220
    neglecta 221
    pseudolavatera 222
    pusilla 220
    sylvestris 220
MALVACEAE 220
Maple, Field 219
Marigold, Corn 390
    Field 401
    Pot 400
Marram 495
Marsh-marigold 116
Marsh-orchid, Southern 433
Marshwort, Lesser 422
Marvel-of-Peru 286
Matricaria chamomilla 391
    discoidea 392
Matthiola incana 239
    longipetala 239
Mayweed, Scented 391
    Scentless 393
    Sea 392
Meadow-grass, Annual 482
    Early 482
    Rough 483
    Smooth 483
    Spreading 483
Meadow-rue, Lesser 123
Meadowsweet 158
Medicago arabica 141
    lupulina 139
    polymorpha 140
    sativa ssp. *sativa* 140
Medick, Black 139
    Spotted 141
    Toothed 140
Melanoselinum decipiens 424
Melilot, Ribbed 139
Melilotus officinalis 139
Melissa officinalis 346
Mentha aquatica 349
    aquatica × *M. spicata* 349
    aquatica × *M. suaveolens* 349
    longifolia × *M. suaveolens* 350
    × *piperita* 349
    pulegium 351
    requienii 351
    × *rotundifolia* 350
    spicata 350
    spicata × *M. suaveolens* 350
    suaveolens 351
    × *suavis* 349
    × *villosa* 350
    × *villosa* var. *alopecuroides* 350
MENYANTHACEAE 359
Menyanthes trifoliata 359, 515
Mercurialis annua 181
Mercury, Annual 181
Michaelmas-daisy, Confused 377
    Late 377
    Narrow-leaved 378
Mignonette, White 223
    Wild 223
Milkwort, Common 153
    Heath 153
Millet, Common 510
    Great 513
Mind-your-own-business 170
Mint, Corsican 351

Hybrid 349
    Round-leaved 351
    Spear 350
    Water 349
Mintweed 352
Mirabilis jalapa 286
Misopates orontium 329
Molinia caerulea 507
Moneywort, Cornish 328
Monk's-hood 514
Montbretia 441
Montia fontana 287
    fontana ssp. *amporitana* 287
    fontana ssp. *chondrosperma* 287
    fontana ssp. *variabilis* 287
MONTIACEAE 286
Moonwort 93
Moor-grass, Purple 507
MORACEAE 168
Morning-glory, Common 514
Mouse-ear, Common 258
    Little 259
    Sea 259
    Sticky 258
Mudwort 339
Muehlenbeckia complexa 248
Mugwort 386
Mullein, Dark 337
    Great 336
    Moth 335
    Orange 336
    Twiggy 336
Muscari comosum 454
Musk-mallow 220
Mustard, Black 234
    Hedge 238
    White 235
Myosotis arvensis 309
    discolor 310
    laxa 309
    laxa subsp. *caespitosa* 309
    ramosissima 310
    scorpioides 308
    secunda 308
Myriophyllum alterniflorum 130
    aquaticum 130
Myrsine africana 289
MYRTACEAE 214
Myrtle, Chilean 217
Naked Ladies 449
Narcissus 451
Nassella trichotoma 475
Nasturtium officinale 228
Nasturtium 222
Navelwort 126
Nerine bowdenii 450
    sarniensis 450
Nettle, Common 169
    Small 169
New Zealand Flax 443
New Zealand Flax, Lesser 444
New Zealand Holly 383
Nicandra physalodes 314
Nicotiana alata 319
    alata × *N. forgetiana* 319
    forgetiana 320
    × *sanderae* 319
Nightshade, Black 316
    Green 316
    Leafy-fruited 317
Nipplewort 364
Nit-grass 495
Nothoscordum borbonicum 448
NYCTAGINACEAE 286
Nymphaea alba 109
NYMPHAEACEAE 109
Oak, Evergreen 173
    Pedunculate 174
    Sessile 173
    Turkey 172

Oat 487
Oat-grass, Downy 485
    False 486
    Yellow 487
Ochagavia carnea 459
Oenanthe crocata 419
    fistulosa 419
    lachenalii 418
Oenothera 212
    biennis 213
    cambrica 212
    × *fallax* 212
    glazioviana 212
OLEACEAE 320
Olearia avicenniifolia 382
    × *haastii* 383
    macrodonta 383
    paniculata 382
    solandri 384
    traversii 384
Oleaster, Spiny 165
ONAGRACEAE 207
Onion, Wild 447
Ononis repens 138
OPHIOGLOSSACEAE 90
Ophioglossum azoricum 91
    lusitanicum 92
    vulgatum 90
Orache, Babington's 273
    Common 274
    Frosted 274
    Garden 272
    Grass-leaved 514
    Hybrid 273
    Shrubby 275
    Spear-leaved 272
Orange-ball-tree 339
Orchid, Fragrant 434
    Pyramidal 434
ORCHIDACEAE 433
Ornithogalum angustifolium 451
    umbellatum 451
    umbellatum ssp. *campestre* 451
Ornithopus perpusillus 132
    pinnatus 133
OROBANCHACEAE 352
Orobanche hederae 355
    minor 355
    minor minor ssp. *maritima* 355
    minor var. *compositarum* 355
Oscularia deltoides 280
Osier 189
    Broad-leaved 189
Osmunda regalis 94
OSMUNDACEAE 94 94
Osteospermum ecklonis 402
    jucundum 401
    jucundum 'Tresco Purple' 401
OXALIDACEAE 177
Oxalis acetosella 179
    articulata 179
    corniculata 177
    corniculata var. *atropurpurea* 177
    debilis 179
    exilis 178
    latifolia 180
    megalorrhiza 178
    pes-caprae 181
    rosea 177
    tetraphylla 180
Oxtongue, Bristly 366
Pampas-grass 507
Panicum miliaceum 510
Pansy, Dwarf 192
    Field 192
    Garden 191
    Wild 191
Papaver arenarium 111
    dubium 111
    pseudoorientale 110

Index to species

rhoeas 111
rhoeas var. *commutatum* 111
somniferum 110
PAPAVERACEAE 110
Parentucellia viscosa 354
Parietaria judaica 170
officinalis 170
Parrot's-feather 130
Parsley, Cow 415
Garden 422
Parsley-piert 163
Slender 163
Parsnip, Wild 424
Passiflora caerulea 515
Passionflower, Blue 515
Pastinaca sativa ssp. *sylvestris* 424
Pea, Field 138
Fodder 138
Sea 137
Pearlwort, Annual 260
Procumbent 260
Sea 261
Pedicularis sylvatica 354
Pelargonium × *hybridum* 205
tomentosum 205
Pellitory-of-the-wall 170
Eastern 170
Penny-cress, Field 238
Pennyroyal 351
Pennywort, Floating 414
Marsh 414
Pentaglottis sempervirens 306
Peppermint 349
Pepperwort, Smith's 230
Pericallis hybrida 398
Periwinkle, Greater 303
Intermediate 515
Lesser 302
Persicaria amphibia 242
capitata 515
hydropiper 244
lapathifolia 243
maculosa 243
Persicaria, Pale 243
Pink-headed 515
Petasites fragans 400
Petroselinum crispum 422
Petunia × *hybrida* 515
Petunia 515
Phacelia 311
Phacelia tanacetifolia 311
Phalaris arundinacea 490
canariensis 491
minor 491
Phleum bertolonii 498
pratense 498
Phormium cookianum 444
Phormium tenax 443
Phragmites australis 508
Phymatosorus diversifolius 107
Physalis peruviana 315
Pickerelweed 457
Pigmyweed, Scilly 125
Pignut 417
Pigweed, White 277
Pilosella officinarum 371
Pimpernel, Bog 292
Scarlet 293
Yellow 290
Pimpinella saxifraga 417
PINACEAE 108
Pine, Lodgepole 108
Maritime 108
Monterey 108
Pineappleweed 392
Pink 268
Pink-sorrel 179
Annual 177
Four-leaved 180
Garden 180

Large-flowered 179
Pinus contorta 108
pinaster 108
radiata 108
Pisum sativum var. *arvense* 138
PITTOSPORACEAE 410
Pittosporum crassifolium 410
tenuifolium 411
Pittosporum 410
PLANTAGINACEAE 331
Plantago coronopus 331
lanceolata 332
major 332
maritima 331
Plantain, Buck's-horn 331
Greater 332
Ribwort 332
Sea 331
Plecostachys serpyllifolia 375
Plum, Cherry 154
Wild 155
PLUMBAGINACEAE 242
Poa annua 482
humilis 483
infirma 482
pratensis 483
trivialis 483
POACEAE 475
POLEMONIACEAE 289
Polycarpon tetraphyllum 261
tetraphyllum var. *diphyllum* 261
Polygala serpyllifolia 153
vulgaris 153
POLYGALACEAE 153
POLYGONACEAE 242
Polygonum arenastrum 246
aviculare 246
maritimum 245
oxyspermum 245
oxyspermum ssp. *raii* 245
POLYPODIACEAE 105
Polypodies 105
Polypodium 105
interjectum 106
interjectum × *P. vulgare* 107
× *mantoniae* 107
vulgare 106
Polypody 106
Hybrid 107
Intermediate 106
Polypogon viridis 496
Polystichum setiferum 102
Pond-sedge, Greater 471
Pondweed, Bog 430
Broad-leaved 430
Curled 431
Fennel 431
Lesser 431
Perfoliate 430
Pontederia cordata 457
PONTEDERIACEAE 457
Poplar, Grey 187
White 186
Poppy, Californian 514
Common 111
Flanders 111
Ladybird 111
Long-headed 111
Mexican 514
Opium 110
Oriental 110
Yellow-horned 112
Populus alba 186
balsamifera 188
× *canadensis* 187
× *canadensis* 'Serotina' 187
× *canescens* 187
× *jackii* 187
Portulaca oleracea 288
PORTULACACEAE 288

Potamogeton crispus 431
natans 430
pectinatus 431
perfoliatus 430
polygonifolius 430
pusillus 431
POTAMOGETONACEAE 430
Potato 318
Potentilla anglica 162
anserina 161
erecta 162
reptans 162
Poterium sanguisorba ssp. *balearicum* 162
Primrose 290
Primula vulgaris 290
PRIMULACEAE 289
Privet, Garden 321
Wild 321
Prunella vulgaris 346
Prunus avium 155
cerasifera 154
domestica 155
domestica ssp. *insititia* 155
spinosa 155
Pseudofumaria lutea 515
Pseudosasa japonica 475
PTERIDACEAE 96
Pteridium aquilinum 95
Pteris cretica 96
Puccinellia maritima 480
Pulicaria dysenterica 375
Purple-loosestrife 206
Purslane, Common 288
Pink 287
Quaking-grass, Greater 481
Lesser 481
Quercus cerris 172
ilex 173
petraea 173
robur 174
Quince, Japanese 514
Quinoa 271
Radiola linoides 194
Radish, Garden 237
Sea 237
Wild 236
Ragged Robin 267
Ragweed 402
Ragwort, Common 394
Hedge 399
Monro's 399
Oxford 396
Purple 395
Silver 394
Woad-leaved 395
Ramping-fumitory, Common 114
Purple 115
Tall 114
Western 113
White 113
RANUNCULACEAE 116
Ranunculus acris 117
arvensis 119
baudotii 122
bulbosus 118
flammula 121
hederaceus 122
marginatus var. *trachycarpus* 119
muricatus 120
parviflorus 120
repens 117
sardous 118
sceleratus 121
trichophyllus 123
tripartitus 121
Rape, Oil-seed 233
Raphanus raphanistrum ssp. *maritimum* 237
raphanistrum ssp. *raphanistrum* 236
sativus 237
Red-hot-poker 442

Greater 443
Redshank 243
Reed, Common 508
Reseda alba 223
   lutea
RESEDACEAE 223
Restharrow, Common 138
Rhododendron ponticum 295
Rhododendron 295
Rhodostachys 458
   Tresco 459
Rhus typhina 217
Rocket, Eastern 238
   Sea 235
Rorippa palustris 227
   sylvestris 227
Rosa canina 164
   micrantha 165
   multiflora var. *cathayensis* 164
   rubiginosa 165
   rugosa 164
ROSACEAE 154
Rose, Dog 164
   Japanese 164
   Multi-flowered 164
Rosemary 352
Rose-of-heaven 266
Rosmarinus officinalis 352
Rostraria cristata 487
Rowan 156
Rubia peregrina 301
RUBIACEAE 298
Rubus microspecies 159
   angusticuspis 160
   caesius 159
   daveyi 160
   dumetorum var. *ferox* 161
   dumnoniensis 160
   fruticosus agg. 159
   hastiformis 161
   intensior 161
   iricus 160
   leyanus 160
   mollissimus 160
   newbouldianus 160
   peninsulae 160
   polyanthemus 160
   prolongatus 160
   pydarensis 160
   riddelsdellii 160
   rilstonei 160
   rubritinctus 160
   sprengelii 160
   transmarinus 161
   tuberculatus 161
   ulmifolius 160
   venetorum 161
   viridescens 160
Rumex × *abortivus* 254
   acetosa 249
   acetosa ssp. *hibernicus* 249
   acetosella 249
   conglomeratus 251
   conglomeratus × *R. obtusifolius* 254
   conglomeratus × *R. pulcher* 254
   crispus 250
   crispus ssp. *crispus* 250
   crispus ssp. *littoreus* 250
   crispus × *R. obtusifolius* 254
   crispus × *R. pulcher* 254
   crispus × *R. rupestris* 254
   × *dufftii* 254
   frutescens 250
   hydrolapathum 250
   × *muretii* 254
   obtusifolius 253
   obtusifolius × *R. pulcher* 254
   obtusifolius × *R. sanguineus* 254
   × *ogulinensis* 254
   × *pratensis* 254

   × *pseudopulcher* 254
   pulcher 253
   pulcher × *R. rupestris* 254
   rupestris 252
   sanguineus 251
   × *trimenii* 254
Ruppia maritima 432
RUPPIACEAE 432
Ruschia caroli 280
Ruscus aculeatus 454
Rush, Bulbous 461
   Compact 464
   Hard 463
   Jointed 459
   Sea 461
   Sharp-flowered 460
   Slender 461
   Toad 462
Russian-vine 247
Rustyback 103
Rye-grass, Italian 477
   Perennial 476
Sage, Wood 344
Sagina apetala 260
   apetala ssp. *apetala* 260
   apetala ssp. *erecta* 260
   maritima 261
   procumbens 260
SALICACEAE 186
Salix alba L. 188
   caprea × *S. cinerea* 189
   caprea 189
   cinerea ssp. *oleifolia* 189
   fragilis 188
   × *multinervis* 189
   × *reichardtii* 189
   × *rubens* 188
   × *smithiana* 189
   triandra 188
   viminalis 189
Sally-my-handsome 283
Salsola kali ssp. *kali* 276
Saltmarsh Rush 462
Saltmarsh-grass, Common 480
Saltwort, Prickly 276
Salvia reflexa 352
SALVINIACEAE 94
Sambucus nigra 407
Samolus valerandi 294
Samphire, Rock 418
Sandwort, Sea 255
   Thyme-leaved 254
Santolina chamaecyparissus 388
SAPINDACEAE 218
Saponaria officinalis 267
Sasa veitchii 475
SAXIFRAGACEAE 125
Saxifrage, Burnet 417
Scandix pecten-veneris 416
Schedonorus arundinaceus 476
Scilla peruviana 452
   verna 452
Scorpion-vetch, Shrubby 133
Scorzoneroides autumnalis 365
Scrophularia auriculata 338
   nodosa 337
   scorodonia 338
SCROPHULARIACEAE 335
Scurvygrass, Common 240
   Danish 240
Scutellaria galericulata 344
Sea-blite, Annual 276
Sea-buckthorn 165
Sea-fig, Angular 285
   Lesser 283
Sea-holly 414
Sea-kale 236
Sea-milkwort 292
Sea-spurrey, Greek 264
   Lesser 263

Rock 262
Sedge, Carnation 473
   Common 474
   Cyperus 471
   Glaucous 473
   Green-ribbed 473
   Grey 470
   Pendulous 472
   Pill 474
   Prickly 470
   Remote 471
   Sand 470
   Smooth-stalked 473
   Star 471
Sedum acre 128
   album 129
   anglicum 129
   confusum 128
   kimnachii 128
Selaginella kraussiana 90
SELAGINELLACEAE 90
Selfheal 346
Senecio cineraria 394
   glastifolius 395
   grandiflorus 395
   jacobaea 394
   minimus 397
   squalidus 396
   sylvaticus 397
   viscosus 397
   vulgaris 396
Setaria pumila 511
   verticillata 511
   viridis 512
Shaggy-soldier 404
Shallon 297
Sheep's-bit 358
Sheep's-fescue 478
Shepherd's-needle 416
Shepherd's-purse 225
Sherardia arvensis 298
Shield-fern, Soft 102
Shoreweed 333
Sibthorpia europaea 328
Silene coeli-rosa 266
   dioica 266
   flos-cuculi 267
   gallica 266
   gallica var. *anglica* 266
   gallica var. *gallica* 266
   gallica var. *quinquevulnera* 266
   latifolia ssp. *alba* 265
   noctiflora 266
   uniflora 265
   vulgaris 264
   vulgaris ssp. *vulgaris* 264
Silverweed 161
Sinapis alba 235
   arvensis 234
Sisymbrium officinale 238
   orientale 238
Sisyrinchium californicum 436
Skullcap 344
Small-reed, Wood 494
Smyrnium olusatrum 416
Snowflake, Summer 450
Snow-in-cerastium 257
Soapwort 267
Soft-brome 500
   Lesser 500
   Slender 501
Soft-grass, Creeping 488
Soft-rush 463
SOLANACEAE 314
Solanum dulcamara 317
   dulcamara var. *marinum* 317
   laciniatum 319
   lycopersicum 318
   nigrum 316
   physalifolium 316

Index to species

physalifolium var. *nitidibaccatum* 316
sarachoides 317
tuberosum 318
Soleirolia soleirolii 170
Solidago canadensis 376
virgaurea 376
Sonchus arvensis 367
asper 368
oleraceus 367
Sorbus aucuparia 156
Sorghum bicolor 513
Sorrel, Common 249
Sheep's 249
Sowthistle, Perennial 367
Prickly 368
Smooth 367
Spanish-dagger 455
Sparaxis grandiflora 440
Sparganium erectum 457
Spartium junceum 151
Spearwort, Lesser 121
Speedwell, Germander 326
Heath 322
Ivy-leaved 324
Thyme-leaved 324
Wall 327
Wood 323
Spergula arvensis 262
Spergularia bocconei 264
marina 263
rubra 263
rupicola 262
Spike-rush, Common 466
Many-stalked 467
Slender 466
Spinach, New Zealand 285
Spindle, Japanese 176
Spiraea douglasii 154
× *pseudosalicifolia* 154
salicifolia 154
Spiranthes spiralis 433
Spleenwort, Black 97
Lanceolate 98
Maidenhair 99
Sea 99
Spotted-orchid, Common 434
Springbeauty 286
Spurge, Caper 183
Cypress 185
Honey 182
Mediterranean 514
Petty 183
Portland 184
Sea 184
Sun 182
Wood 185
Spurge-laurel 514
Spurrey, Corn 262
Sand 263
Squill, Portuguese 452
Spring 452
St John's-wort, Marsh 197
Perforate 196
Slender 197
Trailing 196
Stachys arvensis 341
byzantina 515
palustris 340
sylvatica 340
Starflower, Spring 449
Star-of-Bethlehem 451
Steeple-bush 154
Stellaria alsine 257
holostea 256
media 255
neglecta 256
pallida 256
Stink-grass 508
Stinkweed 233
Stitchwort, Bog 257

Greater 256
Stock, Hoary 239
Night-scented 239
Stonecrop, Biting 128
English 129
Mexican 128
Shrubby 126
White 129
Stork's-bill, Common 204
Musk 203
Sea 203
Sticky 204
Strawberry, Wild 514
Strawflower 374
Suaeda maritima 276
Sumach, Stag's-horn 217
Sunflower 403
Lesser 403
Perennial 403
Thin-leaved 403
Sweet-briar
Small-flowered 165
Sweet-grass, Floating 498
Hybrid 498
Plicate 499
Small 499
Sweet-William 268
Swine-cress 231
Lesser 231
Sycamore 218
Symphytum officinale 305
× *uplandicum* 305
TAMARICACEAE 241
Tamarisk 241
African 241
Tamarix africana 241
gallica 241
Tanacetum parthenium 385
vulgare 386
Tansy 386
Taraxacum 369
laevigatum 369
'Obliqua' 369
Tare, Hairy 134
Slender 515
Smooth 135
Tasmanian-fuchsia 219
Tasselweed, Beaked 432
Teasel, Wild 410
Tea-tree, Broom 214
Woolly 214
Tetragonia tetragonioides 285
Teucrium scorodonia 344
Thalictrum minus 123
Thimble-flower, Blue 289, 514
Thistle, Carline 359
Creeping 362
Marsh 361
Slender 360
Spear 361
Thlaspi arvense 238
Thorn-apple 315
Thorow-wax, False 420, 514
Thrift 242
Throatwort 358
Thyme, Lemon 348
Wild 347
Thymus × *citriodorus* 348
polytrichus 347
Timothy 498
Toadflax, Common 330
Ivy-leaved 329
Tobacco 319
Red 320
Sweet 319
Tomato 318
Torilis japonica 425
nodosa 426
Tormentil 162
Trailing 162

Trachelium caeruleum 358
Tradescantia zebrina 515
Traveller's-joy 116
Treacle-mustard 224
Treasureflower 372
Plain 514
Tree of Love 130
Tree-fern, Australian 95
Tree-mallow 221
Smaller 222
Trefoil, Hop 145
Lesser 145
Slender 146
Trifolium arvense 149
campestre 145
dubium 145
fragiferum 144
glomeratum 144
hybridum 143
incarnatum ssp. *incarnatum* 147
medium 147
micranthum 146
occidentale 143
ornithopodioides 141
pratense 146
pratense var. *sativum* 146
repens 142
repens var. *townsendii* 142
resupinatum 515
scabrum 148
striatum 148
subterraneum 149
suffocatum 144
Tripleurospermum inodorum 393
maritimum 392
Trisetum flavescens 487
Tristagma uniflorum 449
Triticum aestivum 506
TROPAEOLACEAE 222
Tropaeolum majus 222
Tulip, Cretan 432
Tulipa saxatilis 432
saxatilis var. *townsendii* 432
Turnip 233
Turnip-rape 233
Tussilago farfara 399
Tussock, Serrated 475
Tussock-sedge, Greater 469
Tutsan 195
Stinking 196
Tall 195
Typha latifolia 458
TYPHACEAE 457
Ugni molinae 216
Ulex europaeus 151
europaeus × *U. gallii* 152
gallii 152
ULMACEAE 166
Ulmus 166
glabra 167
× *hollandica* 166
minor 167
minor ssp. *angustifolia* 167
minor ssp. *sarniensis* 167
procera 167
Umbilicus rupestris 126
Urospermum dalechampii 515
Urtica dioica 169
urens 169
URTICACEAE 169
Valerian, Common 409
Marsh 409
Red 410
Valeriana dioica 409
officinalis 409
VALERIANACEAE 408
Valerianella carinata 408
dentata 408
locusta 408
locusta var. *dunensis* 408

Verbascum blattaria 335
   nigrum 337
   phlomoides 336
   thapsus 336
   virgatum 336
Verbena bonariensis 515
   officinalis 356
VERBENACEAE 356
Vernal-grass, Sweet 490
Veronica agrestis 325
   arvensis 327
   beccabunga 323
   brachysiphon 327
   chamaedrys 326
   dieffenbachii 327
   elliptica × *V. salicifolia* 327
   elliptica × *V. speciosa* 328
   × *franciscana* 327, 328
   × *franciscana* 'Blue Gem' 328
   hederifolia 324
   hederifolia ssp. *lucorum* 324
   hederifolia ssp. *sublobata* 324
   × *lewisii* 327
   montana 323
   officinalis 322
   persica 326
   polita 325
   salicifolia 327
   serpyllifolia ssp. *serpyllifolia* 324
Veronica, Hedge 328
VERONICACEAE 322
Vervain 356
   Argentinian 515
Vetch, Bithynian 137
   Bush 135
   Common 136
   Kidney 131
   Spring 136
   Tufted 134
Vetchling, Meadow 138
Viburnum opulus 407
   tinus 407
Vicia bithynica 137
   cracca 134
   faba 137
   hirsuta 134
   lathyroides 136
   parviflora 515
   sativa 136
   sativa ssp. *nigra* 136
   sativa ssp. *segetalis* 136
   sepium 135
   tetrasperma 135
Vinca difformis 515
   major 303
   major var. *oxyloba* 303
   minor 302
Viola arvensis 192
   canina 191
   × *contempta* 515
   hirta 190
   kitaibeliana 192
   odorata 190
   reichenbachiana 191
   riviniana 190
   riviniana var. *minor* 190
   tricolor 191
   × *wittrockiana* 191
VIOLACEAE 190
Violet, Hairy 190
   Sweet 190
Viper's-bugloss 303
   Giant 304
   Purple 304
Virgin's-bower 116
Vulpia bromoides 478
   fasciculata 478
   myuros 479
Wahlenbergia hederacea 357
Wallflower 224

Wall-rocket, Annual 233
   Perennial 232
Wall-rue 100
Water-cress 228
Water-crowfoot, Brackish 122
   Thread-leaved 123
Water-dropwort, Hemlock 419
   Parsley 418
   Tubular 419
Water-lily, White 109
Water-milfoil, Alternate 130
Water-parsnip, Lesser 418
Water-pepper 244
Water-plantain, Lesser 429
Water-purslane 207
Water-starwort, Blunt-fruited 334
   Common 333
   Intermediate 335
   Pedunculate 334
   Various-leaved 334
Waterwort, Six-stamened 186
Watsonia borbonica 438
Wattle, Cape 152
   Hickory 152
Weasel's-snout 329
Wheat, Bread 506
Whistling Jacks 439
Wild-oat 486
   Winter 486
Willow, Almond 188
   Goat 189
   Grey 189
   White 188
Willowherb, American 210
   Broad-leaved 208
   Great 207
   Hoary 208
   Marsh 211
   Rosebay 211
   Short-fruited 210
   Spear-leaved 209
   Square-stalked 209
Winter-cress 226
   American 226
Wireplant 248
Wood-rush, Field 464
   Heath 465
Wood-sedge 472
WOODSIACEAE 100
Wood-sorrel 179
Woodwardia radicans 101
Wormwood 387
   Sea 387
Woundwort, Field 341
   Hedge 340
   Marsh 340
XANTHORRHOEACEAE 442
Yard-grass 509
Yarrow 388
Yellow Archangel, Garden 342
Yellow-cress, Creeping 227
   Marsh 227
Yellow-eyed-grass 436
Yellow-sedge, Common 474
Yellow-sorrel, Fleshy 178
   Least 178
   Procumbent 177
Yorkshire-fog 488
Yucca gloriosa 455
Zantedeschia aethiopica 427
Zostera marina 430
ZOSTERACEAE 430

Sand flats